21 Springer Series in Chemical Physics

Edited by J. Peter Toennies

Springer Series in Chemical Physics

Editors: V. I. Goldanskii R. Gomer F. P. Schäfer J. P. Toennies

Dynamics of Gas-Surface Interaction

Proceedings of the International School on Material Science and Technology, Erice, Italy, July 1–15, 1981

Editors: G. Benedek and U. Valbusa

With 132 Figures

Springer-Verlag Berlin Heidelberg GmbH 1982

Professor Dr. Giorgio Benedek
Istituto di Fisica dell'Universita Milano,
Via Celoria 16,
I-20133 Milano, Italy

Professor Dr. Ugo Valbusa
Istituto di Scienze Fisiche, Universita Genova,
Viale Benedetto XV, 5
I-16131 Genova, Italy

Series Editors

ISBN 978-3-642-86457-5 ISBN 978-3-642-86455-1 (eBook)
DOI 10.1007/978-3-642-86455-1

Originally published by Springer-Verlag Berlin Heidelberg in 1982
Softcover reprint of the hardcover 1st edition 1982

2153/3130-543210

Preface

In recent decades surface science has experienced a large growth in connection with the development of various experimental techniques which are able to characterize solid surfaces through the observation of the scattering of ions, electrons, photons or atoms. These methods of investigation, known under different labels such as LEED, AES, XPS, UPS, etc. have been extensively applied in describing the structure, morphology, and chemical and physical properties of crystal surfaces and interfaces of a large variety of materials of interest in solid-state physics, electronics, metallurgy, biophysics, and heterogeneous catalysis. Among these methods we wish to emphasize molecular beam scattering from solid surfaces.

Molecular beam scattering has gone through a large development in the last ten years. In this decade a large number of laboratories have used this method to study various clean and adsorbate-covered surfaces. The technique is nonetheless quite old. It dates back to the beginning of the thirties, when Estermann and Stern performed the first atom diffraction experiment proving the wave nature of atoms. In the following years the entire subject of gas-surface interaction was considered a branch of rarefied gas dynamics and developed in connection with aerospace research. Attention was then given to the integral properties of gas-solid interactions (sticking and energy accomodation, mean momentum transfer) rather than to atom-surface scattering from well-characterized surfaces. Meanwhile the molecular beam technique was greatly improved by the development of supersonic beams, high-sensitivity detectors, and UHV facilities, which allowed high-resolution diffraction experiments from clean and reproducible solid surfaces. In the seventies the field was booming. Molecular beams rapidly began competing with other probes because of several of their advantages: a lack of penetration into the solid; a heightened sensitivity to surface coverage; and direct information on the atom-surface potential energy and on inelastic atom-surface interaction.

The aim of the present book is to review our present understanding of atom-surface scattering and to compare the capabilities of this method with others currently used in studying clean surfaces and adsorbates. The book originated from the lectures held at the third course of the International School on Material Science and Technology on "Dynamics of Gas-Surface Interaction" held at the Center for Scientific Culture "Ettore Majorana" in Erice, Italy, July 1-15, 1981.

The book is divided in four parts. Each part contains a few extended lectures introducing the subject followed by the advanced seminars. In the first part, the theory of atom-surface scattering is presented together with a description of the recent advances in experimentation. Emphasis is placed on diffractive scattering and the characterization of clean surfaces. In view of the traditional interest in ionic crystals, the authors review the most significant experiments carried out on semiconductor and metal surfaces.

In the second part the molecular beam method is compared with neutron diffraction, X-ray diffraction, LEED, and Raman spectroscopy as far as the characterization of adsorbed phases and desorption is concerned.

The second half of the book is devoted to the spectroscopy of surface phonons, which has registered spectacular progress in recent years. Part 3 deals with the theory of elementary surface excitations and their experimental study by means of optical methods and electron energy loss spectroscopy (EELS). The lectures by Professor H. Ibach on EELS experiments are not included since this subject has been extensively treated by him in two recent books. Finally, part 4 illustrates the state of the art in time-of-flight spectroscopy of atom scattering and the studies of inelastic resonances. Both of these topics are put into perspective by theoretical lectures on surface dynamics with an explicit application to inelastic scattering.

We wish to thank the lecturers and participants who have contributed so much to make this School successful. We are deeply grateful to the Director, Professor Zichichi; to the secretary, Miss Pinola Savelli; to Dr. A. Gabriele; and to the staff of the Majorana Center for their excellent organization, hospitality, and assistance. We thank IBM Italy, the Gruppo Nazionale di Struttura della Materia, and the Technological Committee of CNR for their sponsorship and financial support, as well as Professor M. Balkanski, Director of the International School on Material Science and Technology, and Professor B. Feuerbacher, Chairman of EPS Surface and Interface Section, for their collaboration and advice during the organization of the course.

Milano and Genova,
Italy, April 1982

G. Benedek
U. Valbusa

Contents

Part III. **Spectroscopy of Surface Optical Excitations**

Part IV. **Surface Phonon Spectroscopy by Atom Scattering**

Part I

Scattering of Atoms from Solid Surfaces

Theory of Atom-Surface Scattering

V. Celli* and D. Evans

Department of Physics, University of Virginia
Charlottesville, VA 22901, USA

Introduction

These lectures describe the application of the quantum theory of scattering to atom-surface interaction processes. The treatment will be introductory, self-contained and fairly inclusive. A complete list of references is not given, however; these can be found in recent reviews [1-6].

We start with the simplest cases and proceed to increasing complexity, both in the physics of the system and in the theory. The goal is to provide a framework for the interpretation of the data obtained, mostly, by reflection of nozzle beams of light species (isotopes of He, H, and H_2) from surfaces with negligible sticking probability. Typical beam parameters are:

	Definition, Symbol	Typical size
Wavelength	λ	1 Å
Wave number	$k = 2\pi/\lambda$	5 $Å^{-1}$
Particle mass	m	4 amu
	$m[g] = 1.6606\cdot 10^{-24} m[amu]$	
Energy	$E = \hbar^2 k^2/2m$	20 meV
	$E[meV] = 2.090k^2[Å^{-2}]/m[amu]$	
Temperature	$T = E/2k_{Boltzmann}$	100 K
	$T[K] = 5.802E[meV]$	

The other notation and conventions are basically those of the book by GOODMAN and WACHMAN [7]: the surface is, in the mean, parallel to the xy plane; the molecular beam is incident from above ($z > 0$); capital letters denote vectors in the xy plane. Thus the position of an incident atom is denoted by $\vec{r}$ and also by $(\vec{R}, z)$; its wave vector $\vec{k}$ is also denoted by $(\vec{K}, k_z)$, with $k_z < 0$; the specularly reflected wave vector is $\vec{k}_o = (\vec{K}_o, k_{oz})$, with $\vec{K}_o = \vec{K}$ and $k_{oz} = -k_z > 0$. Often $\vec{k}$ will be called the momentum, but the true momentum is of course $\hbar\vec{k}$.

Much of the theory, although often rediscovered, is an adaptation of the methods developed in atomic and nuclear physics [8,9] or in the classical theory of diffraction, mostly in optics [10,11]. Some elementary considerations on length scales are helpful to connect alternative descriptions of the same phenomena.

---

* Parts of these lectures were prepared at IGV-KFA, Jülich, Germany, with the support of a Humboldt Award, which is gratefully acknowledged.

The sample may be larger or smaller than the incoming beam; in either case let $L(\gg\lambda)$ be a typical size of the illuminated area, say L = 1 mm. The specular and Bragg-diffracted beams, if any, have an angular diffraction width λ/L. (We neglect the divergence of the incoming beam, although it is usually larger than λ/L.) Coherent scattering between different parts of the surface is responsible for the narrow angular definition of these beams; incoherently scattered particles, on the other hand, are smoothly distributed over angles.

At a distance r, the linear diffraction width of a coherent beam is $r\lambda/L$; it becomes equal to L at $r = L^2/\lambda$. It follows that such a beam is well described by a plane wave for $r \ll L^2/\lambda$ and by a divergent spherical wave (within the small angle λ/L) for $r \gg L^2/\lambda$. On the other hand, the incoherently scattered particles are well described by plane waves for $z \ll L$ and by a divergent spherical wave for $r \gg L$. Typically, for atomic beams, L^2/λ = 10 km for L = 1 mm, λ = 1 Å.

We then distinguish three zones:

- the Fresnel diffraction zone, for $z \ll L$: all outgoing particles described by plane waves;
- the Fraunhofer diffraction zone, for $L \ll r \ll L^2/\lambda$: coherent particle beams described by plane waves, incoherently scattered particles described by an angle-modulated spherical wave;
- the scattering zone: all outgoing particles described by an angle-modulated spherical wave.

Molecular beam detectors are set up in the Fraunhofer zone (typically at r = 40 cm). Theory, however, is often done either for $z \ll L$ (using the Fresnel diffraction language) or for $r \gg L^2/\lambda$ (using the language of scattering theory).

All this is true for an ideal surface. A real surface, however, consists of many facets that scatter incoherently. In effect, L must be taken as the size of a facet, which is typically 100 Å. Then the Fraunhofer zone extends only to 1μ, and the typical angular width of the beam is about 1°. This is comparable to the angle subtended by the detector, a fact that can cause problems in data interpretation, as has been discussed by LAPUJOULADE [12]. An ideal measurement can distinguish the diffracted beams from the incoherent background, and this corresponds to the Fraunhofer description.

Lecture 1

In this lecture we review the numerical methods that have been used in atom-surface scattering. Special emphasis is given to the calculation of phase shifts, which are relevant to the theory of resonant scattering, to be presented in Lecture 4.

1.1 Scattering from a Periodic Potential

The simplest problem is to solve for the scattering of a monoenergetic beam from a fixed periodic potential $V = V(\vec{R},z)$. The surface periodicity in $\vec{R}$ is described by the two basic lattice vectors $\vec{A}_1$ and $\vec{A}_2$ (not necessarily orthogonal to each other), or by the two basic reciprocal lattice vectors $\vec{G}_1$ and $\vec{G}_2$ (with $\vec{G}_1\cdot\vec{A}_1 = \vec{G}_2\cdot\vec{A}_2 = 2\pi$ and $\vec{G}_1\cdot\vec{A}_2 = \vec{G}_2\cdot\vec{A}_1 = 0$).

It can be assumed that V is infinitely repulsive for z less than some z_0. For large z, V is of the form $-C_3/z^3$, where C_3 is of the order of 500 meV $Å^3$, due to the van der Waals attraction.

For $z > z_0$, V can be Fourier expanded

$$V(\vec{r}) = \sum_G e^{i\vec{G}\cdot\vec{R}} V_G(z) \tag{1.1}$$

where $\vec{G}$ is a combination of $\vec{G}_1$ and $\vec{G}_2$ with integer coefficients. $V_0(z)$ is called the laterally averaged potential and has the shape shown in Fig. 4.2; the well depth D is of the order of 10 meV. The other Fourier components, $V_G(z)$, are usually assumed to be purely repulsive, and grow faster than V_0 for decreasing z. Care must be taken then to check that the representation (1.1), especially when truncated to a few terms, does not oscillate wildly for small z; this is one good reason to set $V = \infty$ for $z < z_0$.

By Bloch's theorem, there is a complete set of eigenfunctions of the form

$$\psi_K(\vec{r}) = \sum_G e^{i(\vec{K}+\vec{G})\cdot\vec{R}} \psi_G(z) \tag{1.2}$$

and the Schrödinger equation for ψ_K is equivalent to the set of coupled ordinary differential equations

$$-\frac{d^2\psi_G(z)}{dz^2} + \frac{2m}{\hbar^2} \sum_{G'} V_{G-G'}(z)\, \psi_{G'}(z) = k_{Gz}^2\, \psi_G(z) \tag{1.3}$$

where k_{Gz}^2 is related to the energy $E = \hbar^2 k^2/2m$ by

$$k_{Gz}^2 = k^2 - (\vec{K}+\vec{G})^2 \ . \tag{1.4}$$

For a scattering state, k^2 is positive and $K \leq k$; in fact,

$$K = k \sin\theta \tag{1.5}$$

defines the incident angle θ. Some k_{Gz}^2 are positive and give open scattering channels; the corresponding $\vec{G}$ vectors, denoted by $\vec{F}$ (for final), fall within a circle of radius k centered at $-\vec{K}$, as shown by the Ewald construction for the graphic solution of (1.4). Negative values of k_{Gz}^2 give closed scattering channels; the corresponding $\vec{G}$ vectors, denoted by $\vec{E}$ (for evanescent), fall outside the Ewald circle.

The asymptotic ($z \to +\infty$) conditions on a scattering state are:

$$\psi_0(z) \sim e^{ik_z z} + A_0 e^{ik_{0z} z} \quad , \quad k_z = -k_{0z} \tag{1.6}$$

(incoming beam and specularly reflected beam); and for $G \neq 0$

$$\psi_G(z) \sim A_G\, e^{ik_{Gz} z} \tag{1.7}$$

(outgoing diffracted beams for $\vec{G} = \vec{F}$, k_{Fz} non-negative real; evanescent beams for $\vec{G} = \vec{E}$, k_{Ez} positive imaginary).

A diffracted intensity is given by the ratio of fluxes across a plane z = const. at large positive z: the incoming flux is $|k_z|/m$, the flux in beam $\vec{F}$ is $k_{Fz}|A_F|^2/m$; hence the diffracted intensities are given by $|S_{Fo}|^2$, with

$$S_{Fo} = -(k_{Fz}/k_{oz})^{1/2} A_F \quad . \tag{1.8}$$

S_{Fo} is an element of the S matrix, about which more is said in Lecture 2. Conservation of flux implies that

$$\sum_F |S_{Fo}|^2 = 1 \quad . \tag{1.9}$$

The size of the diffracting area, L, appears nowhere in these formulas, which are obtained by flux balance in the Fresnel zone ($z \ll L$). The actual measurements are performed in the Fraunhofer zone ($r \gg L$), where the diffracted beams diverge radially from each other. The ratio of total flux in the $\vec{F}$ beam to incident flux is still $|S_{FO}|^2$, however. One can also approach the problem in a way more similar to ordinary potential scattering theory, by assuming that the incident beam is wider than the sample, which has area L^2. One then computes differential cross sections, integrates over the angular width of beam F, and normalizes by the total cross section for scattering in the region $z > 0$, which is $L^2\cos\theta$ by simple geometry. The result is again $|S_{Fo}|^2$.

An incident beam with momentum $(\vec{K} + \vec{F}'\text{-}k_{F'z})$ also diffracts into the set of beams $(\vec{K} + \vec{F}, k_{Fz})$, and we can call the corresponding intensities $|S_{FF'}|^2$. Flux conservation applied to this case gives the analog of (1.9), with $0 \to F'$; more generally, flux conservation also holds when two (or more) beams are simultaneously incident, and it implies that

$$\sum_F S^*_{FF'} S_{FF''} = \delta_{F'F''} \quad , \tag{1.10}$$

i.e., the S matrix is unitary. It can then be diagonalized, and the eigenvalues are of the form exp $(2i\delta_\nu)$, with δ_ν real. Knowledge of the phases δ_ν, and of the eigenvectors of $\tilde{S}$, as a function of incident energy, would enable us to reconstruct $V(\vec{r})$ directly. Unfortunately, as usual in scattering problems, only $|S_{Fo}|^2$ is experimentally determined.

1.2 Numerical Methods and Phase Shifts

In order to solve (1.2) numerically, the summation over $\vec{G}'$ is truncated after N terms. Given the starting values of ψ_G and $\psi_{G'}$ ($= d\psi_G/dz$), integration is straightforward, except for stability difficulties to be discussed below; in reality, however, we are given instead the boundary conditions: (1.6) and (1.7) for $z \to \infty$, and $\psi_G = 0$ for $z = z_0$. Thus we have a problem.

Recall first how this is handled for a single equation: choose arbitrarily $\psi'(z_0)$ (= 1, for instance), integrate out to a large z_f, and compare with the desired asymptotic form

$$\tilde{\psi}(z) = e^{-ikz} - e^{2i\delta} e^{ikz} \quad . \tag{1.11}$$

Because the solution that vanishes at z_0 is unique, up to a factor Q, it must be true that $\psi = Q\tilde{\psi}$ and $\psi' = Q\tilde{\psi}'$ at $z = z_f$. From these two equations, eliminate Q, which is irrelevant, and solve for exp $(2i\delta)$, which in this case is the only element of the S matrix. Find

$$S\, e^{2ikz_f} = e^{2i(kz_f+\delta)} = (\psi' + ik\psi)/(\psi' - ik\psi) \tag{1.12}$$

The point is that δ occurs naturally in the course of the numerical solution procedure.

Numerical integration is straightforward by the Numerov method [14]. The method is said to fail for large values of E-V, because ψ oscillates with a wavelength inversely proportional to $(E-V)^{1/2}$, and one requires a fixed number of grid points per oscillation. Actually, the Numerov method always works well near the classical turning point (where $E = V$); the solution can then be joined, as described below, to the simplest semiclassical approximation, which is valid for large E-V.

A more elaborate alternative is to use Gordon's method [19]: V is approximated by a piecewise linear potential, the solutions of which are known in terms of Airy functions, Ai and Bi. Explicitly, $V(z)$ is approximated in the interval (z_1, z_2) by the expression

$$V(z_3) + (z-z_3)\, V'\,(z_3) \tag{1.13}$$

where $z_3 = (z_1 + z_2)/2$; the wave function is then

$$\psi(z) = a\mathrm{Ai}(\gamma(z+\delta)) + b\mathrm{Bi}(\gamma(z+\delta)) \tag{1.14}$$

where $\gamma = [2mV'(z_3)/\hbar^2]^{1/3}$, $\delta + z_3 = (V(z_3)-E)/V'(z_3)$.

The constants a and b are determined by matching ψ and ψ' to the solution previously found for $z < z_1$ (or setting $\psi = 0, \psi'$ arbitrary, if $z_1 = z_0$). Accurate and convenient expressions for the Airy functions have been developed for use with this method [14].

Whatever numerical method one uses, it is often better to proceed somewhat differently, especially to find the bound state wave functions: integrate upward from z_0 and downward from z_f, then match at an intermediate point z_m. This is advantageous because the solution started at one end becomes numerically unstable at the other end, in the classically forbidden region beyond the turning point. Let $\psi_>$ be the solution started at z_f with $\psi_> = \exp(-\beta z_f)$ and $\psi_>' = -\beta\exp(-\beta z_f)$, putting $k = i\beta$ for a bound state; let $\psi_<$ be the solution ψ that was found above, starting from z_0. Then the matching condition at z_m can be written $W(\psi_>, \psi_<) = 0$, where W is the Wronskian $W(\phi,\psi) = \phi\psi' - \psi\phi'$. The choice of z_m does not matter because W does not depend on z. The condition $W = 0$ determines the eigenvalues.

A catalog of wave functions may help. For $E < 0$, $\psi_<$ and $\psi_>$ are in general distinct and neither is a well behaved solution of Schrödinger's equation. An exception occurs when E is equal to a bound state energy E_n: then $\psi_<$ and $\psi_>$ are both proportional to the normalized eigenfunction. For $E > 0$, $\psi_<$ is the only well behaved solution of Schrödinger's equation, up to a factor; $\psi_>$ and $\psi_>^*$ are not well behaved at small z, but one can form the well behaved complex combination $\psi_+ = \psi_>^* - e^{2i\delta}\psi_>$. For some purposes (see Section 1.4) it is better to use the real standing wave combination, ψ_s:

$$i\psi_s = e^{-i\delta}\psi_+ = e^{-i\delta}\psi_>^* - e^{i\delta}\psi_> \ . \tag{1.15}$$

The bound state eigenfunctions and the ψ_s, normalized in a box, are collectively designated by ψ_n in (1.19) below.

Having computed $\psi_<$ and $\psi_>$, we can immediately construct the Green's function, $G^+(z,z';E)$. G^+ is defined as the solution of the equation

$$\{E + \frac{\hbar^2}{2m}\frac{d^2}{dz^2} - V(z)\}\, G = \delta(z-z') \tag{1.16}$$

with outgoing wave boundary conditions, i.e. as a function of z (and also of z') G^+ behaves as $\psi_<$ at $-\infty$ and as $\psi_>$ at $+\infty$. It is also useful at times to consider a solution G^- of (1.16) that, at $z \to +\infty$, behaves as $\exp(-\beta z)$ for $E < 0$, and as $\exp(-ikz)$ for $E > 0$; actually G^- is the complex conjugate of G^+. When no superscript is shown, G will stand for G^+ in the following. It is easy to check that, for $z > z'$, (1.16) is satisfied by

$$G = \frac{-2m}{\hbar^2 W(\psi_>,\psi_<)} \quad \psi_>(z)\ \psi_<(z'); \tag{1.17}$$

for $z < z'$, z and z' are interchanged in (1.17). One can regard (1.16) as an operator equation, which defines G as the inverse of E-H, with $H = (-\hbar^2/2m)(d^2/dz^2) + V$. Taking matrix elements between the normalized eigenfunctions of H, ψ_n, with eigenvalues E_n, one obtains

$$G_{nn'}(E) = \delta_{nn'}/(E-E_n + i\eta) \quad , \tag{1.18}$$

$$G(z,z';E) = \sum_n \frac{\psi_n(z)\ \psi_n(z')}{E-E_n + i\eta} \quad . \tag{1.19}$$

The infinitesimal $+ i\eta$ assures that G obeys outgoing wave boundary conditions. Although (1.19) is useful for the formal theory, (1.17) is in general the practical way to compute G.

The relations between phase shifts, Green's functions, and Wronskians take many forms, not always obvious. For instance, δ is essentially the phase of a Wronskian, or, up to a multiple of π,

$$\delta = \mathrm{Im}\ \ell n\ W\ (\psi_>^*,\psi_<) \quad . \tag{1.20}$$

To derive (1.20), which is interesting only for $E > 0$, proceed as in the derivation of (1.12), but now match $\psi_<$ to $\psi_>^*-S\psi_>$ at any point z_m, rather than to the asymptotic form ψ at large $z > 0$; S is then obtained as the ratio of two Wronskians, as in (1.12),

$$S = e^{2i\delta} = W(\psi_<,\psi_>^*)/W(\psi_<,\psi_>) \quad . \tag{1.21}$$

The two Wronskians are the complex conjugate of each other, however, because $\psi_<$ is real; hence (1.20) follows.

We can use (1.20) or (1.21) to find the effect on δ of a change in V that affects only $z < z_m$; only $\psi_<$ needs to be recomputed. Typically $\psi_>$ is given well enough by the semiclassical approximation

$$\psi_>(z) = (k/k(z))^{1/2}\, e^{i\phi_{cl}(z)} \quad , \tag{1.22}$$

$$k(z) = [(2m/\hbar^2)\,(E-V(z))]^{1/2} \quad , \tag{1.22'}$$

$$\phi_{cl}(z) = kz_f - \int_z^{z_f} k(z')dz' \quad , \tag{1.22''}$$

It follows then from (1.20), or (1.21), that

$$\delta = -\phi_{cl}(z_m) + \mathrm{Im}\,\ln\,(\psi_<'(z_m) + ik(z_m)\psi_<(z_m)) . \tag{1.23}$$

According to (1.12), the second term on the r.h.s. of (1.23) can be regarded as $-k(z_m)z_m + \delta_<(z_m)$, where $\delta_<(z_m)$ is a "hard-wall" phase shift, due to scattering of a particle of momentum $k(z_m)$ from the repulsive part of the potential. Thus, in this simple case, the phase shifts from different parts of the potential are essentially additive.

The complete semiclassical expression for the phase shift is similar to (1.23),

$$\delta = \int_{z_t}^{z_f} k(z)dz - kz_f + \pi/4 = -\phi_{cl}(z_t) + \pi/4 \tag{1.24}$$

where z_t is the classical turning point, i.e., $E = V(z_t)$.

Taking a derivative of (1.24) with respect to E, it is seen that the time taken by a particle to travel from z_f, to the turning point, and back to z_f is

$$2\int_{z_t}^{z_f} \frac{dz}{v(z)} = 2\left\{ z_f \frac{m}{\hbar k} + \hbar \frac{d\delta}{dE} \right\} \tag{1.25}$$

where $v(z) = \hbar k(z)/m$ is the classical speed at point z. Thus, $\hbar d\delta/dE$ gives the extra time delay due to the fact that reflection happens (classically) at $z = z_t$, rather than at $z = 0$, and that the speed of the particle changes as it travels. Actually, the interpretation of $d\delta/dE$ as a time delay is not just semiclassical; when the exact δ is used, quantum effects due to penetration of the potential, and to resonances, are also given correctly.

1.3 Close Coupling Calculations

We now discuss the numerical methods for the solution of the coupled-channel equations (1.3) and their relation to the S matrix. Formally, we can regard ψ_G as an N-component vector ψ, $V_{G-G'}$ as a matrix V, and k^2_{Gz} as the diagonal elements $\alpha^2{}_G$ of a diagonal matrix α^2.

The Numerov method can be applied essentially as it is. If δz is the chosen increment and $\psi(n)$ denotes $\psi(z_0 + n\delta z)$, the formula is

$$\psi(n+1) = 2\psi(n) - \psi(n-1) + \frac{1}{12}(\delta z)^2 \{F(n+1) + 10F(n) + F(n-1)\} , \qquad (1.26)$$

$$F(n) = \{(2m/\hbar^2)V(z_0+n\delta z) - \alpha^2\} \psi(n) .$$

It is better to use the method as a predictor-corrector procedure, replacing F(n+1) by F(n) the first time, and inserting the computed $\psi(n+1)$ on the next iteration. Gordon's method can also be applied [16], but further elaboration is needed to handle the fact that V is a matrix with non-negligible off-diagonal elements. The basic idea is to first make $\alpha^2-V(z)$ approximately diagonal: for each strip (z_1,z_2), centered at z_3, one can find a unitary matrix $M(z_3)$ that diagonalizes $\alpha^2-V(z_3)$. This amounts to solving the two-dimensional band structure problem of finding the eigenstates of $V(\vec{R},z_3)$ by the plane wave method. The new potential, $MV(z)M^{-1}$, is now expanded linearly near z_3, and only the diagonal elements of $MV'(z_3)M^{-1}$ are kept in the approximate hamiltonian to be solved exactly, while the off-diagonal elements are treated as a perturbation. The result is a set of decoupled equations, with solutions of the type (1.13); these solutions give $M(z_3)\psi$, however, and one must still undo the effect of the transformation $M(z_3)$ when matching ψ and ψ' with the adjacent strips.

In principle, then, we need only construct N linearly independent solutions ψ_{Gn} that vanish at z_0 (and differ in the derivative there), then take a linear combination that satisfies (1.6) and (1.7), at $z = z_f$, i.e.

$$\Sigma_n \psi_{Gn}(z)C_n = e^{-i\alpha_G z} \delta_{Go} + A_G e^{i\alpha_G z} , \qquad (1.27a)$$

$$\Sigma_n \psi'_{Gn}(z)C_n = -i\alpha_G e^{-i\alpha_G z} \delta_{Go} + i\alpha_G A_G e^{i\alpha_G z} . \qquad (1.27b)$$

These are 2N linear equations in the 2N unknowns C_n, A_G. However, N equations for the C_n alone are obtained by multiplying (a) by $i\alpha_G$ and subtracting (b) side by side. The solution of this system of N equations is the crucial step and solves the boundary-value problem. Once the C_n are known, A_G is obtained directly. Numerical instabilities can occur because the $\psi_{Gn}(z)$ for the evanescent beams contain also the exponentially growing part. There are several ways to avoid this problem [17]; an approach useful for further developments is to start out solutions at z_f of the type $\psi_>$ and match at an intermediate z_m to the solutions of type $\psi_<$, constructed starting from z_0. In particular let $\psi^>_{GG'}$ be the solution that behaves as $\delta_{GG'}e^{i\alpha_{G'}z}$ at $+\infty$. The matching at z_m gives equations similar to (1.27),

$$\Sigma_n \psi^<_{Gn}C_n = \psi^>_{Go} + \Sigma_{G'}\psi^>_{GG'}A_{G'} , \qquad (1.28a)$$

$$\Sigma_n(d\psi^<_{Gn}/dz)C'_n = (d\psi^>_{Go}/dz) + \Sigma_{G'}(d\psi^>_{GG'}/dz)A_{G'} . \qquad (1.28b)$$

The analogy of this procedure to the one-dimensional procedure leading to (1.17) should be obvious. It is also possible to generalize the procedure to construct G(z,z';E), which is now a matrix.

It will be noted in (1.28) that $\psi^>_{GG'}$, for instance, can be regarded as a square matrix; in fact all the equations can be conveniently written in matrix form, especially for formal manipulations. From this point of view, A_G is to be regarded as A_{GO}, the amplitude for the transition $G \leftarrow 0$. It is possible to consider simultaneously all the transition amplitudes, simply by changing 0 to one of the $\vec{F}$ vectors in (1.27) and (1.28).

Thus, for example, the solution of (1.27) gives the S matrix

$$S e^{2i\alpha z_f} = \alpha^{1/2}(\psi'-i\alpha\psi)\ (\psi'+i\alpha\psi)^{-1}\alpha^{-1/2} \quad , \tag{1.29}$$

where the fractional powers of α come in according to (1.8).

From the practical point of view, the size N x N of the matrices to be inverted, and not the z-integration procedure, sets a limit on the applicability of the close-coupling methods. For simple, unreconstructed surfaces there are no difficulties, however; the problem, of course, is that one does not know the potential, and one needs to develop a reasonable picture of it from the atom-scattering data themselves, and other input, before doing lengthy calculations.

1.4 The Distorted Wave Born Approximation

A useful approximation is obtained when the Fourier coefficients $V_G(z)$ are small in the region where $\psi_G(z)$ is appreciable. Then to zero order we have a one-dimensional problem with potential $V_0(z)$, to which the considerations of Section 1.2 apply: the wave function depends just on z and the Green's function is given by (1.17). Let us denote $\hbar^2\alpha_G^2/2m$ by E_G and let $\chi^+(z',E_G)$ be the eigenfunction of $V_0(z)$ with energy E_G. We can rewrite (1.3) as an integral equation of the Lippmann-Schwinger type (See Lect. 2),

$$\psi_G^+(z) = \chi^+(z,E_0)\delta_{G0} + \int dz'\ G(z,z';\ E_G) \sum_{G'\neq 0} V_{G-G'}(z')\ \psi_{G'}^+(z') \tag{1.30}$$

and replace, to first order, $\psi_{G'}^+$ by $\chi^+(z';E_0)\ \delta_{G'0}$ on the right hand side. We obtain then, for G ≠ 0, using also (1.15)

$$\psi_G^+(z) = e^{i\delta(E_0)} \int dz'\ G(z,z';E_G)\ V_G(z')\ \chi_s\ (z';E_0) \quad . \tag{1.31}$$

The amplitude A_G is given by the form of $\psi_G^+(z)$ for large z. Using (1.17) we have the answer at once: because $\chi_>$ becomes $\exp(i\alpha_G z)$ and $\chi_<$ can be identified with $\chi_s(z,E_G)$ (up to an irrelevant constant), we find

$$A_G = \left(\frac{2m}{\hbar^2 W}\right)e^{i\delta(E_0)} \int dz'\ \chi_s(z',E_G)\ V_G(z')\ \chi_s(z',E_0) \quad . \tag{1.32}$$

The Wronskian $W = W(\chi_s,\ \chi_>)$ can be conveniently evaluated at large z, and is just $i\alpha_G \exp(-i\delta(E_G))$. The final result for the S matrix elements is then, by (1.8), for $\vec{F} \neq 0$

$$S_{F0} = e^{i(\delta(E_F)+\delta(E_0))} \frac{2mi}{\hbar^2(\alpha_F\alpha_0)^{1/2}} \int dz\chi_s(z,E_F)V_{F-0}(z)\chi_s(z,E_0) \tag{1.33}$$

where, explicitly

$$V_G(z) = \int \frac{d^2R}{A}\ e^{-i\vec{G}\cdot\vec{R}}\ \ V(\vec{R};z) \tag{1.34}$$

and A is the area of the unit cell. On the other hand, S_{00} is still $\exp(2i\delta(E_0))$, because the specular wave function is unchanged to first order.

This is the distorted wave Born approximation (DWBA), so called because the matrix elements are not taken between plane waves, but between eigenstates of the "large" part of the potential. It will be rederived in lecture 2 from a more formal point of view. The DWBA is not useful, as such, to compute the specular intensity; however, it can be combined with the exact unitarity condition (1.9) to give

$$|S_{00}|^2 = 1 - \sum_{F \neq 0} |S_{F0}|^2 \qquad (1.35)$$

where then (1.33) is inserted. This shows that the change of specular intensity is of the second order in the matrix elements of V; a second order DWBA formula gives, of course, a result consistent with (1.35) (see (3.9)).

The computation of the matrix elements in (1.33) is numerically trivial, once χ is obtained by the methods of Sect. 1.2. Simple general formulas can be given when $V(\vec{R},z)$ is of the general form $V(z-\zeta(\vec{R}))$ and one can further expand to first order

$$V(\vec{R},z) = V_0(z) - V_0'(z)\zeta(\vec{R}) \quad . \qquad (1.36)$$

Then the integral in (1.33) is just the product of the matrix element

$$\int dz \chi_s(z,E_F)\, V_0'(z)\, \chi_s(z,E_0) \qquad (1.37)$$

and the Fourier component ζ_{F-0} of the corrugation $\zeta(\vec{R})$.

When $V_0(z)$ is a step of height V_0, $V_0'(z)$ is $-V_0\delta(z)$ and (1.37) gives $(2\hbar^2\alpha_F\alpha_0/m)$, independent of V_0 (for $E_F < V_0$, $E_0 > V_0$). It is then exactly the same result as for a hard corrugated surface, which corresponds to $V_0 \to \infty$. The S matrix depends on V_0 only through the factors $\exp(i\delta)$, and the intensities are

$$|S_{F0}|^2 = 4\alpha_F\alpha_0|\zeta_{F-0}|^2 \qquad (1.38)$$

(incidentally, it is essential to keep V_0 finite until the final step, where it cancels, or incorrect results are obtained).

The result (1.38) actually applies to the scattering from any potential of the form (1.34), in the limit as $\alpha_F \to \alpha_0$.

A proof of this result is instructive. To simplify the notation, let χ_F and χ_0 be the eigenfunctions of $V_0(z)$. Take the derivative of Schrödinger's equation for χ_0, multiply by χ_F and subtract Schrödinger's equation for χ_F, multiplied by $d\chi_0/dz = \chi_0'$. The result is

$$(\chi_F\chi_0''' - \chi_0'\chi_F'') - (2m/\hbar^2)\chi_F V'\chi_0 = (\alpha_0^2-\alpha_F^2)\chi_F\chi_0' \quad . \qquad (1.39a)$$

Now integrate between z_0 and z_f. The first term, after integrations by parts, gives

$$[\chi_F\chi_0'' - \chi_F'\chi_0']_{z_0}^{z_f} \quad . \qquad (1.39b)$$

For $z_f \to \infty$, this vanishes if either χ_F or χ_0 is a bound state, and gives simply $-4\alpha_0^2$, from the upper limit, if both are continuum states. Thus we conclude that the matrix element (1.37) approaches $(2\hbar^2\alpha_0^2/m)$ for $\alpha_F \to \alpha_0$.

It is clear that the DWBA fails when it gives off-specular intensities that add up to $\Sigma > 1$ (and thus a negative specular, $1 - \Sigma$, according to (1.35)). A simple cure for this is to divide all the first order DWBA intensities by the normalization $1 + \Sigma$ (and thus the specular is now $1 - \Sigma/(1 + \Sigma) = 1/(1 + \Sigma)$). GOODMAN, MANSON and collaborators have used instead the CCGM [18] prescription, which for a $V_0(z)$ without bound states is to divide the off-specular intensities by $(1 + \Sigma/4)^2$, with the specular then given by $(1 - \Sigma/4)^2/(1 + \Sigma/4)^2$. Neither prescription includes all the next order corrections in the DWBA series, which are obtained by resubstituting (1.31) in (1.32); in particular, the intensities which are zero in the lowest order DWBA are still zero in the normalized formulae, but can in fact be large. Higher order DWBA formulae are discussed by WEARE et al. [19], and the entire DWB series has actually been summed by ARMAND and MANSON [20] for an exponential repulsive potential, by a procedure which is equivalent to solving (1.30) by successive iteration. Comparison of their exact results with the DWBA shows that the unitárized formulae discussed above are only qualitatively beyond the range of validity of the DWBA. In principle, CCGM should be better than simple normalization, because it is derived from a completely unitary theory which includes all the multiple scattering processes with energy conservation in the intermediate stages; in practice, however, the processes "off the energy shell" give comparable contributions and cannot be safely neglected.

When the scattering is weak and the DWBA applies, deviations from the simple predictions of (1.38) indicate that the potential does not shift rigidly according to (1.36), or that $V_0(z)$ is "soft", or both. The effect of softness can be seen in the well known DWBA result for an exponential, $V_0(z) = \exp(-2z/a)$: the matrix element (1.37) is given by the hard wall result, $2\hbar\alpha_F\alpha_0/m$, times [Ref. 7, p. 180; also Ref. 20]

$$\frac{(p-q)/2}{\sinh((p-q)/2)}\;\frac{(p+q)/2}{\sinh((p+q)/2)}\;\left(\frac{\sinh p\,\sinh q}{pq}\right)^{1/2} \qquad (1.40)$$

where $p = \pi a\alpha_0$, $q = \pi a\alpha_F$. This correction factor is plotted in Fig. 1.1. It is seen that the hard wall does not give a good approximation, except for $|p-q| < 1$ (remember that the result must be squared to obtain the intensity) and that the DWBA works better when the potential is soft; however the stronger beams usually correspond to small values of $|p-q|$, where the hard

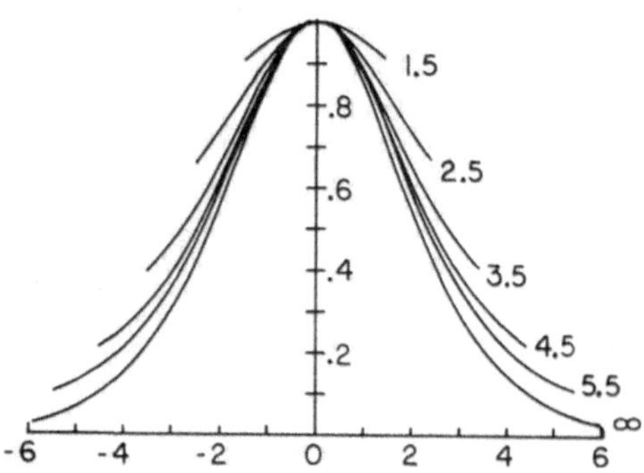

Fig.1.1 Graph of the correction factor (1.40) as a function of $x = p - q$ for several different values of $p + q$. The values of $p + q$ are shown on the right hand side of the plot.

wall result is better. Examination of a typical atom-surface potential (Fig. 4.2) shows that only the short-range part varies as a function of $\vec{R}$. A simple prescription, which has been used with some success, is to say that the repulsive part of $V_0(z)$, to the left of the minimum at z_m, shifts rigidly, while the attractive part, to the right of z_m, is constant and simply accelerates the atom, increasing its perpendicular energy by the well depth D. If the repulsive part is effectively a hard wall, one finds that the DWBA result is still (1.38) where now α_0 and α_F are measured with respect to the bottom of the well, i.e.

$$\alpha_0^2 = k_{0z}^2 + 2mD/\hbar^2 \quad . \tag{1.41}$$

One can again show, by the argument leading to (1.39), that this result is valid in general when the collision is specular; deviations from this result are still described, qualitatively, by the plots of Fig. 1.1.

Different separations of V_0 into repulsive and attractive parts give quite different results, however. For instance, for the Morse potential

$$V_0(z) = D(e^{-2z/a} - 2e^{-z/a}), \tag{1.42}$$

it is natural to take $De^{-2z/a}$ as the repulsive short range part, and shift it rigidly with corrugation $\zeta(\vec{R})$. We can consider the general potential

$$V_0(z) + F(z)\, \zeta(\vec{R}) \tag{1.43}$$

and compare, when V_0 is Morse, the three choices: (i) $F = -V_0'$ for all z, (ii) $F = -V_0'$ for $z < z_m$ only, (iii) $F = (2D/a)\exp(-2z/a)$. It turns out that for (i) the results are similar to those of the simple exponential (1.40), that (ii) is well described by replacing α with the value (1.41), but that (iii) gives considerably larger intensities than (ii). We will come back to these considerations in Lecture 5 when discussing the "Beeby correction" to the Debye exponent.

Lecture 2

In this lecture are first summarized some results of scattering theory [9]. The aim of the presentation is to reduce everything to algebra. Then the formal methods are applied to surface problems.

2.1 Formal Scattering Theory

Let E be the total energy, H_0 the kinetic energy of the incident molecule plus the internal Hamiltonians of the solid and of the molecule; let $H = H_0 + V$, where V is the molecule-solid interaction. Define ϕ and ψ by the Schrödinger equations

$$H_0\phi = E\phi \quad , \tag{2.1}$$
$$H\psi = E\psi \quad . \tag{2.2}$$

Let G_0 and G be the Green's functions defined by the equations

$$(E-H_0)G_0 = 1 = G_0\,(E-H_0) \quad , \tag{2.3}$$
$$(E-H)G = 1 = G(E-H) \quad . \tag{2.4}$$

The solution of these equations depends also on the boundary conditions. Let a superscript + denote the solution corresponding to outgoing wave boundary conditions, which were discussed in Lecture 1; let - denote the solution corresponding to incoming wave boundary conditions. Unless otherwise noted, a superscript + will be understood. Formally,

$$G^{\pm} = (E-H \pm i\eta)^{-1} \quad . \tag{2.5}$$

It is convenient to convert the Schrödinger equation into an integral equation that already incorporates the boundary conditions. This is the Lippmann-Schwinger equation:

$$\psi^+ = \phi + G_0^+ V \psi^+ \quad , \tag{2.6}$$

$$\psi^- = \phi + G_0^- V \psi^- \quad . \tag{2.7}$$

(Multiply both sides of (2.6), or (2.7), by (E-K) and use (2.3) and (2.1) to obtain (2.2).)

Let T stand for the T matrix, defined by $T\phi = V\psi$, or more explicitly,

$$T_{fi} = \langle \phi_f | V | \psi_i^+ \rangle \quad , \tag{2.8}$$

where f and i stand for final and initial.

The basic equation for T follows from (2.6) and (2.8):

$$T = V + V\, G_0\, T \quad . \tag{2.9}$$

In the standard time-dependent version of scattering theory, it is shown that the transition rates are related to the square moduli of the T-matrix elements; formulae for the cross section, or for the reflection coefficients, are then easily derived. It is also possible to obtain such formulae by the procedure of Sect. 1.4. A more formal procedure is followed here, by explicitly introducing the S matrix

$$S_{fi} = \langle \psi_f^- | \psi_i^+ \rangle \quad . \tag{2.10}$$

Far from the surface, ψ_i^+ consists of the incoming wave and all the outgoing waves, while ψ_f^- consists of a single outgoing wave, and all the incoming waves needed to make it up. Far from the surface, the general expansion

$$\psi_i^+ = \Sigma_f\, S_{fi}\, \psi_f^- \tag{2.11}$$

becomes an expansion of ψ_i^+ in free particle states; and, by the properties of ψ_f^-, the coefficient of the outgoing wave ϕ_f is just S_{fi}. All we need then, for the reflection coefficients, is a relation between S and T.

For this purpose, and for other formal developments, we list many equivalent relations involving G_0, G, V, and T. If we multiply both sides of (2.4) on the left by G_0 in the first equality, and on the right in the second equality, we get

$$G = G_0 + G_0 V G = G_0 + G V G_0 \quad . \tag{2.12}$$

If we multiply (2.9) on the left by G_0, multiply (2.12) on the right by V, and compare results, we see that G_0T and GV obey the same equation, so

$$GV = G_0 T \tag{2.13}$$

and similarly we find

$$VG = TG_0 \ . \tag{2.14}$$

If we substitute (2.13) into (2.9), we see that

$$T = V + VGV \ . \tag{2.15}$$

If we substitute (2.14) into (2.15) and (2.12), we get respectively

$$T = V + TG_0V \quad , \tag{2.16}$$

$$G = G_0 + G_0TG_0 \ . \tag{2.17}$$

We can write a formal solution of the Lippmann-Schwinger equation by using the definition $V\psi = T\phi$, and also (2.13):

$$\psi = \phi + G_0T\phi = \phi + GV\phi \ . \tag{2.18}$$

This shows that the operator $1 + G_0T$, or $1 + GV$, converts ϕ into ψ.

From all these formal relations we can see that knowledge of G, or T, or ψ gives all the necessary information. All the equations are of the same type and lead to a geometric expansion, e.g.

$$T = V + VG_0V + VG_0VG_0V + \dots \ . \tag{2.19}$$

Further relations follow from the orthogonality and completeness of the sets $\{\psi^+\}$ and $\{\psi^-\}$ for $E > 0$. From

$$\delta_{fi} = \langle\psi_f^+|\psi_i^+\rangle = \sum_k \langle\psi_f^+|\psi_k^-\rangle \langle\psi_k^-|\psi_i^+\rangle = \sum_k S_{fk}^* \, S_{ki} \tag{2.20}$$

follows the unitarity of the S matrix. From (2.18) and (2.5):

$$G^- - G^+ = 2\pi i\delta(E-H) \ , \tag{2.21}$$

$$\psi_f^- - \psi_f^+ = 2\pi i\delta(E_f-H)V\phi_f \ . \tag{2.22}$$

Using (2.22) in (2.10), we find the desired relation between S and T:

$$S_{fi} = \langle\psi_f^+|\psi_i^+\rangle - 2\pi i\langle\phi_f|V\delta(E_f-H)\psi_i^+\rangle = \delta_{fi} - 2\pi i\delta(E_f-E_i)T_{fi} \ . \tag{2.23}$$

It is seen that the S matrix connects only states that are "on the energy shell" and is quite singular. With energy conservation understood, S gives the distribution of amplitudes between the scattered beams. Formally, one of the summations over quantum numbers of the final state, f, is carried out explicitly, with the result that the energy delta function is replaced by a density of states, $N(E_f)$. In the case of surface scattering, the summation over perpendicular momentum is carried out, leaving only a summation over parallel momentum, as in (1.2), and over the internal quantum numbers of the solid (phonons) and of the molecule (rotations and vibrations), if any.

The amplitudes of the reflected "on shell" beams are then given by

$$A_F = -2\pi i\ N(E_f)\ T_{fi} \ , \tag{2.24}$$

$$N(E) = (dE/dk_z)^{-1} = m/h^2|k_z| \ . \tag{2.25}$$

In Lecture 1 we introduced

$$S_{F0} = -(k_{Fz}/k_{0z})^{1/2}A_F = -(N(E_i)/N(E_f))^{1/2}A_F \tag{2.26}$$

which was also called an element of the S matrix. The actual relation to the formal S matrix used here is:

$$(N(E_f)N(E_i))^{1/2}S_{fi} = -\delta(E_f-E_i)\ S_{F0} \tag{2.27}$$

(the change of sign is to make $S_{00} = 1$ for a flat hard surface.)

The replacement of the energy delta function with a geometric average of the densities of states is just what is needed to convert the unitarity condition (2.20), where the intermediate sum is over all quantum numbers, to (1.10), where the sum is only on the energy shell. It should also be noted that, in surface scattering, the initial states always have negative k_z and the final states have positive k_z.

2.2 Partial Processes

One use of all the formal machinery is to derive in a purely algebraic way some not so obvious relations. One assumes that the entire scattering process can be subdivided into simpler parts, and expresses then the overall S matrix, or T matrix, in terms of those for the partial processes. At the basis of these manipulations lies the geometric kind of expansions such as (2.19).

The first, but not the simplest, derivation is that of the two potential formula already introduced in (1.33). The potential is written as $V = V_0 + v$, the eigenfunctions of V_0 are denoted by χ, the corresponding Green's function by $G(V_0)$. When the Hamiltonian is broken up as $(H_0 + V_0) + v$, there follows an equation entirely analogous to (2.18) (and already encountered in (1.30)):

$$\psi = \chi + G(V_0)\ v\psi \ . \tag{2.28}$$

Then we form $V\psi = V_0\psi + v\psi$, rewrite $V_0\psi$ by use of (2.28) and obtain

$$V\psi = V_0\chi + (1 + V_0G(V_0))\ v\psi \ . \tag{2.29}$$

Now we define new matrices, T_0 and t, by $T_0\phi = V_0\chi$ and $t\chi = v\psi$, by analogy with (2.16). T_0 solves the problem of scattering from V_0 and is presumably easy to determine, while t may be treated by perturbation theory; from (2.28)

$$t = v + v\ G(V_0)t \ . \tag{2.30}$$

Then (2.29) is rewritten as $T\phi = T_0\phi + (1 + V_0G(V_0))t\chi$. Insert $\chi = (1 + G(V_0)V_0)\phi$, which is the analog of (2.9) and find

$$T = T_0 + (1 + V_0G(V_0))t(1 + G(V_0)V_0) \ . \tag{2.31}$$

This is the two potential formula. It will be recognized that $(1 + G(V_0)V_0)$ is the operator that transforms ϕ into χ^+, and that $(1 + V_0G(V_0))$ $(1 + G(V_0)V_0)$ is nothing but the S matrix for the potential V_0. In the application already given in Sect. 1.4, where V_0 is the laterally averaged potential, it is convenient to take matrix elements between standing wave functions χ_S; then the factor $(1 + G(V_0)V_0)$ in (2.31) is responsible for the $\exp(i\delta(E_0))$ that appears in (1.33) and expresses the difference between χ^+ and χ_S for the initial state. Similarly $1 + V_0G(V_0) = (1 + G^-(V_0)V_0)^\dagger$ transforms $(\chi_S)_f$ into $(\chi_f^-)^*$.

Next, suppose instead that $G(V_0)$ can be conveniently divided into two additive parts, $G(V_0) = G_1 + G_2$. Then (2.9) is equivalent to the two equations

$$\tau = v + v\, G_1\, \tau \quad , \tag{2.32}$$

$$t = \tau + \tau\, G_2\, t \quad . \tag{2.33}$$

To show this, substitute (2.32) in (2.33) and write the result as $t = v + v\, G_2 t + v\, G_1\, (\tau + \tau\, G_2\, t)$. The term in parenthesis is again t, by (2.33), and (2.30) follows. Analogous equations can be obtained for the matrix T, because the structure of (2.30) is the same as that of (2.9). A natural way to divide $G(V_0)$, for instance, is to separate away the sum over bound states, or over some bound states, in (1.19). This will be discussed in detail in lecture 4.

Finally, we write explicitly the relation between the matrix S, or the amplitude A, and the matrix t, which follows from (2.24) and (2.31). If V_0 gives only specular scattering, we have

$$S_{F0} = \delta_{F0}\, e^{2i\delta(E_0)} + 2\pi i (N(E_i)N(E_f))^{1/2}\, (\chi_f^-|t|\chi_i^+) \quad . \tag{2.34}$$

In practice, it is more convenient to work with matrix elements of t between stationary states, χ_S. The formula is then rewritten once more as (4.18). One should also keep in mind that the states χ^+ (or χ^-) do not always form a complete set of states, but must be supplemented by the bound states of V_0, if any. Nevertheless, they can be used as complete sets for expansions that involve only states of positive energy, such as (2.11).

Lecture 3

3.1 Semiclassical Methods [8]

Semiclassical methods work amazingly well as an approximation to quantum theory. In principle, the condition for the validity of the semiclassical ideas is that $|\lambda V'/V| << 1$, where V' is a typical derivative of the potential; and even then the semiclassical approximation (SCA) is supposed to be only asymptotic, i.e., it gets better as λ gets smaller, but it is not the first term of a convergent set of better approximations. In practice, however, there are well known cases where the SCA produces exact results, as for instance in predicting the bound states of the harmonic oscillator by the Bohr formula, or in all sorts of properties of the coulomb potential.

Much of optics, of course, is handled semiclassically. By suitable modifications of the theory it is not difficult to treat semiclassically potentials that vary quite rapidly on the scale of the wavelength, for instance a hard wall. There are other limitations to the use of the SCA, however, which will now be discussed. [21-24]

One can derive semiclassical formulas from Schrödinger's equation, but it is better conceptually to view them in the context of the Dirac-Feynman formulation, that is, to find the amplitude of a given process, sum $\exp(iS_{cl}/\hbar)$ over all the possible paths leading to that process. Here S_{cl} in the classical action, $\int p dx$ in the simplest case, and the path is completely arbitrary between the given initial and final conditions. In SCA, the sum is evaluated by the method of stationary phase, and the dominant result comes from the paths which extremize S_{cl}, which are the classical paths. As always in the method of stationary phase, the exponent is expanded to second order around an extremal point, x_0, according to $S_{cl} = S_0 + \frac{1}{2}(\partial^2 S_0/\partial x^2)_0(x-x_0)^2$. One is then left with a Gaussian integral, which gives a result proportional to

$$|\partial^2 S_{cl}/\partial x^2|^{-1/2} \quad . \tag{3.1}$$

In the multidimensional case, the determinant of second derivatives appears in (3.1). [21]

The approximation scheme described above is known as primitive semiclassical (PSC). It is beset by three types of difficulties, somewhat related to each other: (i) a stationary point may occur for a complex value of x, (ii) the second derivative in (3.1) may vanish, or become very small, and (iii) there may be many classical paths, and they may become so entangled that the whole scheme is hopeless.

Difficulties of type (i) can be handled if one is just willing to contemplate complex paths, which correspond to motion in classically forbidden regions. Often, more direct methods yield the correct answer, for instance, the WKB approximation to the one-dimensional Schrödinger equation gives correctly the behavior beyond the turning point, and is adequate to describe tunnelling.

Difficulties of type (ii) are handled in most cases by keeping the third (and sometimes fourth) derivatives of S_{cl}. A typical case, related to (i) occurs when a minimum and a maximum coalesce, then move off into the complex plane. The classical action is then approximated by a cubic, which is the simplest function displaying this type of behavior; the resulting integral is expressible in terms of Airy functions, which we already met in (1.14). The behavior at a classical turning point z_t is a well known example: on one side of z_t there are two classical paths, going left and right, on the other side of z_t there is none. The rainbow, to be discussed below, is another example, quite important to our subject. Ways have been devised to join smoothly the Airy solution to the PSC solution, which is valid away from the turning point (or the rainbow). The result is the uniform semiclassical approximation (USC), which represents a considerable improvement on the PSC [22]. The coalescence of three or four stationary points is also of interest, because it occurs in the "cusps" of surface rainbows [23]. Finally, many or all derivatives may disappear, but then usually the problem simplifies in other ways. This occurs, for example, when a corrugated surface becomes flat.

Difficulties of type (iii) are harder to handle, and in practice limit the effectiveness of all SCA methods. In the simplest case, the number of contributing classical paths is large, and the method simply becomes too expensive. Worse, the ultimate fate of a classical trajectory may depend critically on the initial values of the impact parameter. This actually occurs in atom-surface scattering for trajectories where the impinging particle bounces back and forth several times between the attractive and repulsive parts of the surface potential, or anyhow makes hard-core collisions with the surface atoms more than once. It also happens when the particle has an energy close to the top of a tunnelling barrier, in which case one has a typical bifurcation: the rest of the path depends critically on whether the particle makes it through the barrier. One can picture the flow of a stream of classical particles: if the flow lines do not become tangled, semiclassical methods are effective. In the worst case, however, the flow becomes turbulent, due both to geometrical tangling and to bifurcations. Perhaps recent advances in the theory of turbulence will help with this problem, but at present the practical alternative is to cut up space into strips, in each of which the classical flow is orderly, and match amplitudes at the boundaries between strips. Gordon's method for the Schrödinger equation [14] is a procedure of this type: the Airy functions represent the USC solution in each narrow strip (see (1.14)). More generally, one can separate the overall process into partial processes by the formal methods of Lecture 2, or equivalently by matching procedures such as those leading to (1.26) and (1.28). Semiclassical methods can then be applied to some partial process.

The failures and limitations of semiclassical methods are interesting in themselves, as they provide physical insight into the scattering process. In practice, it is often true that some behavior can be described semiclassically (typically, at near normal incidence and reflection) [24]; valuable inferences can then be made about the surface potential, and full-scale quantum calculations can later be done for the potential so constructed. In this way of proceeding, and in the application of the SCA to compute partial amplitudes, one seldom needs to go beyond the simple PSC formulae (1.22)-(1.27) and the Airy function formulas (1.14), except for the handling of the hard wall problem, which is taken up next.

3.2 Scattering from a Hard Corrugated Surface (HCS)

The HCS simply represents the locus of classical turning points in the potential. It is better to discuss first the exact theory of scattering from a HCS, and then to develop specific semiclassical approximations. The surface will be described by $z = \zeta(\vec{R})$, with $\zeta_{max} > \zeta(\vec{R}) > \zeta_{min}$.

The simplest method of solution goes back to Lord Rayleigh but is still imperfectly understood [25]. One simply takes the asymptotic form of the solution, (1.6) and (1.7) and continues it all the way to the surface, thus

$$\psi = e^{i\vec{K}\cdot\vec{R}} \left(e^{-i\alpha_0 z} + \sum_G A_G \, e^{i\alpha_G z} \, e^{i\vec{G}\cdot\vec{R}} \right) . \qquad (3.2)$$

One then sets $\psi = 0$ on the surface and determines the coefficients A_G. Because $V = 0$ for $z > \zeta(\vec{R})$, each term in (3.2) is a solution of Schrödinger's equation, and the series in (3.2) certainly converges for $z > \zeta_{max}$, because it is just a Fourier expansion of $\psi(z,\vec{R})$ at fixed z. For $\zeta_{max} > z > \zeta_{min}$, (3.2) represents the solution, if the series converges, only where $z > \zeta(\vec{R})$;

for $z < \zeta(\vec{R})$ the true solution vanishes and (3.2) does not, although it is zero at $z = \zeta(\vec{R})$. The convergence of the series depends on the behavior for large $G(\equiv |\vec{G}|)$, where α_G is simply replaced by iG; it is clear then that the convergence becomes poorer as z becomes more negative, and in fact the lack of convergence for a particular $(\vec{R}_c, z_c)$ implies lack of convergence for all $\vec{R}$ at $z < z_c$. The question of convergence is thus well understood, and explicit criteria can be worked out for simple profiles: for the one-dimensional corrugation $D = h \cos(G_0 x)$, the series converges for $G_0 h \lesssim .448$; for $D = \frac{1}{2} h \{\cos(G_0 x) + \cos(G_0 y)\}$, it is expected to converge for $G_0 h \lesssim .592$ [25]. The convergence criterion, however, is simply a statement about the limit of $A_G \exp(G\zeta_{min})$ for $G \to \infty$: the series does not converge if the limit is not zero. In reality, we are interested in finding the coefficient A_F for rather small F using a truncation of the series after N terms, which is in fact a good solution of Schrödinger's equation. The true question is then one of finality of the coefficients A_F, i.e., do they approach a limit as $N \to \infty$, in principle and in practice? (The method may work in principle but not in practice because of numerical problems due to ill-conditioned matrices). It is found in practice that finality is more difficult to achieve as the incident wavelength λ decreases, while the convergence criterion is independent of λ. This can be understood physically as follows. For large λ we have the equivalent of an electrostatic problem, where the incident field is canceled by an accumulation of charge on the tops of the corrugation profile and does not penetrate into deep grooves; it does not really matter whether the Rayleigh assumption is valid in the grooves. For small λ, on the other hand, the waves penetrate in the grooves and incorrect results can be expected if the Rayleigh assumption fails; but then one can use the semiclassical eikonal formula (see Sect. 3.3), which is derived from the Rayleigh equations, as long as multiple hits and shadowing are unimportant; thus the Rayleigh method is applicable once again, in an asymptotic sense.

In practice, there are at least three ways to determine the first N amplitudes A_G [26]. One can directly set $\psi = 0$ at N points on the surface in (3.2); or one can first take Fourier transforms, convert $\psi(\vec{R},\zeta(\vec{R})) = 0$ to

$$0 = (e^{-i\alpha_0\zeta})_G + \sum_{G'} (e^{i\alpha_{G'}\zeta})_{G-G'} A_{G'} \quad , \tag{3.3}$$

and solve the first N equations of the system (3.3). Here we have defined, for any q,

$$(e^{iq\zeta})_G = \int \frac{d^2R}{A^2} e^{-i\vec{G}\cdot\vec{R}} e^{iq\zeta(\vec{R})} \quad . \tag{3.4}$$

A variant of this method is to multiply (3.2) by $e^{i\alpha_0\zeta}$ first. Thus

$$0 = \delta_{G0} + \sum_{G'} (e^{iq_{G'}\zeta})_{G-G'} A_{G'} \tag{3.5}$$

where $q_{G'} = \alpha_0 + \alpha_{G'}$.

For numerical purposes, the Fourier space (GG') methods (3.3) and (3.5) are inferior to the mixed space (GR) method, unless the Fourier coefficients (3.4) are known explicitly, as in the case of a sawtooth or a sinusoidal profile. In either case, the limiting step is the inversion of an N x N matrix, as in the close-coupling calculations [26].

A third method is to regard ζ as a perturbation and solve for A_G to high order of perturbation theory. This can be done systematically with surprising ease, starting from (3.3) or (3.5). Expand

$$(e^{iq\zeta})_G = \delta_{G0} + iq\zeta_G - \frac{1}{2} q^2 \zeta_G^2 - \frac{1}{6} q^3 \zeta_G^3 + \dots \quad , \tag{3.6}$$

$$A_G = A_G^{(o)} + A_G^{(1)} + \frac{1}{2} A_G^{(2)} + \frac{1}{6} A_G^{(3)} + \dots \tag{3.7}$$

where, formally, $A_G^{(n)}$ is of order ζ^n. Insert in (3.3), for instance, and obtain through second order

$$\delta_{G0} + A_0^{(o)} = 0 \quad , \tag{3.8a}$$

$$-\alpha_o\zeta_G + \sum_{G'} \{i\alpha_{G'} \zeta_{G-G'} A_{G'}^{(o)}\} + A_G^{(1)} = 0 \quad , \tag{3.8b}$$

$$-\alpha_o^2 (\zeta^2)_G + \sum_{G'} \{-\alpha_{G'}^2(\zeta^2)_{G-G'} A_{G'}^{(o)} - 2i\alpha_{G'}\zeta_{G-G'} A_{G'}^{(1)}\} + A_G^{(2)} = 0 . \tag{3.8c}$$

To any order n, $A_G^{(n)}$ is given recursively in terms of lower order coefficients, and is formally easy to find, as pointed out by LOPEZ, YNDURAIN and GARCIA [27]. Unfortunately, the recursive relation is numerically unstable: the explicit expression of a high order $A_G^{(n)}$ in powers of ζ involves many terms with large coefficients and alternating signs, which compensate each other, leaving a small result. Nevertheless, excellent results for A_G can be obtained, for corrugations well beyond the limit of validity of the Rayleigh expansion, by judiciously stopping at a high enough order (n = 20, say). It is an open question whether the formal expansion (3.7) is convergent, or only asymptotic. This question is not directly tied, it seems, to the validity of the Rayleigh method, which simply provides the easiest way to proceed; the formal expansion of A_G in powers of ζ is unique and can be derived in other ways.

The explicit expressions to second order are, from (3.8),

$$A_G^{(1)} = 2i\alpha_o \zeta_G \quad , \tag{3.9a}$$

$$A_G^{(2)} = 4\alpha_o \sum_{G'} \alpha_{G'} \zeta_{G-G'} \zeta_{G'} \quad . \tag{3.9b}$$

The S matrix elements are now given by (1.8) and are seen to be symmetrical in α_o and α_G, as they should be. It is not accidental that (3.9a) leads to the same result (1.35) of the first order DWBA, and that (3.9b) for G = 0 is consistent with (a) and with the unitarity condition (1.9). In fact, one can check that the LYG expansion [27] is consistent with the unitary conditions (1.10) through fourth order (and presumably to all orders, but a proof is lacking).

For large corrugations, one must go to the exact Lippmann-Schwinger equation (2.7). Because the HCS potential is singular, one must first write the equation for a finite step height V_0, then let $V_0 \to \infty$. Actually, the finite V_0 case is not much harder than the HCS, and can also be handled by the T-matrix equations (2.8) or (2.27), which become singular for $V_0 \to \infty$. [20] Detailed studies of the effect of a finite V_0, which have been carried out, show that there is little to be gained by keeping this additional parameter in the model potential.

Equivalently, the HCS problem can be viewed as a boundary value problem on the free-space Schrödinger equation. One then starts from Green's theorem, which is familiar in electrostatics, where the potential ψ satisfies $\nabla^2\psi = 0$; exactly the same result holds in our case, and in optics, where it is also known as Huygens' principle or Kirchhoff's integral. [10,11] In fact, the result holds not only for the free-space wave equation $(\nabla^2 + k^2)\psi = 0$, but also when an arbitrary potential V is present (one must know the Green's function of V, however). The theorem is usually stated for a closed surface S which in our case can be taken as a flat box built on the surface of the solid. Green's theorem is that, for points $\vec{r}$ inside the volume bounded by S,

$$\psi(\vec{r}) = \int dS \, [G(\vec{r}-\vec{r}\,') \frac{\partial\psi}{\partial n'} - \frac{\partial G}{\partial n'} (\vec{r}-\vec{r}\,') \, \psi \, (\vec{r}\,')] \tag{3.10}$$

where n is the outward normal direction to S and G is the free-space Green's function, $G(\vec{r}) = e^{ikr}/4\pi r$. The integral over the sides of the box can be neglected, the integral over the flat top just gives the incident wave (the nature of Kirchhoff's integral is such that, when taken over a plane, it picks up only the waves moving in the inward normal direction); the remaining integral is over the surface of the solid, where however $\psi = 0$. Therefore (3.10) gives

$$\psi(\vec{r}) = e^{i\vec{k}\cdot\vec{r}} + \int dS \; G(\vec{r}-\vec{r}\,') \frac{\partial\psi}{\partial n'} (\vec{r}\,') \quad . \tag{3.11}$$

If $\vec{r}$ is also taken to be on the surface, the left side vanishes and (3.11) gives an integral equation for the "source function" $\partial\psi/\partial n$; this is then reinserted into (3.11) to obtain the reflected field. For a periodic surface, the integral in (3.11) can be reduced to a unit cell using the Bloch property of ψ and $\partial\psi/\partial n$. Explicitly, define the periodic source function $f_K(\vec{r})$ by $(\partial\psi/\partial n)dS = \exp(i\vec{K}\cdot\vec{R}) \, f_K \, d^2R$. Then the integral equation becomes

$$0 = e^{-i\alpha_0\zeta(\vec{R})} + \int d^2R' \; G_K(\vec{r}-\vec{r}\,') \; f_K(\vec{R}\,') \tag{3.12}$$

where $\vec{r}\,' = (\vec{R},\zeta(\vec{R}\,'))$, $\vec{r} = (\vec{R},\zeta(\vec{R}))$, the integral is over a unit cell, and G_K is a sum over cells,

$$G_K(\vec{r}) = \sum_{\ell} e^{-i\vec{K}\cdot(\vec{R}-\vec{R}_\ell)} \; G(\vec{r}-\vec{R}_\ell) = i \sum_{G} e^{i\vec{G}\cdot\vec{R}} \, e^{i\alpha_G|z|}/\alpha_G \quad . \tag{3.13}$$

The evaluation of (3.13) involves additional expense, compared to the Rayleigh method; otherwise one is back to a matrix equation, this time all in

$\vec{R}$ space, when (3.12) is evaluated on a grid. [28] In fact, the left hand side of (3.12) must hold for all $z \leq \zeta(\vec{R})$, and not just on the surface; this result is known in optics as the extinction theorem. The equation greatly simplifies if z is taken to be a constant, and smaller than the minimum value of $\zeta(\vec{R})$. Then in the expression for $G_K(\vec{r}-\vec{r}\,')$, as given by (3.13), we can replace $|z-\zeta(\vec{R}\,')|$ by $\zeta(R')-z$, and instead of (3.12) we have

$$0 = \alpha_0 + \int d^2R' \;\; e^{i\alpha_G \zeta(R')} \;\; e^{-i\vec{G}\cdot\vec{R}\,'} \;\; f_K(R') \quad . \tag{3.12'}$$

This is known as the MMM equation in atom-surface scattering. [29] It is obvious that the true f_K is a solution of (3.14), and there is proof that the solution is unique. [30] In practice, however, it is found that this method suffers of numerical limitations similar to those of the Rayleigh method. [31]

3.3 Eikonal and Kirchoff Approximations

These semiclassical formulae for the HCS were first applied to molecular scattering by LEVI and collaborators [32]. They are one of the most useful approximate results, replacing the DWBA for quick data analysis when the effective corrugation is not small. There actually are several formulae of the type

$$A_G = -P\,(e^{-iq\zeta})_G = -P\int \frac{d^2R}{A}\, e^{-i\vec{G}\cdot\vec{R}\,-iq\zeta(\vec{R})} \tag{3.14}$$

for different values of the prefactor P and the momentum transfer q. They work best, often very well, for beams close to the specular, i.e. when

$$|\alpha_G - \alpha_0| \ll \alpha_0 \quad . \tag{3.15}$$

The simplest formula has $P = 1$ and $q = 2\alpha_0$, and is seen to be a solution of (3.5), when $\alpha_{G'}$ is approximated by α_0, consistently with (3.15). A reasonable improvement, still leaving $P = 1$, is to take $q = q_G$, the actual perpendicular momentum transfer in the process under consideration. The exponent is now in agreement with the general prescription of semiclassical theory, and is known in optics as the eikonal. The correct semiclassical formula, known as the Kirchhoff approximation in optics, has $q = q_G$ and

$$P = \frac{k^2 - \vec{k}\cdot\vec{k}_G}{\alpha_G^2 + \alpha_0\alpha_G} = \frac{1 + \cos\theta_0\cos\theta_G - \sin\theta_0\sin\theta_G\sin(\phi_0-\phi_G)}{\cos\theta_G(\cos\theta_0 + \cos\theta_G)} \quad . \tag{3.16}$$

This can be derived from (3.11) when the source function $\partial\psi/\partial n$ is replaced by the value appropriate locally to a flat surface, $2\,\partial\psi_i/\partial n$, where ψ_i is the incident wave; explicitly, make the replacement

$$\frac{\partial\psi}{\partial n}\; dS \rightarrow 2i\; \vec{k}\cdot\vec{\nabla}(z-\zeta(\vec{R}))\; e^{i\vec{k}\cdot\vec{r}}\; d^2R \tag{3.17}$$

and carry out an integration by parts [10,11,32].

The prefactor P is always positive and of order unity, approaching one when (3.15) holds. The important effects are described by the eikonal integral. It may help to display first the explicit value for the corrugation

$$\zeta(\vec{R}) = h_1 \cos(2\pi x/a_1) + h_2 \cos(2\pi y/a_2) \quad (3.18)$$

which includes as particular cases a one-dimensional sinusoid ($a_2 = 0$) and the square corrugation ($a_1 = a_2$, $h_1 = h_2$) that gives a good description of the prototype system, He/LiF(001). With $\vec{G} = (2\pi m/a_1, 2\pi n/a_2)$, one finds

$$(e^{-iq\zeta})_G = i^{|m| + |n|} J_{|m|}(qh_1)\, J_{|n|}(qh_2) \quad . \quad (3.19)$$

The Bessel functions peak, roughly, where the index equals the argument; they are oscillatory where the index is smaller, and decay exponentially where the index is larger than the argument.

This behavior is an example of a general feature of scattering from a HCS, the surface rainbow effect. Take as an example a simple sinusoidal corrugation, as shown in Fig. 4.1. For small deflections, $|m| < qh_1$, there are two classical paths contributing to the amplitude; the two paths coalesce for $|m| = qh_1$, when reflection occurs, classically, from the inflection points of the surface profile; and for $|m| > qh_1$ there is no contributing classical path.

Generally, the classical deflection angle is just twice the angle, at the point of incidence, between the surface normal and the z direction. Thus the angular spread of the diffraction pattern (the rainbow angle) is a direct measure of the maximum slope of the surface profile (which occurs at the inflection points). The rainbow may be masked by double reflections or by shadowing of parts of the surface, and in fact under these conditions the eikonal approximation is not valid, but is otherwise one of the most striking effects in surface scattering.

Explicitly, the condition for specular reflection is $(\vec{k}-\vec{k}_F)\cdot\vec{n} = 0$, where the normal $\vec{n}$ is in the direction of $\nabla(z-\zeta(\vec{R}))$, i.e.

$$\alpha_0 + \alpha_F + \vec{G}\cdot\nabla\zeta = 0 \quad . \quad (3.20)$$

For a one-dimensional profile, the inflection points occur where $\partial^2\zeta/\partial x^2 = 0$ (the condition $|m| = qh_1$ follows from this and (3.20) for a sinusoid); for a two-dimensional profile, there are inflection lines given by $H = 0$, where H (the Hessian) is the determinant of the second derivatives of ζ. In the simplest semiclassical approximation (PSC), as well as classically the scattered intensity is proportional to $1/|H|$, and thus peaks at the rainbow. It is clear physically that the reflectivity is peaked where the curvature vanishes. Further details, including the beautiful relation between the topology of the surface and the rainbow pattern, are given by BERRY [23].

Lecture 4

In this lecture we consider elastic atom-surface resonant scattering and selective adsorption.

4.1 Introduction

The scattering of mono-energetic molecular beams from well-characterized crystal surfaces often shows sharp features that can be ascribed to resonant scattering, both in the elastic and in the inelastic intensities. The most extensively studied resonances are those of ^4He with various simple surfaces, although data are available for other light species such as ^3He, H, D, H_2, and D_2. The (001) plane of alkali halides (primarily LiF and NaF) and the basal plane of graphite have been used frequently as targets, due to the ease with which stable, well-defined surfaces can be obtained [1-7].

In resonant scattering, an impinging molecule skims along the surface for times of the order of 10^{-12} seconds and distances of the order of 10^{-7} cm before rescattering outwards. The probability of inelastic collisions is enhanced by the long interaction time, and there is a corresponding decrease in the total elastic intensity, although not necessarily in that of any particular elastic beam. For this reason, resonance phenomena can also be described, at least in part, as "selective adsorption".

As a simple example, consider the processes leading to specular scattering of an atom from a rigid surface with a one-dimensional corrugation $z = \zeta(x)$. The effective surface corrugation is the locus of turning points of the atom's center of mass in the potential generated by the surface. For atoms incident in the xz plane, the specular trajectories also lie in the xz plane. Some of these trajectories are shown in Figure 4.1 below.

All five of the classical trajectories shown in Fig. 4.1 contribute to the specular intensity, which is computed classically by adding the probabilities of the various allowed processes. The one-hit trajectories are bent by acceleration in the long range potential (a typical example of which is shown in Fig. 4.2), but are generally similar to trajectories corresponding

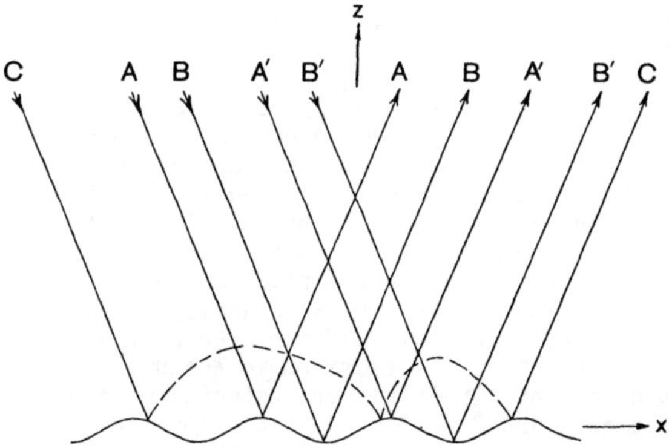

Fig.4.1 Schematic drawing of trajectories that contribute to specular reflection. At resonance, the dashed part of path C corresponds to back and forth motion in a bound state.

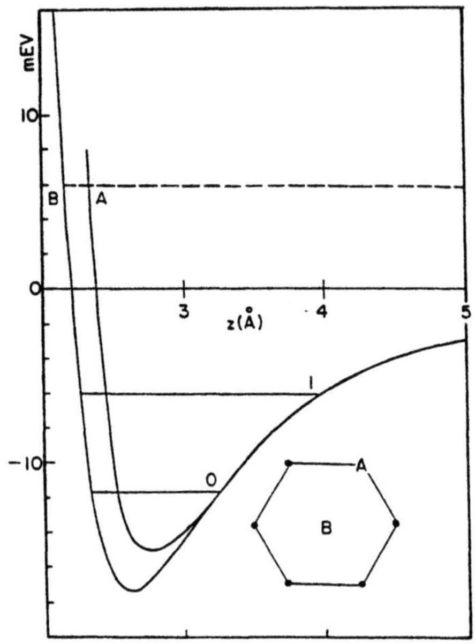

Fig.4.2 Potential energy diagram for the system $^4He/C$. The insert shows the carbon ring; A is the position of an atom, and B is the center of the ring. Curves A and B give the potential energy as a function of the distance, z, above the points A and B. If the dashed line is at the energy corresponding to the incoming perpendicular motion, points A and B mark the turning points for trajectories A and B of Fig. 4.1. The horizontal lines, with labels 0 and 1, are at the energies of the two lowest bound states of $V_0(z)$.

to scattering from a hard corrugated surface. On the other hand, if the particle undergoes many reflections in the surface potential well (path C in Fig. 4.1), the final scattering direction can be an apparently random function of the incident direction.

The quantum mechanical scattering amplitude is found by adding a term $\exp(iS/\hbar)$, where $S = \int \vec{p}\cdot d\vec{x}$ is the classical action, for every conceivable path, even for ones not classically allowed. Evaluating the squared modulus of this amplitude then yields the scattered intensity. The classically permitted trajectories provide the main contribution to the path integral. Quantum interference effects arise from the phase difference, $\Delta S/\hbar$, between different classical paths. In the simple case of an infinite periodic surface, Bragg diffraction results from interference between analogous paths from different unit cells of the surface, such as interference between A and A' in Fig. 4.1. The "quantum rainbow effect" described in lecture 3 is caused by interference between different single-hit trajectories within the same unit cell. The paths labeled A and B in Fig. 4.1 interfere in this fashion. Prominent resonant adsorption features (both peaks and dips), and other fine structure, result from interference with multiple-hit trajectories such as C. Figs. 4.3 and 4.4 show resonant maxima and minima. Many experimental papers contain plots of this type.

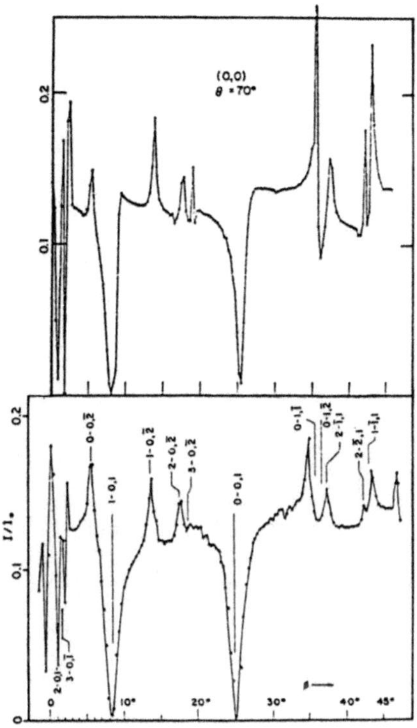

Fig.4.3 Theory (above) and experiment (below) for the azimuthal dependence of the specular intensity for He/LiF. The incident wave vector was 5.76 reciprocal angstroms, and the incident polar angle was 70 degrees. The label 1-0,1, for example, indicates a resonant transition to the first excited state through $\vec{G} = (2\pi/a)(0,1)$. The theoretical model is a hard corrugated wall with an attractive well in front of it. Inelastic effects are not included, except for an overall Debye-Waller scaling of the diffracted intensity.
(Reproduced from Phys. Rev. B19 (1979) 20).

$K_{oy} = k_o \sin\theta \sin\phi$, and $k_{oz} = k_o \cos\theta$. Thus, for any fixed reciprocal surface vector, the surface well acts as a resonant cavity, with eigenfrequencies $\varepsilon_n/\hbar$. The surface corrugation couples the incident atomic wavefunction to the surface well. If the resonances are relatively isolated, it is possible to use (4.1) to find the bound state energies by measuring the diffracted intensities as a function of the incident energy or the incident polar and azimuthal angles. In practice, it is much more convenient to vary the incident angles instead of the incident energy. Various experiments with different isotopes have verified the simple mass dependence

4.2 Kinematics

We assume an infinite periodic surface. Hence, at every intermediate state (for instance, after each hard-wall collision) that atom's parallel momentum is $\hbar(\vec{K} + \vec{G})$, where $\vec{G}$ is a reciprocal vector of the surface. In addition, the scattering amplitude is strongly affected if the frequency of the cyclical

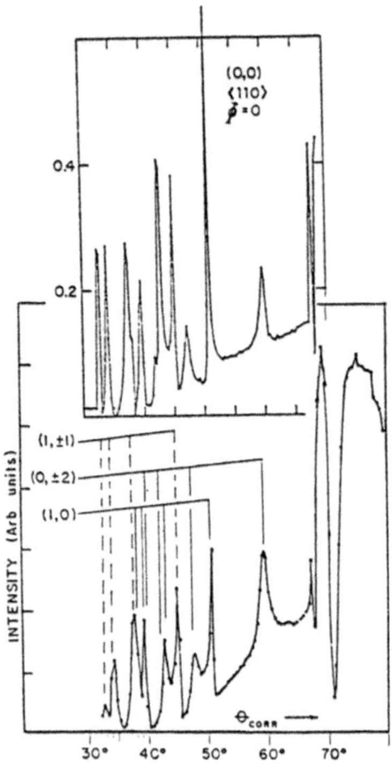

Fig.4.4 Theory (above) and experiment (below) for the polar dependence of the specular intensity for He/LiF, at a fixed azimuthal angle. The theoretical graph is calculated elastically, except for an overall Debye-Waller scaling.
(Reproduced from Phys. Rev. B19 (1979) 20.)

exhibited in (4.1) [33]. In the scattering of light atoms from simple surfaces, the number of bound states is usually between 2 and 6. For example, in the system $^4He/LiF$, there are 4 bound states, with binding energies between 0.21 mev and 5.9 mev.

The energy of a bound state, ε_n, must be between 0 and -D, where D is the depth of the surface potential well. If $\vec{G}$ has components such that $\hbar^2(\vec{K}_0 + \vec{G})^2/(2m)$ is between E_0 and $E_0 + D$, then the associated perpendicular energy will be between 0 and -D, rendering resonant interactions possible. Such reciprocal surface vectors will be said to belong to the class $\vec{N}$. If $\hbar^2(\vec{K}_0 + \vec{G})^2/(2m)$ is less than E_0 (positive perpendicular energy), then $\vec{G}$ is an element of the class $\vec{F}$, consisting of observable vectors. If the perpendicular energy associated with $\vec{G}$ is less than -D, then $\vec{G}$ is an evanescent, class $\vec{E}$, vector. In the quantum regime, k_0^2 is comparable to the squared modulus of one of the smallest reciprocal vectors, and to $(2mD/\hbar^2)$, leading to relatively few vectors belonging to class $\vec{N}$. Hence, resonances will be few and usually well separated. In the classical limit, k_0 is much motion in the surface well is such that the atom's kinetic energy is equal to the binding energy of an adsorbed state. These two statements can be combined in the kinematic equation

$$E_0 = \hbar^2(\vec{K}_0 + \vec{G})^2/(2m) + \varepsilon_n \quad , \tag{4.1}$$

where $E_0 = \hbar^2 k_0^2/(2m)$ is the incident energy, $\hbar\vec{K}_0$ is the component of the incident momentum parallel to the mean surface, m is the mass of the atom, and ε_n is the energy of the adsorbed state. In (4.1), $K_{0x} = k_0 \sin\theta \cos\phi$, larger than the smaller reciprocal vectors, and there are many class $\vec{N}$ vec-

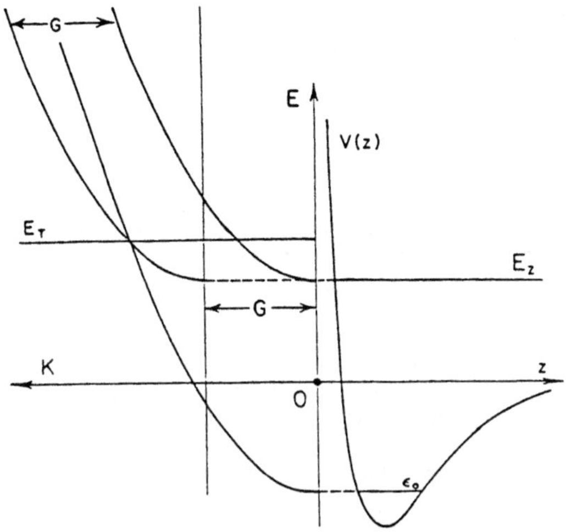

Fig.4.5 Schematic plots of potential energy, perpendicular energy, and total energy. The potential, $V(z)$, is plotted on the right as a function of z, the distance from the surface. The incident perpendicular energy and the energy of the lowest bound state are also shown on the right. The lowest curve on the left is a graph of $\varepsilon_0 + (\hbar k)^2/(2m)$, as a function of the parallel momentum K. The two upper curves on the left are the total incident energy for fixed perpendicular energy and variable K, and the Bragg-diffracted energy $E_z + \hbar^2(\vec{K} - \vec{G})^2/(2m)$. The Bragg-diffracted curve intersects the lower curve at values of the total energy and the parallel momentum that satisfy equation (4.1) for n = 0. The resonant value of the incident energy is indicated on the graph, and is denoted by E_T.

tors, leading to numerous interacting resonances and an apparently random fine structure for the diffracted intensities.

Figure 4.5 shows that the resonant adsorption processes involve transitions to states with positive total energy, and not to permanently trapped states, which have negative total energy. Nevertheless, with the assumption that the lateral motion is nearly free, one can infer the surface-band structure of the negative energy states.

The eigenvalues of the laterally averaged potential $V_0(z)$ are approximately equal to the energies ε_n of the surface adsorbed states. By performing experiments to determine the binding energies of the adsorbed states, it is possible to assign values to any adjustable parameters which appear in an assumed expression for $V_0(z)$, which has a form suggested by atomic and solid-state theory. Far from the surface, the laterally averaged potential is dominated by a $1/z^3$ van der Waals term, but, for reasons of computational ease, some investigators have used other potentials, such as the potential [35]

$$V_0(z) = D((1+(z/a))^{-2\lambda a} - 2(1 + (z/a))^{-\lambda a}) \quad , \tag{4.2}$$

which has three adjustable parameters, namely the well depth D (about 6 to 12 meV), the range a (1 or 2 Å), and λ. This potential approaches the Morse potential for large λ. The qualitatively reasonable potential defined by (4.2) has energy eigenvalues which can be approximated by the expression

$$\varepsilon_n = -D(1 - C(n + 0.5))^r \quad , \tag{4.3}$$

where the adjustable parameters are the well depth D, C, and r, with C and r being explicitly given in terms of a and λ. Essentially, any reasonable long range potential which goes smoothly to zero far from the surface, and has the right well depth and range, can be made to fit the bound state data by introducing an additional adjustable parameter which describes the asymmetry of the well.

In the semiclassical limit, the bound state energies are determined by the Bohr quantization condition, $\phi(\varepsilon_n) = 2n\pi$, where ϕ is the phase accumulated between successive collisions with the hard wall. Much can be learned if ε_n and $\phi(\varepsilon_n)$ are known, even without assuming any explicit expression for the laterally averaged potential. The relation between ϕ and the potential was discussed in Lecture 1.

An incident atom trapped on the surface is influenced by the periodic lateral variation in the surface potential, leading to deviations from Eq. (4.1) [36]. If we neglect transitions out of the bound state, we find a full two-dimensional band structure $E_n(\vec{K})$ for each surface energy eigenvalue. Unlike the simple parabolic form of (4.1), $E_n(\vec{K})$ has energy gaps at the boundaries of the surface Brillouin zones. For a one-dimensional corrugation, these gaps occur whenever 2K is a multiple of the smallest reciprocal vector of the surface. In addition, there are energy gaps within the Brillouin zones, which arise from the mixing of the different bands. Both types of gaps are shown in Fig. 4.6. The band gaps occur when, for some n,j, and $\vec{G}$,

$$\varepsilon_n + \hbar^2K^2/(2m) = \varepsilon_j + h^2(\vec{K} - \vec{G})^2/(2m) \quad . \tag{4.4}$$

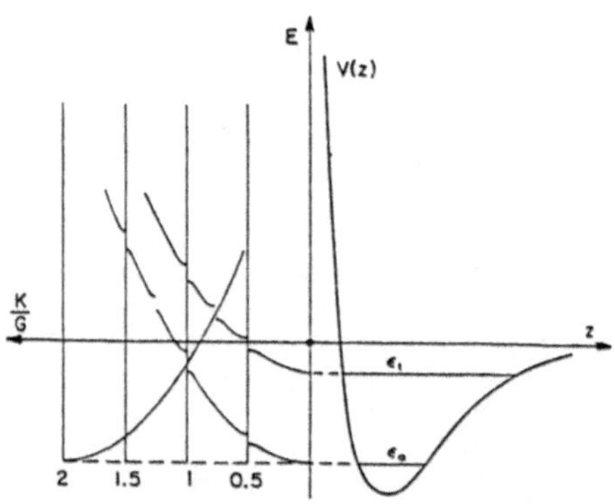

Fig.4.6 Surface energy bands (schematic diagram). The potential, V(z), and the two lowest bound state energies are shown on the right hand side of the graph. On the left side the two lowest bands for the corrugated surface appear. Vertical lines are drawn at the Brillouin zone boundaries. The Bragg diffracted band $\varepsilon_0 + \hbar^2(\vec{K} - 2\vec{G}_0)^2/(2m)$ is also drawn on the left. Additional band gaps arise when this band, or any other Bragg diffracted band, intersects any of the undiffracted bands.

The size of an energy gap is equal to twice the absolute value of the matrix element of the potential V(x,y,z) between the unperturbed eigenstates. An unperturbed eigenfunction will consist of an eigenfunction of the laterally averaged potential multiplied by a plane wave. Hence, knowledge of the band gaps can enable one to determine the lateral variation of the surface potential within the well. For example, investigators used these considerations to ascertain the potential for the ^{4}He/C system, which was shown in Fig. 4.2. [5,31]

Resonant states decay back into continuum states at a rate inversely proportional to the width of the resonances, thus causing a broadening which tends to obscure the gaps; but the gaps are first order in the lateral variation of the potential, while the resonant width is only second order. Hence, the band gaps remain visible.

The simple kinematical considerations described in this section are adequate to determine the locations of the adsorption resonances, but mere kinematical manipulations cannot reveal the strength or width of these resonances. Indeed, nothing mentioned in this chapter would make it possible to know whether a given resonance will be a maximum or a minimum. The close coupling formalism described in section 1.3 permits a full scattering calculation for any specified V(x,y,z), but it does not separate the resonant scattering effects from the "direct" scattering [38]. We adapt the results and techniques of lecture two in order to present a theory which handles explicitly the resonant effects.

4.3 Resonant Scattering Formalism [39,40]

Let v, ψ, χ, and t be defined as they were in Lecture 2. As discussed there, it is convenient to work with matrix elements of t between stationary states χ_s, which are now simply called χ. We then define

$$t_{fi} = (\chi_i|t|\chi_f) = -\exp[-i(\delta(E_i) + \delta(E_f))](\chi_f^-|v|\psi_i^+) \quad . \tag{4.5}$$

Equation (2.30) presents a general relationship involving t, and is here restated as

$$t = v + vG(V_0)t \quad . \tag{4.6}$$

If we let E_j stand for the energy of χ_j, then $G_{ij}(V_0) = \delta_{ij}(E_0-E_j\pm i\eta)^{-1}$, where η is a positive infinitesimal. We can use this expression to write (4.6) in a more explicit fashion

$$t_{fi} = v_{fi} + \sum_j v_{fj}(E_0 - E_j + i\eta)^{-1}t_{ji} \quad , \tag{4.7}$$

where the summation is over intermediate states.

The distored wave Born approximation is, of course, simply $t_{fi} = v_{fi}$. More sophisticated approximations include some of the terms in the summation in (4.7). The intermediate states can be either surface bound states with energy $E_b = \varepsilon_n + \hbar^2(\vec{K} + \vec{G})^2/(2m)$, or continuum states. Resonances happen when E_0 matches E_b for some $\vec{G}$ and some n. Near a resonance, approximations must be made with great care.

Eq. (4.6) is equivalent to (2.32) and (2.33):

$$\tau = v + vG_1\tau \quad , \tag{4.8}$$

$$t = \tau + \tau G_2 t \quad , \tag{4.9}$$

where $G_1 + G_2 = G(V_0)$. We would like to choose G_1 and G_2 in such a way that (4.9) contains most of the essential physics in the problem, and can be solved exactly for a suitable, if approximate, τ. In the present case, G_2 must contain the resonant contributions.

In the simplest nontrivial case, when there is a single surface bound state, we can write (4.9) in the explicit form

$$t_{bi} = \tau_{bi} + \tau_{bb}(E_0 - E_b + i\eta)^{-1} t_{bi} \quad , \tag{4.10}$$

$$t_{fi} = \tau_{fi} + \tau_{fb}(E_0 - E_b + i\eta)^{-1} t_{bi} \tag{4.11}$$

Meanwhile, (4.8) implies that

$$\tau_{bb} = v_{bb} + \sum_{j \neq b} v_{bj}(E_0 - E_j + i\eta)^{-1} \tau_{jb} \ , \tag{4.12a}$$

$$\tau_{bi} = v_{bi} + \sum_{j \neq b} v_{bj}(E_0 - E_j + i\eta)^{-1} \tau_{ji} \quad , \tag{4.12b}$$

with an analogous expression for τ_{fi}.

If we solve (4.10) for t_{bi} and then substitute the result in (4.11), we find that

$$t_{fi} = \tau_{fi} + \tau_{fb}(E_0 - E_b - \tau_{bb})^{-1} \tau_{bi} \quad , \tag{4.13}$$

where the infinitesimal $i\eta$ has been absorbed into the imaginary part of τ_{bb}. The two terms in this equation correspond to resonant and direct scattering, thus exhibiting the resonant effects separately. It is permissible, as a first approximation, to replace τ_{bi}, τ_{fb}, and τ_{fi} by v_{bi}, v_{fb}, and v_{fi}, but the τ_{bb} which appears in the denominator of (4.13) cannot be approximated by v_{bb}, since v_{bb} is zero. Instead, we get an approximate form for τ_{bb} from (4.12), namely

$$\tau_{bb} = \sum_{j \neq b} \delta(E_0 - E_j + i\eta)^{-1} |v_{bj}|^2 \tag{4.14}$$

The real part of τ_{bb} accounts for a shift in the resonant position, whereas the imaginary part of τ_{bb} determines the total transition rate out of the surface bound state. We can define the resonant line width Γ by

$$\Gamma = (-2)\mathrm{Im}(\tau_{bb}) = (2\pi) \sum_{j \neq b} \delta(E_0 - E_j) |\tau_{bj}|^2 \tag{4.15}$$

Eq. (4.15), which is the "optical theorem", is consistent with (4.14) if we let $\tau_{bi} = v_{bi}$ and note that $\mathrm{Im}(E_o - E_j + i\eta)^{-1} = -\pi\delta(E_o - E_j)$.

The formalism represented by Eqs. (4.10) through (4.15) is sufficient to handle an isolated resonance. However, if one is to account for the mixing of resonant states, and the consequences of such mixing, then it becomes necessary to include more than one bound state. This transforms the relatively simple equation (4.10) into a matrix equation. The theory of CABRERA, CELLI, GOODMAN, and MANSON [18] contains in $G(V_o)$ the effects of all the bound states, G_b, plus the imaginary part of the contribution due to the continuum states, G_c. That is, in (4.6), the CCGM theory sets $G(V_o) = G_b + (i)\mathrm{Im}(G_c) = \mathrm{Re}(G_b) + (i)\mathrm{Im}(G(V_o))$. Notice that the CCGM theory handles the imaginary part of the total Green's function in an exact manner, thus insuring that particle flux is conserved (i.e the theory is unitary) and that the optical theorem is valid. On the other hand, in the theory of Wolfe and Weare [41], $G_2 = G_b$ in (4.9), and $G_1 = G_c$ in (4.8). Both of these theories are in agreement with the results presented here for the simple case of an isolated resonance. Moreover, the two theories are essentially equivalent to second order in v [42,43].

An alternative procedure consists of dividing the potential into an attractive part V_a and a repulsive part V_r, instead of dividing the potential into a laterally averaged part $V_o(z)$ and a perturbing potential v. We assume that V_a depends only on z, the distance from the mean surface, while V_r shifts rigidly in the z direction as $\vec{R}$ changes, as shown in Fig. 4.2. The intended procedure consists of solving the scattering problems for V_a and V_r separately, and then matching the two solutions at the bottom of the surface well, which we assume to be located at $z = z_m$. This matching procedure automatically accounts for any complicated series of multiple reflections which might occur in the surface well. In addition, semiclassical approximations can be used in considering the scattering due to V_a and V_r. This theory is an analog of McRae's work on resonances in electron diffraction [44].

In the notation used here, the wave function for points far from the surface will be written in the form

$$\psi = \exp(ik_{oz}z - \vec{K}\cdot\vec{R}) + \sum_G A_G \exp(i(\vec{K}_o + \vec{G})\cdot\vec{R}) \quad , \tag{4.16}$$

where A_G is a scattering amplitude [45]. The normalized diffracted intensity for the transition from momentum $\hbar\vec{k}_o$, where $\vec{k}_o = (\vec{K}_o, k_{oz})$, to the final momentum $\hbar\vec{k}_F$, where $\vec{k}_F = (\vec{K}_o + \vec{F}, k_{Fz})$, is

$$P_F = |S_{FO}|^2 = (k_{Fz}/k_{oz})|A_F|^2 \quad . \tag{4.17}$$

The t-matrix element for this transition is related to A_F by (2.34), rewritten here as

$$-(k_{Fz}/k_{oz})^{1/2}A_F = [\delta_{FO} - (2\pi im)t_{fi}/h^2(k_{oz}k_{Fz})^{1/2}]\exp[i(\delta(E_F)+\delta(E_o))]. \tag{4.18}$$

The wave function at the bottom of the well ($z = z_m$) is written

$$\psi(x,y,z) = \sum_G (B_G^+\exp(izp_G) + B_G^-\exp(-izp_G))\exp(i(\vec{K}+\vec{G})\cdot\vec{R}) \quad , \tag{4.19}$$

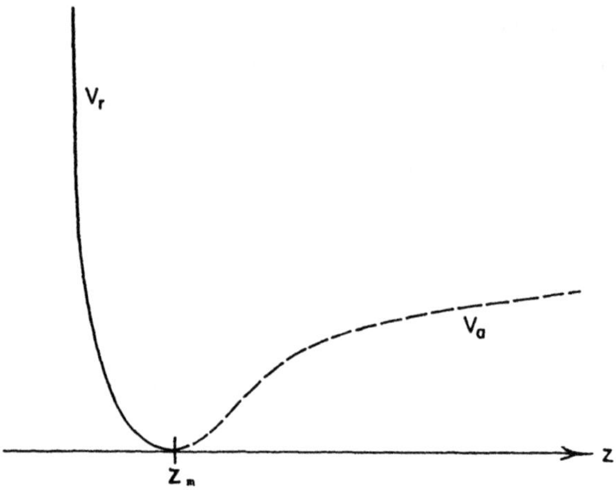

Fig.4.7 Schematic plot of the atom-surface interaction potential $V(\vec{R},z)$ for a fixed value of $\vec{R}$ showing decomposition into an attractive part V_a and a repulsive part V_r (solid line).

where p_G is the perpendicular momentum measured from the bottom of the well, that is,

$$p_G^2 + (\vec{K} + \vec{G})^2 = 2m(E + D)/\hbar^2 \quad . \tag{4.20}$$

It is assumed that the scattering problem for V_r in the absence of V_a has been solved. With respect to V_r alone, B_G^+ is the amplitude of an outgoing wave, and is related to B_G^- by a known matrix M, where

$$B_G^+ = \sum_G M(\vec{G},\vec{G}')B_{G'}^- \quad . \tag{4.21}$$

Analogously, A_F and B_G^- are outgoing amplitudes for the scattering from the attractive potential V_a, and they are related to the incoming amplitudes by the equations

$$B_G^- = T'(\vec{G},\vec{0}) + \sum_N R(\vec{G},\vec{N})B_N^+ \quad , \tag{4.22}$$

$$A_F = R'(\vec{F},\vec{0}) + \sum_G T(\vec{F},\vec{G})B_G^+ \quad , \tag{4.23}$$

where the known matrices R and R' describe reflections from V_a and the known matrices T and T' describe transmission across V_a. With reference to Fig. 4.7, T and R refer to particles incident upon V_a from the left, whereas T' and R' refer to incidence from the right. The matrix M, and the four matrices R, R', T, T', can be computed exactly by separate close coupling calculations, or can be approximated as described below; they are in fact expressible in terms of the matrices $\psi_<$ and $\psi_>$ that appear in (1.28). Notice that if $\vec{G}$ does not belong to the class $\vec{N}$, that is, if $\vec{G}$ is such that p_G^2 is not between $2mD/\hbar^2$ and zero, then $R(\vec{G},\vec{G}')$ can be neglected. This makes it possible for us to truncate the sum in (4.22), restricting it to the finite set of reciprocal vectors belonging to class $\vec{N}$. If we substitute the result of (4.22) into (4.21), we can express the result in the following symbolic fashion:

$$B^+ = MT' + MRB^+ \quad . \tag{4.24}$$

This equation is similar in structure to (4.11), and, like (4.11), it can display separately the resonant effects.

If, as we have assumed, V_a is a function of z only, then the reflection and transmission matrices are diagonal, thus introducing a major simplification into (4.24). Of course, the matrix $M(\vec{G},\vec{G}')$ is still not diagonal.

We can write (4.24) in a more explicit manner, as follows

$$B_G^+ = \sum_{G} M(\vec{G},\vec{G}')T'(\vec{G}',\vec{0}) + \sum_{G',N} M(\vec{G},\vec{G}')R(\vec{G}',\vec{N})B_N^+ \quad . \qquad (4.25)$$

We can simplify (4.25) to a considerable extent by using the following observations. First $T'(\vec{G}',\vec{0})$ can be neglected except for $\vec{G}' = \vec{0}$. Second, the product $M(\vec{G},\vec{G}')R(\vec{G}',\vec{N})$ can also be neglected unless $\vec{G}'$ belongs to the class $\vec{N}$. We can now rewrite (4.25) in the following form:

$$B_G^+ = M(\vec{G},\vec{0})T'(\vec{0},\vec{0}) + \sum_{N',N} M(\vec{G},\vec{N}')R(\vec{N}',\vec{N})B_N^+ \quad . \qquad (4.26)$$

We can describe an isolated resonance by restricting the sum in (4.26) to a single resonant vector $\vec{N}' = \vec{N}$. Let $\vec{G} = \vec{N}$ on the left hand side of (4.26), and then solve the resulting equation for B_N^+. After doing this, let $\vec{G} = \vec{F}$ on the left hand side of (4.26), since the class $\vec{F}$ vectors are the only ones leading to observable effects, and get an expression for B_F. We find that

$$B_F = T'(\vec{0},\vec{0})(M(\vec{F},\vec{0}) + M(\vec{F},\vec{N})R(\vec{N},\vec{N})M(\vec{N},\vec{0})/(1 - R(\vec{N},\vec{N})M(\vec{N},\vec{N}))) \, . \qquad (4.27)$$

One then finds the scattered intensity by using (4.23) and (4.17).

If we set $R(\vec{N},\vec{N})M(\vec{N},\vec{N}) = \rho\exp(i\delta)$, it becomes apparent that the resonant condition is equivalent to $\delta = 2\pi n$, where n must be an integer. If the lateral dependence of V_a is neglected, then the resonances become bound states, and δ is the same as $\phi(\varepsilon)$, which was defined shortly after equation (4.3), at $\varepsilon = (\hbar^2 p_N^2/(2m)) - D$. The difference between δ and ϕ shifts the resonant positions, as this changes the real part of τ_{bb}. Moreover, by performing a Taylor expansion on δ, we can get an expression for the half width, $\Gamma/2$, of the resonance. To be explicit,

$$\Gamma/2 = ((1 - \rho)/\rho)/(d\delta/dE) \quad . \qquad (4.28)$$

In this equation, $1 - \rho^2$ is the total scattering probability during one oscillation in the resonant state, $\hbar(d\delta/dE)$ is the period of this oscillation, and the decay rate is equal to the scattering probability multiplied by the frequency of oscillation.

4.4 Resonance Line Shapes

Define the quantity b to be the ratio of the resonant term to the direct term, evaluated at resonance. The value of b determines the line shape of a resonance, and also determines whether the resonance will be a maximum or a minimum. For an isolated resonance, we can get a simple expression for b from (4.13) and (4.27). For a nonspecular final beam we have

$$b = \tau_{fb}\tau_{bi}/(-i\tau_{fi}\mathrm{Im}(\tau_{bb})) = M(\vec{F},\vec{N})M(\vec{N},\vec{0})\rho/(1 - \rho)M(\vec{F},\vec{0})M(\vec{N},\vec{N}) \quad . \qquad (4.29)$$

For the specular beam, $-i\tau_{fi}$ must be replaced by $-i\tau_{fi} + h^2|k_{oz}|/(2\pi m)$, according to (4.18).

Real values of b lead to Lorentzian line shapes, with maxima occurring for positive b, and minima for b between zero and negative two. If $b = -1$, the minimum intensity goes to zero, as seen in Fig. 4.3. If $b = -2$, the diffracted intensity does not display any resonant behavior. However, at points where the resonant condition is satisfied, the scattered amplitude will vary by 2π over a very narrow range of energies or angles. In effect, there is a "resonance" in the phase, although not in the intensity. This effect appears whenever the incident and specular beams are coupled only to a bound state. In such a case, all of the intensity is specularly reflected, since it cannot go anywhere else. This conclusion depends on the absence of inelastic transitions.

When b is less than negative two, the resonances are all maxima, and the line shape is still Lorentzian. Several such resonant maxima are seen in Figure 4.4, for polar angles between 30 degrees and 68 degrees. The large maximum near 50 degrees has a large negative b value. In this graph, the direct specular intensity is weak, because of destructive interference between direct paths, such as A and B in Fig. 4.1.

The value of b need not be real. Some typical line shapes for complex values of b are shown in Fig. 4.8. Turning the drawing upside down, or exchanging left and right would also result in permissible line shapes [46].

First order perturbation theory predicts that b will be real and negative for resonances in the specular beam. To be precise,

$$b = -2|v_{bo}|^2/\Gamma \tag{4.30}$$

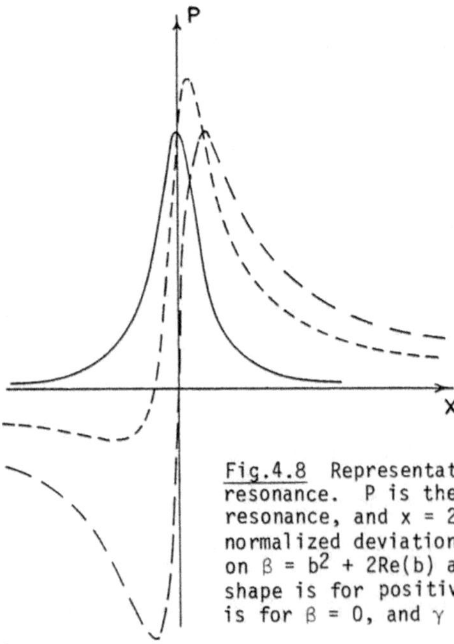

Fig.4.8 Representative line shapes for an isolated resonance. P is the excess scattered intensity due to the resonance, and $x = 2((\hbar^2 p_N^2/2m) - \varepsilon)$, which stands for the normalized deviation from resonance. The line shape depends on $\beta = b^2 + 2\mathrm{Re}(b)$ and on $\gamma = \mathrm{Im}(b)$. The Lorentzian shape is for positive real b. The antisymmetric shape is for $\beta = 0$, and γ positive. The third curve has $\beta = 2\gamma$.

If v_{bo} vanishes, then the second order contribution to b will have a positive real part and a negligible (or zero) imaginary part. Thus we have the simple rules of WEARE and co-workers [47]: For a weak corrugation and in the absence of close channel effects (i.e. for $|b|$ less than 2), first order resonances give minima, while second order resonances give maxima in the elastic intensity. Perturbation theory, applied to the calculation of the matrix elements of τ, also makes predictions for resonance shapes in the diffracted beams.

If V_r is represented by a hard corrugated wall, as is often done, the Kirchhoff approximation can be used to predict the phase of $M(\vec{G},\vec{G}')$ and the shape of the resonances. As a simple example, consider the corrugation profile

$$z = \zeta_1 \cos(2\pi x/a_1) + \zeta_2 \cos(2\pi y/a_2) \quad , \tag{4.31}$$

which could represent an alkali halide surface. Define m and n by $\vec{G} - \vec{G}' = (2\pi m/a_1, 2\pi n/a_2)$. According to (3.19), the Kirchhoff approximation then yields [32]

$$M(\vec{G},\vec{G}') = P i^{|m| + |n|} J_{|m|}((p_G + p_{G'})\zeta_1) J_{|n|}((p_G + p_{G'})\zeta_2) \quad , \tag{4.32}$$

where P is a positive factor of order unity, and J is a cylindrical Bessel function of integral order. It turns out then that only maxima or minima can occur [42]. Complex values of b can result from higher order Fourier components in the corrugation, or from the mixing of different resonances.

The observed resonances for He/LiF, which are shown in Fig. 4.2 and Fig. 4.3, are consistent with the corrugation defined in (4.31), with $a_1 = a_2$ and $\zeta_1 = \zeta_2$. When helium atoms are scattered from a (1x2) adsorbate on a Ni(110) surface, asymmetric resonances may be present [48]. This surface has a corrugation which is nearly one dimensional (ζ_1 far exceeds ζ_2), but the shape is much more complex than a sine wave.

In the case of two interacting resonances, (4.26) can still yield an explicit result for B_F^+ and the scattered intensity. In this case, there will be a gap in the surface-band structure given by

$$\mathrm{Re}(\tau_{bb'}) = \mathrm{Im}(M(\vec{N},\vec{N}'))/((d\delta/d\varepsilon)(d\delta'/d\varepsilon))^{1/2} \quad . \tag{4.33}$$

In using (4.26), it is not necessary to restrict consideration to problems involving isolated resonances or dual resonances, since (4.26) is suitable for computer calculations. Inelastic effects must be included, however, as discussed elsewhere (see, e.g., Cantini's lecture on bound-state inelastic resonances). Here we remark only that, as the surface temperature increases quantum effects (diffraction and resonances) may still be visible.

References

1. F.O.Goodman: Critical Reviews in Solid State and Materials Sciences, 7, 33 (1977)
2. M.W.Cole and D.R.Frankl: Surf. Sci. 20, 585 (1978)
3. V.Celli: Proc. of the 12th Intern. Rarefied Gas Dynamics Conf., AIAA (1980), p. 50; F.O.Goodman: ibid., p. 1

4. H.Hoinkes: Rev. Mod. Phys. 52, 933 (1980)
5. M.W.Cole and D.R.Frankl: Rev. Mod. Phys. 53, 199 (1981)
6. M.J.Cardillo: Annual Rev. of Phys. Chem. 32, 331 (1981)
7. F.O.Goodman and H.Wachman: Dynamics of Gas-Surface Scattering, Academic Press (1976)
8. See, for instance, the books on Quantum Mechanics by L.D.Landau and E.M.Lifshitz, Pergamon (1958), or by L.I.Schiff, McGraw Hill (1968)
9. L.S.Rodberg and R.M.Thaler: Introduction to the Quantum Theory of Scattering, Academic Press (1967)
10. M.Born and E.Wolf: Principles of Optics, Pergamon (1980), is the classical reference for optics. See also J.D.Jackson: Classical Electrodynamics, John Wiley and Sons (1975)
11. P.Beckmann and A.Spizzichino: The Scattering of Electromagnetic Waves from Rough Surfaces, Macmillan (1963)
12. J.Lapujoulade, Y.LeJay and G.Armand: Surf. Sci. 95, 107 (1980)
13. F.Scheid: Theory and Problems of Numerical Analysis, McGraw-Hill (1968), p. 232
14. R.G.Gordon: Methods of Computational Phys. 10, 81 (1971)
15. A.T.Yinnon, S.Bosanac, and R.B.Gerber: Chem. Phys. Letters 58, 364 (1978)
16. G.Wolken: J. Chem. Phys. 58, 3047 (1973)
17. R.G.Gordon: J. Chem. Phys. 51, 14 (1969)
18. N.Cabrera, V.Celli, F.O.Goodman, and J.R.Manson: Surf. Sci. 19, 93 (1970)
19. D.J.Malik and J.H.Weare: Chem. Phys. 2, 313 (1973)
20. G.Armand and J.R.Manson: Phys. Rev. B20, 5020 (1979); and G.Armand: J. Physique 41, 1975 (1980)
21. The general theory is reviewed by W.H.Miller: Adv. Chem. Phys. 25, 69 (1974). For applications to surface problems, see Ref. 24
22. See, for instance, H.Kriek, R.L.Ellis, and R.A.Marcus: J. Chem. Phys. 61, 4540 (1974), and references therein
23. M.V.Berry: J. Phys. A8, 566 (1975)
24. J.D.Doll: Chem. Phys. 3, 257 (1974), gives the general semiclassical theory from atom-surface scattering, and an application to He/LiF
25. N.R.Hill and V.Celli: Phys. Rev. B17, 17 (1978). This paper contains references to earlier work, especially by Petit and Cadilhac, and by Millar
26. N.Garcia, T.Ibanez, T.Solana, and N.Cabrera: Surf. Sci. 60, 385 (1976) and N.Garcia: J. Chem. Phys. 67, 897 (1977)
27. C.Lopez, F.J.Yndurain, and N.Garcia: Phys. Rev. B18, 970 (1978). This method is used independently by Weare and collaborators (see Ref. 41)
28. N.Garcia and N.Cabrera: Phys. Rev. B18, 576 (1978)
29. R.I.Masel, R.P.Merrill, and H.W.Miller: Phys. Rev. B12, 5545 (1975)
30. F.Toigo, A.Marvin, V.Celli, and N.R.Hill: Phys. Rev. B15, 5618 (1977)
31. N.Garcia, V.Celli, N.R.Hill, and N.Cabrera: Phys. Rev. B18, 5184 (1978); G.Armand and J.R.Manson: Phys. Rev. B19, 4091 (1979)
32. U.Garibaldi, A.C.Levi, R.Spadacini, and G.E.Tommei: Surf. Sci. 48, 649 (1975). For a critical discussion of this and later work, see N.R.Hill and V.Celli: Phys. Rev. B17, 17 (1978)
33. D.R.O'Keefe, J.N.Smith,Jr., R.L.Palmer and H.Saltsburg: J. Chem. Phys. 55, 3220 (1971)
34. H.U.Finzel, H.Frank, H.Hoinkes, M.Luschke, H.Nahr, H.Wilsch, and U.Wonka: Surf. Sci. 49, 577 (1975)
35. L.Mattera, C.Salvo, S.Terreni, and F.Tommasini: Surf. Sci. 97, 158 (1980)
36. H.Chow and E.D.Thompson: Surf. Sci. 59, 225 (1976)
37. W.Carlos and M.W.Cole: Surf. Sci. 91, 339 (1980)
38. G.Wolken: J. Chem. Phys. 58, 3047 (1973)
39. R.Manson and V.Celli: Surf. Sci. 24, 495 (1971)
40. C.E.Harvie and J.H.Weare: Phys. Rev. Letters 40, 187 (1978)
41. K.Wolfe and J.Weare: Phys. Rev. Letters 41, 1663 (1978)

42. V.Celli, N.Garcia, and J.Hutchison: Surf. Sci. 87, 112 (1979)
43. F.O.Goodman: Surf. Sci. 94, 581 (1980)
44. E.G.McRae: J. Chem. Phys. 45, 3258 (1966)
45. J.S.Hutchison: Phys. Rev. B22, 5671 (1980)
46. Y.Hamauzu: J. Phys. Soc. of Japan 42, 961 (1977)
47. K.L.Wolfe, C.H.Harvie, and J.H.Weare: Solid State Commun. 27, 1293 (1978)
48. J.M.Soler, V.Celli, N.Garcia, K.H.Rieder, and T.Engel: Surf. Sci. 108, 1 (1981)

He Diffraction from Semiconductor Surfaces.

Lecture I: Si (100)

M.J. Cardillo

Bell Laboratories
Murray Hill, NJ 07974, USA

1. Introduction

The outermost layers of many single crystal materials undergo reconstruction, i.e., the geometry at the surface differs from that expected for an ideal termination of the bulk. Most of these reconstructions result in a change in the size of the repeating surface unit cell and therefore can immediately be identified by the simple observation of the diffraction spot pattern in a low-energy electron diffraction (LEED) apparatus. In particular the common semiconductors, which are primarily covalent materials with strong directional bonding, all undergo surface reconstruction with some faces exhibiting one or more metastable surface structures. These may be considered to be driven by the high potential energy of the half-filled orbitals generated when the bulk bonds are broken to make surface atoms. The resulting periodicities of these surfaces in many cases have been known from LEED for over twenty years. Despite numerous experimental and theoretical studies, the geometrical configurations of almost all these surfaces are still to be resolved. In Table 1 are listed a selection of the reconstructions observed for some low index faces of common semiconductors. Out of this list, the geometries of GaAs(110) and Si(111)2×1 are the only ones generally accepted as well determined.

There are a variety of new and exciting probes of surface structure which are presently at early stages of development, such as Surface Extended X-Ray Absorption Fine Structure (SEXAFS), glancing incidence X-ray diffraction, ion scattering, and atom diffraction which is the topic of these lectures. Each of these probes may play a key role in the determination of surface structures. However, the principal and most productive tool for surface structural studies has been LEED, and therefore it serves as a standard with which to compare atom diffraction.

Table 1

Si(111)	-	(7×7), (1×1)?
Si(100)	-	(2×1)↔c(2×4) disordered?
Si(110)	-	(5×1), c(1×2), (5×4),...
Ge(111)	-	(2×8), (1×1)?
Ge(100)	-	c(2×4)
Ge(110)	-	(2×2),...
GaAs(111)	-	(2×2), ($\sqrt{19}\times\sqrt{19}$) R 23.9°, 3×3
GaAs(100)	-	(2×8), (2×4), (1×6),...
GaAs(110)	-	(1×1) "relaxed"

The information contained in the array of LEED diffraction beams generally observed and their variation with incident beam voltage is truly comprehensive in terms of structural details. However, a theoretical computation of these beam intensities is fraught with difficulties, particularly in the case of semiconductors. One of the basic problems in LEED structural analysis is the penetration and subsequent backscattering of the electron beam from depths approaching 10Å or 5 layers into the bulk. Strong multiple scattering occurs between and within each layer. In addition the semiconductor crystals may not become periodic in the direction perpendicular to the surface until 5 layers into the bulk. Thus a dynamic calculation of the LEED intensities must include a sufficiently accurate description for the electron-solid potential, the complexities of multiple scattering, and possibly different geometries for 3-5 layers.

The analysis of the diffraction of thermal energy light neutral particles, in particular He atoms, appears to be an attractive complement. At the low energies employed, typically 10-100meV, the atoms strictly do not penetrate with the result that the observed diffraction pattern is due to the periodic potential of the outermost exposed layer. Only a single geometry is in question. However, in contrast to 100eV electrons, for which the scattering is dominated by the atomic cores, thermal He atoms reflect from the surface potential at distances in the range of van der Waals diameters. Although experiments and theory have shown that the diffraction pattern is extremely sensitive to the details of this weak interaction potential, the information about nuclear positions is quite indirect. Thus the analytical aspects of the development of He diffraction into a quantitative surface structural tool may be divided into two substantial problems. The first is the derivation of a satisfactory description of the surface scattering potential from the diffraction beams. The second is the relation of that potential to the nuclear positions.

The semiconductor surfaces offer an exciting prospect for this development. There is a continuing challenge to resolve the geometries of these surfaces. He diffraction patterns have shown the semiconductor potentials to be sufficiently corrugated at the van der Waals distances of closest approach that there are large phase shift variations over the unit mesh. This implies sufficient structural sensitivity that a solution of the He scattering problem will give rise to useful geometric information. Strongly corrugated surfaces, however, present somewhat of a theoretical challenge. As will be discussed in later lectures, exciting progress has been made in both theoretical aspects of the problem.

For the remainder of this and the following lectures, progress in the study of He diffraction from semiconductor surfaces will be reviewed. Experimental results for the disordered Si(100)(2×1) surface will be discussed first along with a simple but informative analytical description of the scattering. Subsequently, the study of a known structure, the GaAs(110) surface, will be described. There we have treated the theoretical aspects of atom diffraction from strongly corrugated surfaces as unknowns. Finally recent results from the Si(111)7×7 surface will be described, from which a new type of reconstruction model has been proposed.

2. Si(100): Disordered Dimer Array

2.1 Si(100) Periodicity

The Si(100)(2×1) reconstruction was first observed with LEED by Schlier and Farnsworth [1] in 1959. They proposed the dimerization of adjacent rows of Si atoms in order to saturate the adjacent half-filled orbitals on the surface. For an accurately cut and annealed crystal (±1/2° in orientation) there are generally monoatomic steps spaced 300-500Å so that a surface diffraction pattern is the superposition of the scattering from each single terrace region. For Si(100) alternate layers are rotated 90° so that the two-domain Si(100)2×1 presents a square array of diffraction beams in reciprocal space. Many LEED studies of the Si(100) surface have been carried out and with two exceptions all have observed just the two-domain (2×1) pattern. Lander and Morrison [2] observed a c(2×4) periodicity with intensity streaks extending from the 1/4 order spots. Poppendiek, Ngoc, and Webb [3] reported a sharp c(2×4) after careful annealing by reducing the temperature from 13000 at 1-2°/sec. The 1/4 order beams they observe amount to ∿1% of the integer beams, i.e., they are sharp but weak.

In the atom diffraction study [4] care was also taken in high temperature annealing and a sharp two domain (2×1) reciprocal net was observed with additional disordered intensity suggesting a secondary reconstruction. The additional intensity forms a diffuse cross (Figs. 1 and 2) which would appear to include patches of c(2×4) and p(2×2) or c(2×2). In Figs. 1 and 2 the azimuth dependence of the diffraction patterns for different wavelengths are plotted vs. the parallel momentum transfer ($\Delta K_{||} = (2\pi/\lambda)(\sin\theta_r - \sin\theta_i)$) from which the reciprocal net illustrated above is derived. Note in Fig. 2 the detector acceptance angle is substantially smaller so that both the background and the disordered diffraction intensities are considerably

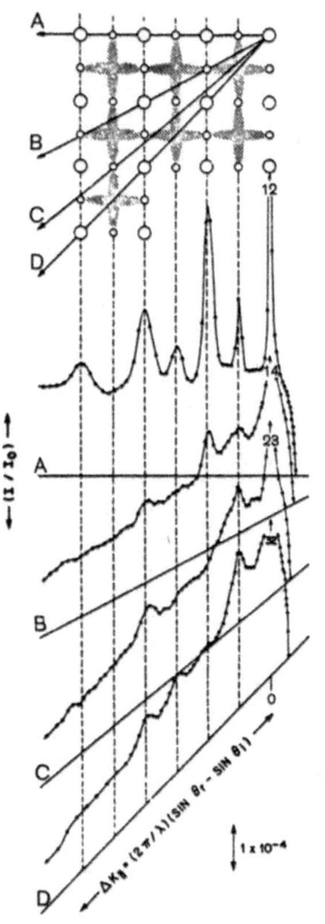

Fig.1 Diffraction scans plotted vs $\Delta K_{||}$, the reciprocal space coordinate. Each scan is rotated to the appropriate azimuth ϕ so that a vertical correspondence exists between the diffraction data and the reciprocal net above. The wavelength λ = 0.57Å and the acceptance angle $\Delta\theta_r$ = 1.5°. Incidence angle θ_i = 70°.

reduced. However, the "extra" intensity is clearly observable and in addition it appears structured, i.e. an intensity minimum is actually observed along the axis of the diffuse cross. This is a striking example of the sensitivity of atom diffraction. The structure observed here lies within a diffusely scattered intensity which is generally unobservable in display LEED systems. The observation of directional disorder based on He diffraction is discussed [4] in terms of the distorted dimer model based on recent total energy calculations of Chadi.[5] In Fig. 3 the buckled dimer model and the possible periodicities involved are illustrated. Chadi's calculations strongly favor the basic dimerization model over several others tested and in particular indicate that a slight tilting of the dimer bond with respect to the surface plane leaves the surface in a (2×1) while it lowers the surface energy on the order of 0.3eV/pair and removes a metallic dispersion in the valence band. An alteration of the tilting so that every other dimer in a row is different reduces the energy somewhat more (.1eV/pair). The total energy calculations cannot distinguish between the p(2×2), c(2×4), and c(2×2) unit meshes, all of which were considered possible based on the atom diffraction results. We interpret the streak (cross in two domains) observed with atom diffraction as a second order reconstruction perpendicular to the dimer direction which is of substantially reduced energy compared to the dimerization, consistent with the Chadi

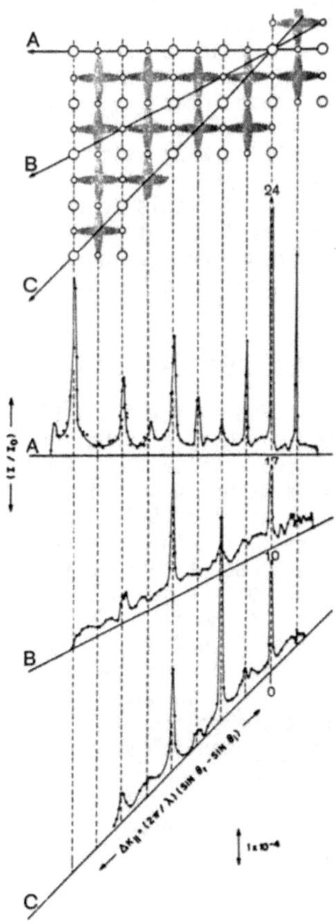

Fig.2 Diffraction scans plotted as in Fig.1 but for a wavelength λ = 0.98Å and $\Delta\Theta_r$ = 0.55°. Incidence angle θ_i = 50°.

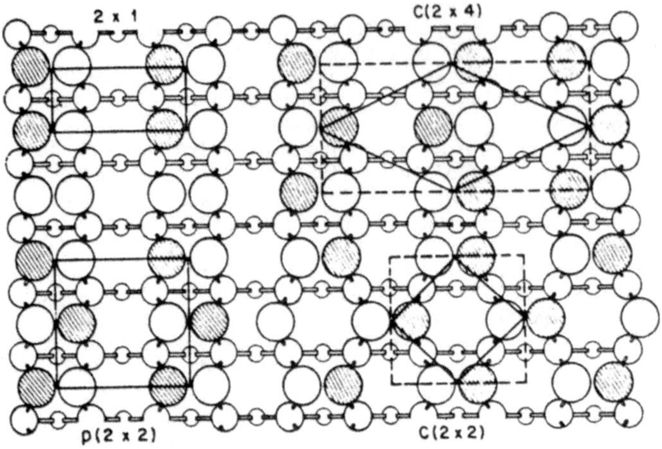

Fig.3 Illustration of the buckled dimer model in various periodic arrays according to the registry of the buckling rows.

buckling or a twist. However because of the weak forces (Jahn-Teller like) involved in the buckling, adjacent rows are not strongly coupled and do not register well at room temperature. Thus the equilibrium surface at 300K has domains of registered rows of buckled (or twisted) dimers which may be ∿25Å in length and give rise to the splitting in the diffuse intensity. The domain edges can actually be small regions of p(2×2) which can be easily converted to c(2×4). Alternatively there may be regions of c(2×2) which would be favored at step edges making a significant angle with the [10] direction and would not be easily converted to c(2×4) since it would require a row of bond breaking. In summary, the disorder in the registry of adjacent dimer rows is interpreted in terms of the weak coupling arising from the tilting perturbation to symmetric dimer. This gives rise to the diffuse cross observed in the He diffraction experiment.

2.2 Diffraction Scans and a Qualitative Feature of the Si(100) Surface

Diffraction scans for a wavelength $\lambda \approx 1\text{Å}$ are plotted in Fig. 4 for the [$\bar{1}$0] azimuth. Several interesting features can be observed. A sharp alternation in the intensity of adjacent beams can be observed in some regions ($\Theta_i = 20°$) and in other data of Ref. 4. This arises from the two domains -- one of which contributes only to integer peaks and the other of which contributes to both the integer and half-order peaks in this direction. In the spectra at $\Theta_i = 70°$ and 60°, a broad maxima in the intensity envelope of both half and integer order beams can be observed displaced ∿ 70° from the specular beam. We can associate this maxima with scattering from one domain, i.e. in the direction of the strong second order reconstruction. In Fig. 3 this corresponds to scattering across the dimer rows. This type of envelope maxima is termed a rainbow maxima and has a classical analogue in which it can be directly related to the slope of the potential at the points of inflection. For hard wall scattering from an oscillatory potential shape function $\xi(x)$ the cross section varies as the inverse of the second derivative of the potential, i.e. $\xi''(x)^{-1}$. This gives rise to an integrable singularity displaced an angle $\Delta\Theta_{rb}$ from specular, which is the rainbow maximum,

$$\Delta\Theta_{rb} \approx -2\tan^{-1}\xi''(x) .$$

If the first Fourier component of the potential dominates the corrugation such that $\xi(x) \sim \frac{1}{2} d_{\perp} \cos\left(\frac{2\pi X}{L}\right)$, then

$$d_{\perp} \sim (L/\pi) \tan(\Delta\Theta_{rb}/2) \; .$$

A crude qualitative analysis of the Si(100) surface can be made using this relation. Referring to Fig. 4 the value $\Delta\Theta_{rb} \sim 70°$ is obtained. However this angular displacement includes the effect of refraction due to the attractive portion of the He-surface potential. In Ref. 4 diffraction scans are plotted for the [0$\bar{1}$] direction where a strong rainbow maximum can be observed at $\Delta\Theta_{rb} \sim 60°$. At this higher energy the attractive potential, expected to be less than 10meV, is likely to have little refractive effect. Using $\Delta\Theta_{rb} = 60°$, L = 2×3.84Å = 7.68Å, an average corrugation across the dimerized rows is estimated at $d_{\perp} \sim 1.4$Å. Despite the simplistic nature of this analysis, it serves to illustrate an important point, the relation between strongly corrugated surfaces and angular distributions of diffracted beams. Simple approximate parameters, such as a principal corrugation of $d_{\perp} \sim 1.4$Å, can be helpful in comparing models of the surface which have substantially different gross contours, for example the missing row model [3] compared to

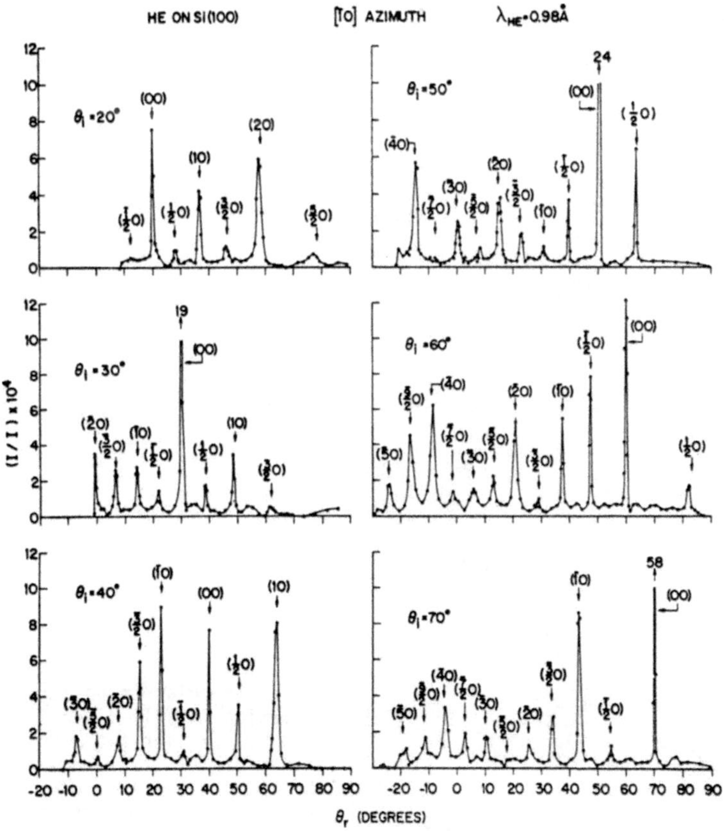

Fig. 4 Diffraction scans in the [10] direction for λ = 0.98Å and $\Delta\theta_r$ = 0.55°.

dimerized rows [1]. Other features in the data of Fig. 4 and in Ref. 4 can be similarly analyzed to give an approximate picture of the He/Si(100) surface potential. A reliable analysis to derive a full description of the potential will entail a full quantum calculation including at least the many open channels. Furthermore, it is not clear that such an effort is worthwhile until the surface can be well ordered.

2.3 Specular Intensities

Structure in specular intensities is often associated with surface resonances. However, in the case of Si(100) these resonance features are generally weak and broad. This is likely the result of a lack of detailed long range order. However there are broad structural features in the specular intensity scans which are due to interference in $\Delta k_\perp$. These features may be observed in the results plotted in Fig. 5. Semiclassical scattering analysis indicates that the specular intensities are dominated by the flat regions of the potential, so that the envelope structure observed is principally related to the peak-to-peak corrugation amplitudes. The inter-

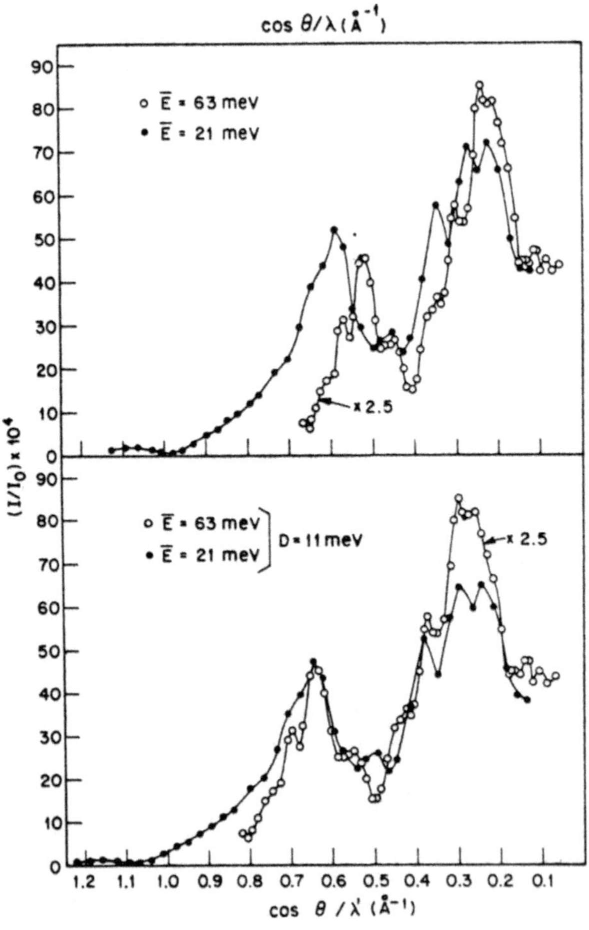

Fig.5 Specular intensity variation in [11] direction with cos Θ/λ for two energies. In the lower curve the abscissa has been adjusted for a wavelength shift but not for refraction (see Ref. 4).

ference oscillations for simple oscillatory potentials vary as the Bessel function $J_0^4(c)$, where $c = dk_i \cos\Theta$. This expression is very close to Bragg's Law, i.e.

$$d_\perp = n\lambda/(2\cos\Theta_i)\ .$$

As in Sec. 2.2, if we consider only the 63meV (λ = 0.57Å) data to reduce the effects of refraction by the attractive potential, the two principal maxima in Fig. 5 yield $d_\perp \sim 1.4$Å. The scan at 21meV appears shifted but similar in features to the 63meV results and is consistent with the interpretation in terms of $\Delta k_\perp$ interference. It is unlikely that the agreement between $d_\perp$ obtained in this simple analysis and that obtained from the classical rainbow analysis is fortuitous. Similar agreement has been obtained for GaAs(110) and subsequently confirmed by rigorous diffraction calculations. In general the diffraction of He from semiconductor surfaces has proven interference analysis of specular intensities to be a useful and nearly independent probe of corrugated surfaces, which complements the full diffraction analysis. Further illustration of its utility will be shown in subsequent lectures.

3. Structural Models for Si(100)

Although a theoretical fit of the Si(100) diffraction data is not yet available, some comments about structural models can be made. All the data for the Si(100) surface are consistant with the distorted dimer model of the surface. The observation of a stable sharp (2×1) reconstruction with a weak second order perturbation in the perpendicular direction, which does not produce long range registry of adjacent rows, is in accord with the alternating buckled dimers proposed by Chadi (or alternating twisted dimers). In addition a simple classical diffraction analysis suggests a principal corrugation in the dimer direction around 1.4Å, i.e. a slope at the point of inflection of ~30°. A second corrugation distance in the perpendicular direction of ~0.4A is also indicated by the data. Using a hard ball model of the Si atom of diameter ~3.8Å (justified in later lectures), a one-dimensional diagram of the tilted dimer is drawn in Fig. 6 in which the appropriate charge contour which accounts for these distances is sketched. An alternative model [3], consisting of missing alternate rows of Si atoms can be easily shown in such a diagram to have a principal corrugation twice that derived here and can therefore be ruled out.

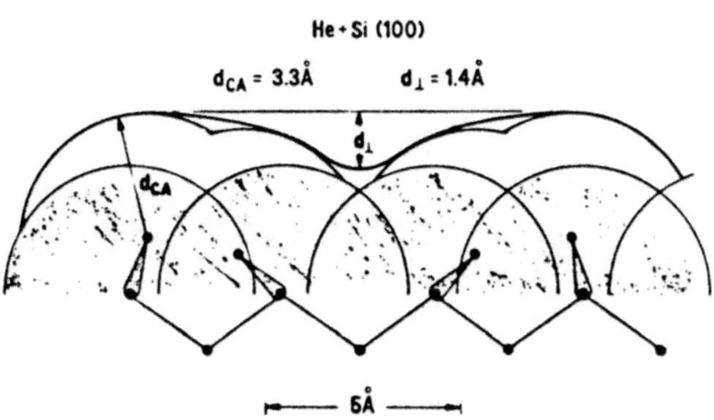

Fig.6 Schematic illustration of a charge contour appropriate for thermal He scattering based on the tilted dimer model and an interaction diameter of 3.5A.

4. Acknowledgement

The author acknowledges the collaboration of G. E. Becker in obtaining the data and J. C. Tully and E. K. Grimmelman for the results of semi-classical calculations and helpful discussions.

5. References

1. R. E. Schlier and H. E. Farnsworth, J. Chem. Phys. 30, 4 (1959).
2. J. J. Lander and J. Morrison, J. Chem. Phys. 37, 729 (1962).
3. T. D. Poppendieck, T. C. Ngoc, and M. B. Webb, Surf. Sci. 75, 287 (1978).
4. M. J. Cardillo and G. E. Becker, Phys. Rev. B21, 1497 (1980).
5. D. J. Chadi, Phys. Rev. Lett. 43, 43 (1979).

He Diffraction from Semiconductor Surfaces.

Lecture II: GaAs (110): Calibration of the Atom-Diffraction Technique

M.J. Cardillo

Bell Laboratories
Murray Hill, NJ 07974, USA

1. Introduction

The study of the Si(100) surface has proven to be an interesting start to the development of He diffraction into a quantitative structural tool for semiconductor surfaces. Several aspects of the problem have become better defined. For corrugated surfaces, simple scattering theories such as the eikonal or sudden approximation are not adequate, in particular in their treatment of multiple scattering, i.e. the diffraction probabilities are strongly coupled. However, they do allow qualitative features of the potential to be described. An accurate computation of the diffraction intensities for realistic potentials is a substantial undertaking. The attractive potential must be well understood as the phase shift variation over the unit mesh is large and the potential energy may also vary significantly over the unit mesh. A problem specific to the Si(100) is the lack of long range order in the secondary reconstruction which makes accurate comparison with theory difficult.

In order to progress further with strongly corrugated surfaces, we have studied He diffraction from a known semiconductor surface, GaAs(110) [1]. By treating the diffraction calculation and the theory of the potential

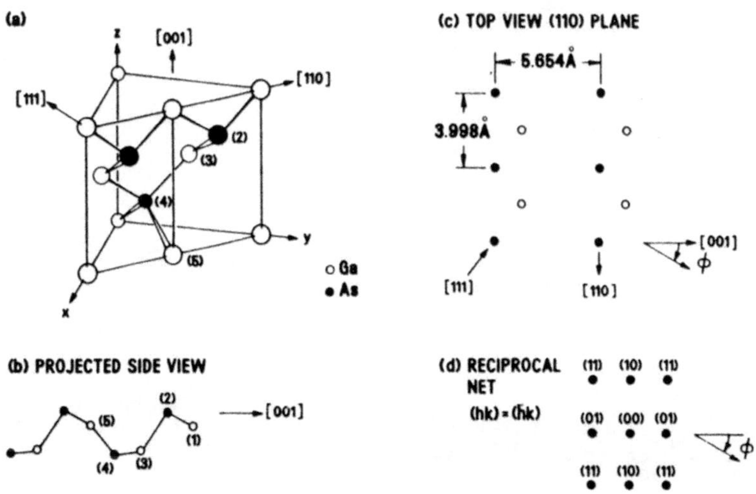

Fig.1 GaAs crystal structure showing a) (110) planar face, b) projected side view of the (110) plane, c) top view of outermost layer, d) reciprocal net indexing and definition of φ.

as unknowns, we wish to develop a suitable description of atom diffraction to enable a quantitative structural study of other corrugated surfaces. The surface of GaAs(110) is illustrated in Fig. 1 along with our conventions for angles and diffraction beams used in this paper. The surface may be thought of as ridges and troughs of gallium-arsenic bonded chains. The surface is "rough" in the [01] direction across the troughs and relatively "smooth" in the [10] direction along the troughs. The surface reconstructs by tilting the surface Ga-As bond by $\sim 27°$ but does not change its periodicity from what is termed 1×1 [2].

2. Diffraction Scans

In Fig. 2 are shown a selected set of diffraction scans with $\lambda = 0.98$Å ($E \sim 21$meV) for a series of Θ_i and for $\phi = 0$ and $180°$. Note that the two sets of scans have gross similarities in their envelopes but that there is a clear asymmetry in the diffraction pattern which can be ascribed to the tilting of the surface Ga-As bond (Ga and As are sufficiently large and close in atomic number that this difference can be ruled out as the cause of the asymmetry). These and many other scans for different ϕ and λ have been analyzed at various levels of theory. Simple one dimensional classical scattering analysis of rainbow maxima, which are clearly seen in the data for example at $\Theta_i = 55°$, assuming the surface to have glide plane symmetry, indicates a gross corrugation across the troughs of $\xi_x \sim 1.0$Å and a corrugation of $\xi_y \sim 0.2$Å along the troughs. The hardwall eikonal wave approximation, which is not expected to be accurate for this surface, has also been used for a qualitative comparison to the classical scattering analysis. It yields values of $\xi_x \sim 1.2$Å and $\xi_y \sim .36$Å, similar to the classical analysis. In these simple analyses the role of the attractive

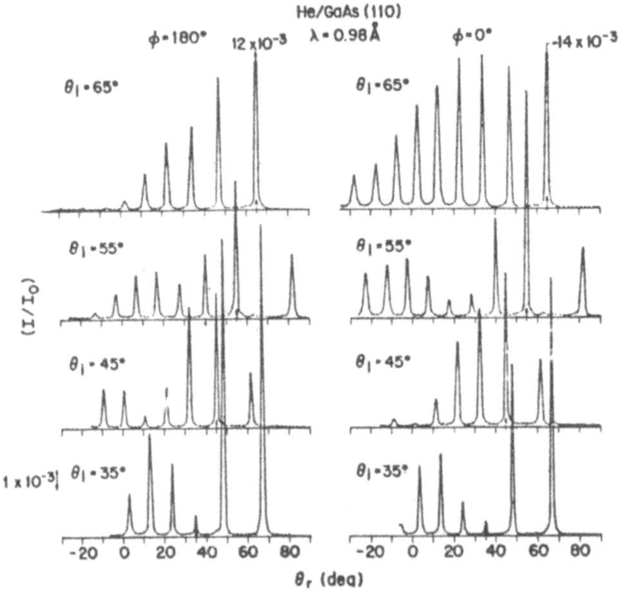

Fig.2 Diffraction intensities for P = 0° and 180° at $\lambda = 0.98$Å. Θ_i is received from the surface normal.

potential has been ignored. However the angular displacements of the rainbow maxima for high energy data ($\bar{E}$ = 63meV, λ = 0.57Å) are similar to those of Fig. 1. This suggests that the attractive potential plays a small role, even at $\bar{E}$ = 21meV, λ = 0.98Å, i.e. the laterally averaged well depth is likely to be less than 10 meV.

3. Specular Intensity Scans

A number of specular intensity vs. polar angle (Θ_i) scans were recorded for various azimuths (ϕ) and wavelengths. In contrast to the disordered Si(100), these scans show sharp resonance oscillations. In Fig. 3 one set of scans in the vicinity of ϕ = 0 at λ = 0.98Å are shown. The sharp features are associated with bound state resonances. For specific incident kinematic conditions the incoming particle can enter a scattering state associated with the attractive potential so that it is effectively bound in the direction normal to the surface. If the surface is not strongly corrugated a useful assumption is that the particle is moving nearly freely so that the conservation condition for parallel momentum K and energy, E_i can be written

$$(K+G)^2 = (2m/\hbar^2)(E_i+|E_j|) \tag{3.1}$$

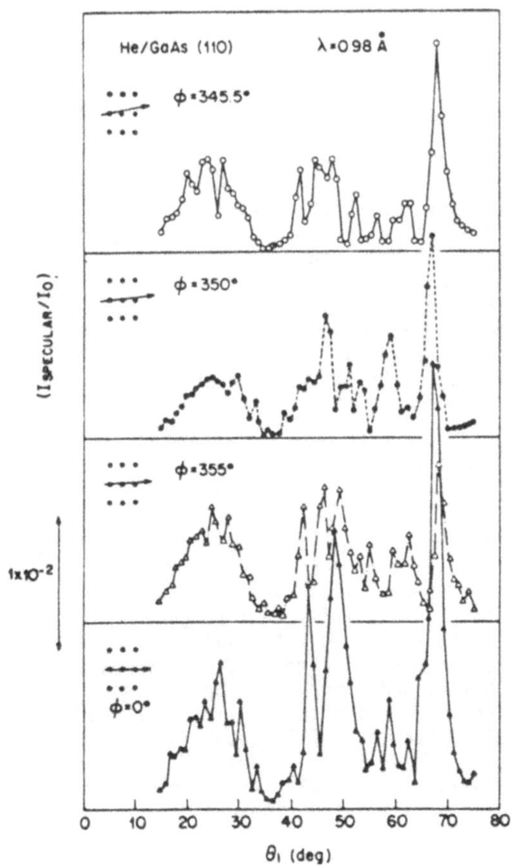

Fig.3 Specular intensity vs. Θ_i for a series of azimuths near ϕ = 0°.

where G is a reciprocal net vector and E_j is a bound state energy level of the laterally averaged potential for motion normal to the surface. Note that Eqn. 3.1 is the equation of a circle in the K_i plane centered at various G. A plot of the positions in the K plane at which the resonances occur should allow the appropriate G for each circle to be indexed. The radii of the circles then determine the various values of E_j. This procedure has been successfully employed to obtain the He surface potential for several weakly corrugated surfaces.

For moderately corrugated surfaces, many more beams can strongly couple to resonances with the result that the density of transitions is high. In the variety of intersections for the multitude of possible circles, there would be free particle state degeneracies which are lifted by splitting these states to form gaps in the bands of allowed energies. Where these splittings can be identified, the Fourier components of the surface potential can be derived.

For the case of GaAs(110) a plot of the resonance positions in the K_i plane yields a high density of points but no readily identifiable circles. The coupling between beams is sufficiently strong that there is substantial splitting all over the K plane and no kinematic analysis is likely to be successful. However, information about the potential has been derived from a comprehensive analysis of the diffraction patterns [3].

In addition to sharp oscillations in the data of Fig. 3, there are broad envelopes similar to those observed for Si(100) specular scans. These features again represent Bragg-like interference in the specular beam and can be analyzed for structural information about vertical displacements. The most consistent feature apparent in all the $\lambda = 0.98$Å data is the maximum observed near $\Theta_i = 25°$. If the effect of the attractive potential is again ignored, the relation $2d_\perp \cos\Theta = n\lambda$ yields $d_\perp/n = 0.54$Å. For $n = 2$, $d_\perp \sim 1.1$Å in agreement with the diffraction analysis. The $n = 1$ maxima should then appear around $\Theta_i = 64°$. Adding a uniform well depth of $D \leq 5$meV brings the assignments into agreement with the data. For the recurrent maxima at $\Theta_i = 45°$, $d_\perp/n = .75$Å. This distance corresponds to the difference between the two corrugations obtained in the diffraction analysis, i.e. $d_x \sim 1.0 - 1.1$Å, $d_y \sim 0.3$Å. Similar analyses with the eikonal approximation as well have been done for a variety of other specular intensity scans with comparable parameters for the corrugation being deduced. In general the results of a Bragg's law or eikonal analysis of the specular intensity are consistent with the diffraction analysis, i.e. the principal corrugation across the troughs is $\xi_x \sim 1.0$-1.1Å and along the troughs $\xi_y \sim 0.3$Å.

4. Rigorous Calculation of Diffraction Intensities

The diffraction intensities for GaAs(110) have been extensively analyzed theoretically by Laughlin. [3] A Green's function technique for carrying out dynamic close-coupled calculations for realistic strongly corrugated potentials was developed. The calculation is accomplished through a recursion relation starting well inside the repulsive potential and integrated to the vacuum. It can routinely handle potentials of any shape or corrugation with good numerical stability and convergence properties. Unitarity is identically satisfied. The diffraction intensities were found to be described by a realistic soft-wall potential. This potential could be roughly described as having an effective peak-trough sinusoidal corrugation of ~ 1.0Å, confirming the qualitative results of the simple approximate analysis described above. In addition the rigorous theory demonstrates that the requirement of the potential to be soft is necessary in order to reproduce accurately the rainbow collapse into the specular beam at grazing incidence as shown in Fig. 2. The average attractive potential well depth is also sensitive to this rainbow collapse and a good fit of the complete set of data indicate $\bar{D} \sim 7$meV.

5. The Origin of the He/GaAs Potential

Having obtained a reasonable description of the effective He/GaAs(110) potential, Laughlin [3] has developed a semi-empirical connection between He-surface potentials and the positions of the atomic cores which is based on the charge density of the surface. Noting the general similarity in form of the energy terms in the Gordon-Kim theory [4] an expression for the repulsive interaction is derived

$$E = T \rho_t \qquad (5.1)$$

where ρ_t is the charge density of the target and $T = 9\times10^4$ meV-Å^3. Correcting empirically for the improper accounting of exchange and correlation and adding the r^{-6} attractive term, the following expression is obtained:

$$E = T\rho_t - X \rho_t^{1/3} - \Sigma C_6/r^6 \qquad (5.2)$$

where $X = 40$meV-Å^3.

The first term is equivalent to a result recently derived by Esbjerg and Norskov [5] for the He-surface repulsion based on a local density calculation for embedding He in a uniform electron gas. Both results quantitatively associate the diffractive scattering of He with the surface

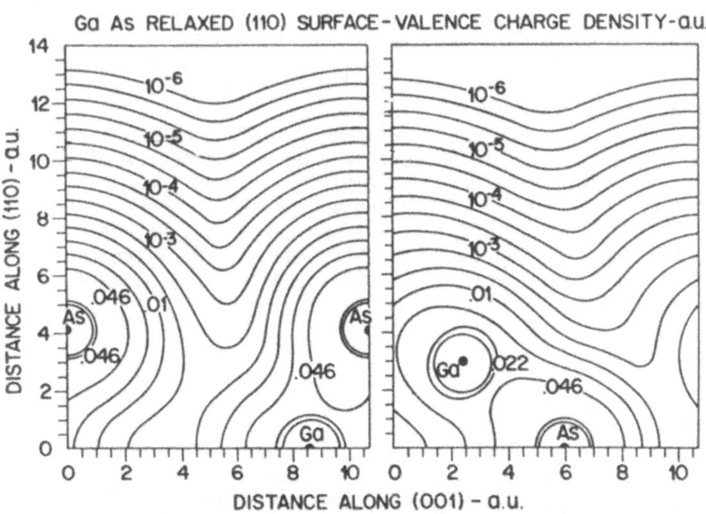

Fig.4 Charge densities for the relaxed (1×1) surface of GaAs(110) calculated by Hamann. The key contour at 2×10^{-5} is appropriate for the estimation of the corrugations observed in He diffraction.

change density but at an extremely low value (10^{-5} au). Thus for the He surface potential to be amenable to direct calculation based on these results, charge densities must become available at an unprecedented level of accuracy.

6. Computation of Rarified Charge Densities

When the specific form of the relation between the He-surface potential and charge densities was proposed, Hamann decided to test the result [6] against the GaAs(110) diffraction study using a surface linear augmented plane wave (SLAPW) method he had recently developed. The exponentially decaying charge densities can be treated accurately and self-consistently with this technique with the only approximation being the local exchange and correlation potential.

The results for the Hamann calculation are shown in Fig. 4. Charge densities are plotted for two cuts across the troughs. The first cut is through the As nuclei and the second through the Ga nuclei. The principal corrugation across the troughs is found to be $\xi_x \sim .9$Å and along the troughs $\xi_y \sim .3$A in remarkable agreement with the analysis of the diffraction results. As a test of the structural sensitivity, the surface was "unrelaxed" to the ideal bulk termination geometry. The principal corrugation parameters dropped to $\zeta_x = 0.6$Å and $\zeta_y \sim 0.1$Å so that the relaxed and unrelaxed surfaces are clearly distinguished. In addition to GaAs(110), Hamann has also shown these calculations to be appropriate for the H on Ni(110) experiment of Rieder and Engel [7]. In that case he was able to determine the specific binding geometry of the hydrogen atom on the Ni(110) surface.

7. Summary

The calibration experiment of He diffraction from a known structure, GaAs(110), must be termed successful in that it has stimulated important progress. It has provided a basis for dramatic theoretical steps forward in the quantitative development of He diffraction into a useful structural probe. A detailed theoretical investigation of scattering from GaAs(110), and strongly corrugated surfaces in general, has been carried out. In this study a new computational technique was developed, a failure of the hard-wall approximation documented, and a characterization of the nature of the potential was discussed. This has led to an empirical expression which quantitatively confirms the proposal of Esbjerg and Norskov. Based on this relationship Hamann has extended SLAPW calculations out to the appropriate charge densities. His results show that the charge density profiles and He-diffraction results are consistent and sensitive to the LEED geometry for the GaAs(110) surface.

8. References

1. M. J. Cardillo, G. E. Becker, S. J. Sibener, D. R. Miller, Surface Sci. 107, (1981).
2. C. B. Duke, et al., J. Vacuum Sci. Technol. 16, 1252 (1979).
3. R. B. Laughlin (to be published).
4. R. G. Gordon and Y. S. Kim, J. Chem. Phys. 56, 3122 (1972).
5. N. Esbjerg and J. K. Norskov, Phys. Rev. Lett. 45, 807 (1980).
6. D. R. Hamann, Phys. Rev. Lett. 46, 1227 (1981).
7. K. H. Rieder and T. Engel, Phys. Rev. Lett. 45, 824 (1980).

He Diffraction from Semiconductor Surfaces.
Lecture III: Si (111) 7 x 7

M.J. Cardillo

Bell Laboratories
Murray Hill, NJ 07974, USA

1. Introduction

The reconstruction of the Si(111) surface to form a (7×7) unit mesh has been known for 20 years [1]. Despite extensive study, no convincing evidence to support any structural model has emerged. In part this is due to the size of the unit mesh which contains 98 top and nearly co-planer second layer atoms. This makes even a kinematic analysis of a core scattering probe, such as LEED, a substantial task and hardly amenable to testing a large range of structural models. In addition dynamic affects associated with the penetration of the electron beam to several layers and the electron-solid potential render kinematic analyses of dubious accuracy.

The diffraction pattern of He atoms reflecting only from the outermost exposed layers is an interesting alternative. Although it is unlikely that a complete diffraction calculation for a detailed structural model will prove much less of a task, it is very possible that key features of the diffraction pattern will yield to the simplifying aspects of approximate analyses in terms of characteristic shapes and distances.

2. Diffraction Scans

In Fig. 1 are plotted a series of recently reported [2] diffraction scans vs $\Delta K_{||}$ in the $[0\bar{1}]$ direction for a wavelength $\lambda = 0.98$Å ($\bar{E} = 21$meV), an acceptance angle $\Delta\Theta_r = 0.56°$, for the values of the incident angle Θ_i indicated. For the 7×7 surface a high resolution was required in order to discriminate against the nearest out of plane beams. The overall intensity distribution in these scans is in contrast to GaAs(110) in that there are no smooth features in the envelope of the diffraction pattern indicative of rainbow scattering. Yet there is substantial scattering to large angles characteristic of strongly corrugated surfaces. We may estimate the expected corrugation for the unreconstructed Si(111) for comparison. In Lecture II the value of the charge density appropriate for the He distance of closest approach was discussed ($\rho \sim \bar{E}$ (meV)/10^5Å^{3}meV). For silicon one may estimate that the effective distance of closest approach for a charge density of 2×10^{-4}Å^{-3} is approximately 3.5Å. Drawing hard overlapping spheres of this radius generates a relatively smooth surface with small cusps which are vertically separated by $\sim$ 0.5Å from the tops. Undoubtedly in this dilute region of charge there will be smoothing of the cusped regions so that one may expect a local corrugation within the 7×7 unit mesh in the vicinity of 0.2-0.3Å. Thus the large angle scattering can not be associated with the local slope of the potential of the undisturbed Si(111) surface. Furthermore, if the reconstruction is of the "smooth" type such as long wavelength undulations or buckling, one would expect to observe intense rainbow maxima characteristic of extended regions of a slowly varying slope of the surface potential.

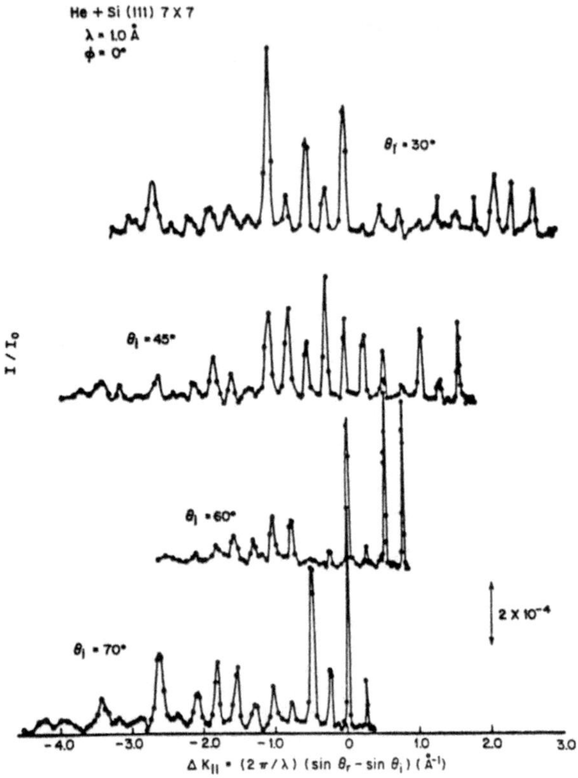

Fig.1 Normalized scattered helium intensity vs parallel momentum transfer for scans in the [01]*($\phi = 0^\circ$) direction. The grating relation allows diffraction peaks at multiples of $0.27Å^{-1}$.

3. Specular Intensity Interference

An important indication as to the nature of the (7×7) reconstruction was inferred recently [3] from a He specular intensity scan taken at a wavelength of $\lambda = 0.57Å$ and a detector acceptance angle $\Delta\Theta_r \sim 1.1°$. The data are reproduced in Fig. 2. There a series of rapid broad oscillations may be observed which generally can be associated with interference in the normal momentum, $\Delta K_\perp$, similar to that for Si(100) and GaAs(110).

The first four oscillations from $\Theta_i = 25°$ to $\Theta_i = 52°$ can be immediately fit to a vertical distance by the simple Bragg relation $2d_\perp \cos\Theta = n\lambda$ and a value $d_\perp \sim 3.2Å$ is obtained. Using this value a series of maxima can be generated assuming that two large flat regions separated by a verticle spacing of 3.2Å exist within the 7×7 unit mesh. They are indicated on Fig. 2 by dark arrows. One can pursue this analysis and postulate a second vertical distance based on the shape of these peaks, i.e. at 44° the oscillation is sharp with a slight shoulder, neighboring peaks appear broader, and the peak at 60° appears split as if two similar distances are beating. This leads to a secondary distance of 2.9Å which along with a Debye-Waller factor accounts reasonably well for the entire shape of the scan. There is some argument as to whether two distances this close can show independent oscillations or if a single oscillation for the average should be observed. This point is not relevent to the rest of the discussion concerning the model of the 7×7. However, it is worth pointing

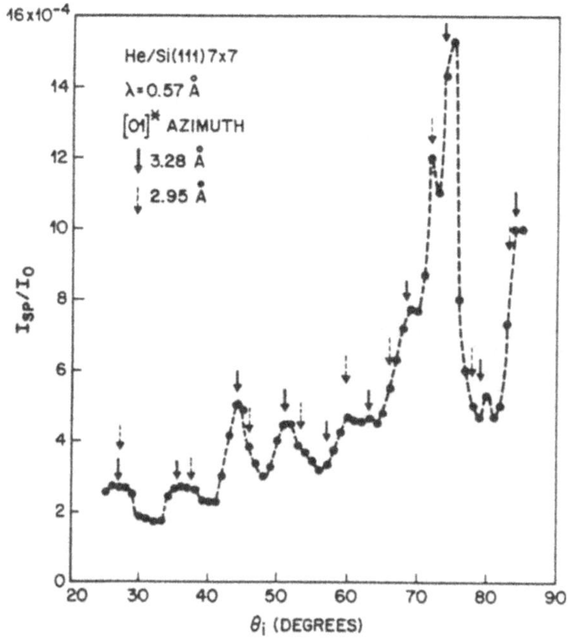

Fig.2 Specular intensity scan for E_i = .063eV He beam. The dark arrows show expected maxima for a vertical spacing of 3.2Å. The dashed arrows correspond to a secondary distance of 2.9Å. Note the difference of these differences is 0.3Å.

out that the theory of specular intensities, even in the rigorous close coupling form, does not fit the data for corrugated surfaces well. The calculations generally show a more "coherent" result for the scattering than the data indicate. At this point it seems reasonable to simply indicate that the data suggest the occurance of two distances and at future studies, in particular in regard to the effect of surface motion on the diffraction scans, resolve this point. Both of these distances are close to, and their average equals, the double layer spacing of the Si(111) crystal. Since the He atom is a completely nonpenetrating probe, these distances are appropriate to the outermost exposed layer. In particular, the angular resolution is $\Delta\Theta_r > 1°$, so that the sensitivity to coherent structural effects is limited to distances < 100Å on the surface. Therefore, these spacings must lie within the (7×7) unit mesh and are not due to random steps. Thus, it can be concluded that the third layer (second double layer) of the Si(111) surface is exposed to the He atom due to the 7×7 reconstruction.

A simple trajectory analysis at Θ_i = 60° in which the distance of closest approach to the Si atoms at a charge density of 2×10^{-4}Å^{-3} is estimated to be ∿ 3.3Å, is illustrated in Fig. 3 to show that the extent of the exposed region is large -- of the order of half the unit mesh dimension. Furthermore the correspondence of the predicted Bragg maxima with the oscillations of Fig. 2 is obtained without the inclusion of the attractive potential. Unless the potential well depth is particularly small (D $\lesssim$ 3 meV), this implies that the area of the interfering regions are large compared to the range of the attractive potential. Adding 2A to the repulsive distance of closest approach (3.3Å) yields a radius of 5.5Å or an area of ∿90Å^2 as the minimum exposed region of the third layer. This confirms the trajectory argument for large open areas.

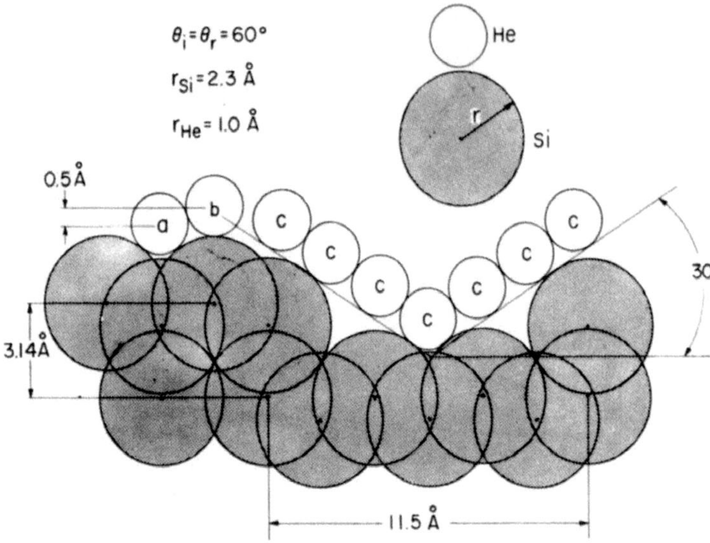

Fig.3 Specular trajectory illustration for $\theta_i = 60°$ showing an estimate of the extent of the exposed third layer required for the He to scatter out.

4. A Model of the Si(111)7×7

In Fig. 4 the implications of these results in terms of a structural model are illustrated. In Fig. 4a the Lander model [4] of the (7×7) is drawn. The ordered array of vacancies were postulated to result in a strong local relaxation (benzene-type of electronic delocalization) which returns more energy than the cost of the vacancies. The analysis presented here, however, requires both top and third layer exposure to the extent drawn in Fig. 4b. This shallow etched region of exposed third layer is the minimum area required to satisfy the data. For this more extended region to form as an exothermic unit, a more complex relaxation or redistribution of charge is required. It would seem that a region of the size drawn in Fig. 4b would generate a (5×5) reconstruction. Since the Si(111) periodicity is (7×7), the open area has been increased in Fig. 4c to occupy one half the unit mesh. At this point the 7×7 units can be thought of as strongly interacting.

It is clear that the data of Fig. 3 do not indicate any of the details of the model in Fig. 4c -- just the extent of the exposed third layer. However, a model of this form does generate interesting possibilities as to the nature of the driving forces for the reconstruction. Some arguments in this regard are presented here as this model is significantly different than previous proposals for the nature of the 7×7 reconstruction [5]. One approach to estimating the energetics of this surface is to consider it a maximum packing density of steps. The implication is that the formation of a step may actually be exothermic. The surface consists of 3 bonded atoms at the top, along the edge, and at the bottom of each N×N triangle. The number of each is $3\times2(N-1)$, $1/2(N-3)(N-2)$, and $1/2(N-2)(N-1)$ respectively. Calling the energy to create a 3-bonded atom E_3, and allowing the edge atoms a relaxation energy R, the total energy of the surface unit can be written $E_T(N) = 2(N-1)(E_3-3R)$ and the energy per unit area $E_A = E_T(N)/N^2$. The issue of the step formation being exothermic depends on the term $(E_3 - 3R)$. If $R > E_3/3$ then the model may be considered as the optimization of the closest packing of steps and the short range

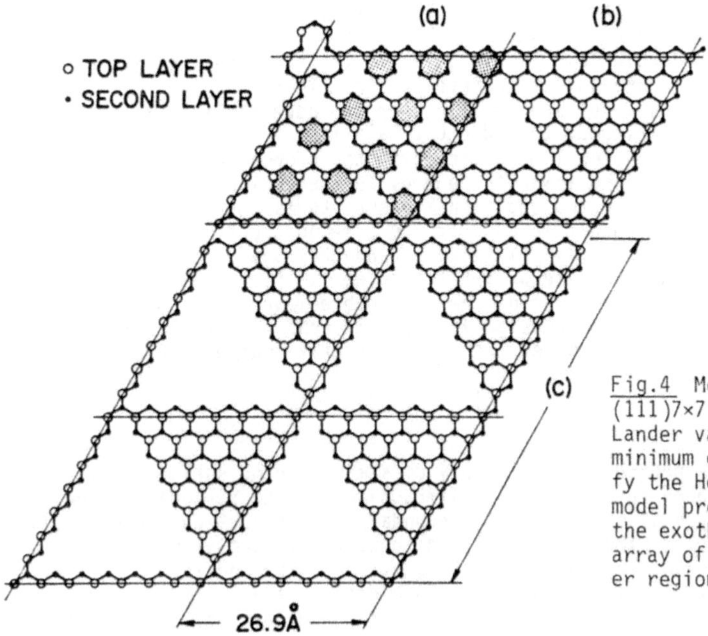

Fig.4 Models for the Si (111)7×7 reconstruction a) Lander vacancy model b) minimum open area to satisfy the He specular data c) model proposed to justify the exothermicity of the array of exposed third layer regions.

step-step repulsion. Note that E_3 should be close to 2.3eV, half the cohesive energy of silicon. A schematic energy diagram (Fig. 5) illustrates this point. (E_3 - 3R) also determines the formation energy of a single linear step so that one would expect Si(111) cleaves to show spontaneous step development if it is negative. The cleaving experiments of Henzler [6] do not show this development. Total energy calculations by Chadi [7] also indicate the $E_3 > 3R$ for the three-bonded step, i.e. that the formation energy is somewhat positive.

There is, however, a significant difference between an infinite linear step, with a positive energy of formation, and the array of Fig. 4c, which

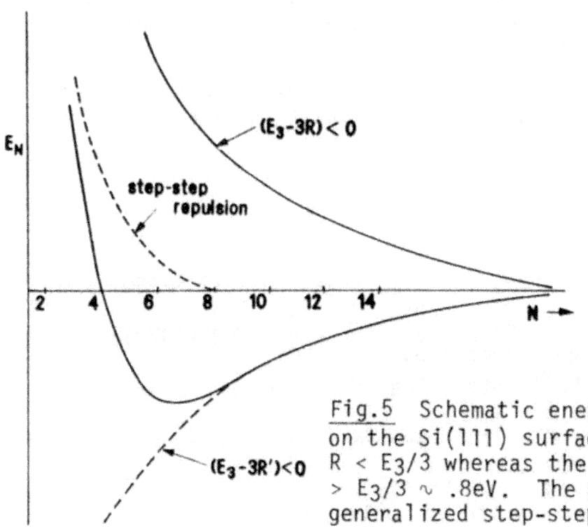

Fig.5 Schematic energy diagram for packing of steps on the Si(111) surface. The repulsive curve is for $R < E_3/3$ whereas the attractive curve is for $R' > E_3/3 \sim .8$eV. The minimum is associated with the generalized step-step repulsion.

should strongly affect the value of R. Each triangular unit has an effective perimeter which allows for relaxation of the inner top layer region. This top double-layer can flatten in the electronic direction of an sp^2 bonding configuration to relieve strain. In relieving this strain it will also be pushing out and rolling over the edge atoms so as to have a strong parallel overlap of the adjacent half-filled orbitals. Note that the long one-dimensional metallic chain thus formed will have a square-root dispersion and lower the energy further. Thus for the triangular unit there are two additional energy terms compared to the linear step which can result in an exothermic reconstruction. The extent of atomic motion to relieve strain is limited and this may be a significant factor in restricting the size of the triangle. In the simplest terms, however, the near island nature of the top double-layer units can increase R through strain relief and by significantly increasing the coupling of the edge orbitals. If $R \gtrsim 0.8$eV then the attractive curve of Fig. 5 is applicable.

5. Summary

In these lectures progress in the study of semiconductor surfaces using He diffraction has been reviewed. The diffraction for the Si(100) surface has revealed a lack of ordering in a secondary reconstruction perpendicular to the strong (2×1). The periodicity and qualitative analysis of both the diffraction scans and specular intensity scans are consistent with a distorted dimer model of the surface. In addition, the Si(100) diffraction has led to some understanding of the analytical problems associated with strongly corrugated surfaces.

A calibration experiment for the theory of He diffraction from corrugated surfaces was undertaken using the GaAs(110) surface for which the structure is well established. This study has stimulated major theoretical advances in the computation of diffraction from rough surfaces, understanding the nature of the He-surface potential, and served as an important and successful test for the direct calculation of that potential from the very low density charge profiles involved.

Finally, experiments and some inciteful results for the Si(111)7×7 were described. Although this surface has yet to receive detailed theoretical attention, a new model has been proposed based on a specular intensity scan. This has led to interesting speculation about the nature of the dominant forces involved on the (7×7) surface and will stimulate much interest in future experiments and theory.

If this rate of progress continues, one may look forward to significant quantitative information becoming available in the next three years based on the technique of He diffraction.

6. References

1. R. E. Schlier and H. E. Farnsworth, J. Chem. Phys. 30, 917 (1959) and J. J. Lander and J. Morrison, J. Chem. Phys. 37, 729 (1962).
2. M. J. Cardillo and G. E. Becker, Phys. Rev. Lett. 42, 508 (1979).
3. M. J. Cardillo, Phys. Rev. BR23, 4279 (1981).
4. J. J. Lander, in Progress in Solid State Chemistry, ed. H. Reiss (Pergamon, Oxford, 1965) Vol. 2.
5. The author acknowledges T. M. Rice for key ideas and clarifications in the development of these arguments.
6. M. Henzler, Surface Sci. 36, 109 (1973).
7. D. J. Chadi and J. R. Chelikowsky (to be published).

Helium Scattering from Clean and Adsorbate-Covered Metal Surfaces

K.H. Rieder

IBM Zurich Research Laboratory
CH-8803 Rüschlikon-ZH, Switzerland

1. Introduction

These Lecture Notes are being written only a few months after the author finished an extensive review on the use of particle diffraction as a surface crystallographic tool together with T. ENGEL [1]. It comprised detailed discussions of the particle-surface interaction potential, of the mathematical methods for calculating diffraction intensities under inclusion of resonant effects as well as of the experimental requirements and present possibilities for performing diffraction investigations. Finally, a broad survey is presented on all results obtained by the end of 1980 with clean ionic, homopolar and metallic surfaces and with adsorbate systems. Some of these topics are also discussed by other authors in the present Lecture Series. Therefore, it is certainly not appropriate to present here just another short review. It seems of much more value to treat the above-mentioned topics only briefly, but nevertheless to such an extent that an understanding without further reading is possible, and to elaborate some special points not dealt with previously in sufficient detail or for which, in the meantime, progress has been such that a comprehensive discussion becomes worthwhile. With this point of view, these Notes can be taken as both a self-contained introductory text and as a first supplement to [1], especially in connection with the contents of Chapters 2, 3 and 5.

2. The He-Surface Interaction Potential and the Crystallographic Information Contained in the Corrugation Function

Diffraction experiments give information on both the dimensions of the unit cell via the angular location of Bragg peaks and the distribution of the scattering centers within the unit cell via the intensities of diffracted beams. As He atoms with thermal energies cannot penetrate the solid, they pick up the periodicity of the outermost atoms.

To understand the nature of the scattering centers in the case of He scattering, we have to discuss the He-surface interaction potential [2,3]. At distances not too far away from the surface, the impinging atoms feel an attraction due to Van der Waal's forces. Closer to the surface, the incoming particles will be repelled according to the Pauli exclusion principle due to the overlap of their electron densities with that of the solid surface. This causes the steeply rising repulsive part of the interaction potential. In general, the classical turning points of the particles will be farther away on top of the surface ions rather than between the ions. This gives rise to a periodic modulation of the repulsive part of the potential parallel to the surface, which follows a contour of constant electron density, as recently shown in theoretical work by ESBJERG and NORSKOV [4]. Every point of the density contour constitutes a scattering center, and the resulting scattering surface is called the corrugation function $\zeta(\underline{R})$. With higher energies, the He atoms penetrate more deeply into the sea of valence electrons, and the corrugation function corresponds to a contour of larger electron density. ESBJERG and NORSKOV [4] established an almost linear relation between the He energy $E_{He}(r)$ (in eV) and the charge density ρ_0 (in atomic units, i.e., electrons per cubic Bohr radius), which for low energies reads

$$E_{He}(\underline{r}) = 750\ \rho_0 \qquad (\text{for } E_{He} \le 100\ \text{meV})\ , \tag{1}$$

and for higher energies, is given by

$$E_{He}(\underline{r}) = 375\ \rho_0 + 0.065 \qquad (\text{for } E_{He} \ge 100\ \text{meV})\ . \tag{1a}$$

With experimentally available He energies between 20 and 300 meV, charge densities between 2.7×10^{-5} and 6.3×10^{-4} au can thus be covered.

The corrugation function very often reflects directly the geometrical arrangement of the surface atoms. Determination of the energy dependence of the corrugation function and the softness of the repulsive potential can shed light onto the bonding characteristics of the surface atoms. A recent attempt by the author to extract such information from diffraction data for Ni(110) taken for a large range of He energies appears very interesting and promises surprising results [5]. Successful first-principle calculations of surface charge densities, necessary for comparison with experimental data, have been recently reported by HAMANN [6].

The periodicity of the corrugation function is responsible for the diffraction of the particle beam. The energy levels allowed in the attractive well can give rise to resonant scattering of the incoming particles into bound states, often referred to as selective adsorption [7]. Although resonance line shapes can be used to remove ambiguities in the determination of corrugation functions from experimental data [8], we shall neglect the existence of the attractive part of the potential entirely in the present discussion. This is justified by the fact that the depth of the attractive potential is always small for He atoms ($D \lesssim 10$ meV) compared to the incoming He energies ($E = 25$-300 meV).

3. Data Analysis

Diffraction Geometry

The Laue condition for diffraction from a two-dimensional periodic array relates the incoming wavevector $\underline{k}_i = (\underline{K}, k_{iz}) = (k_i \sin\theta_i, -k_i \cos\theta_i)$ with the outgoing $\underline{k}_{\underline{G}} = (\underline{K}_{\underline{G}}, k_{\underline{G}z})$ via

$$\underline{K} - \underline{K}_{-\underline{G}} = \underline{G}\ . \tag{2}$$

θ_i denotes the angle of incidence of the incoming beam as measured from the surface normal. The energy of the particles remains unchanged during diffraction, so that $k_i^2 = k_{\underline{G}}^2$.

The reciprocal lattice vectors $\underline{G} = j\underline{b}_1 + l\underline{b}_2$ are specified by the following relationship between the direct ($\underline{a}_i, i = 1,2$) and the reciprocal ($\underline{b}_i, i = 1,2$) unit-cell vectors:

$$\underline{a}_i \cdot \underline{b}_k = 2\pi\, \delta_{ik} . \tag{3}$$

This implies that $\underline{b}_1$ is normal to $\underline{a}_2$, and $\underline{b}_2$ is normal to $\underline{a}_1$. The lengths of b_1 and b_2 are determined by

$$b_1 = \frac{2\pi}{a_1 \sin\beta} \qquad b_2 = \frac{2\pi}{a_2 \sin\beta} , \tag{4}$$

where β denotes the angle between $\underline{a}_1$ and $\underline{a}_2$; $\underline{a}_1$ is usually chosen to be the shorter unit-cell vector. Figure 1 illustrates the relation between the direct and reciprocal lattices.

The Laue condition can be graphically represented by the Ewald construction, shown schematically in Fig.2. A projection of this construction is also shown in Fig.1b. Both figures are exhibited with the purpose of illustrating the derivation of expressions for the diffraction angles for the most general case of an oblique lattice and an arbitrary angle of incidence γ with respect to $\underline{a}_1$ (see Fig.1b). The diffraction angles are measured with a normal goniometer: θ_G is the angle which the detector has to be rotated from the surface normal in the scattering plane (defined by the wavevector $\underline{k}_i$ and the surface normal). The angle through which the detector has to be moved out of plane to reach the reciprocal lattice rod $\underline{G} = (j,l)$ is given by ϕ_G. With the relations $\sin\phi_G = A/k_i$ and $\sin\theta_G = D/(k_i\cos\phi_G)$, which can be obtained

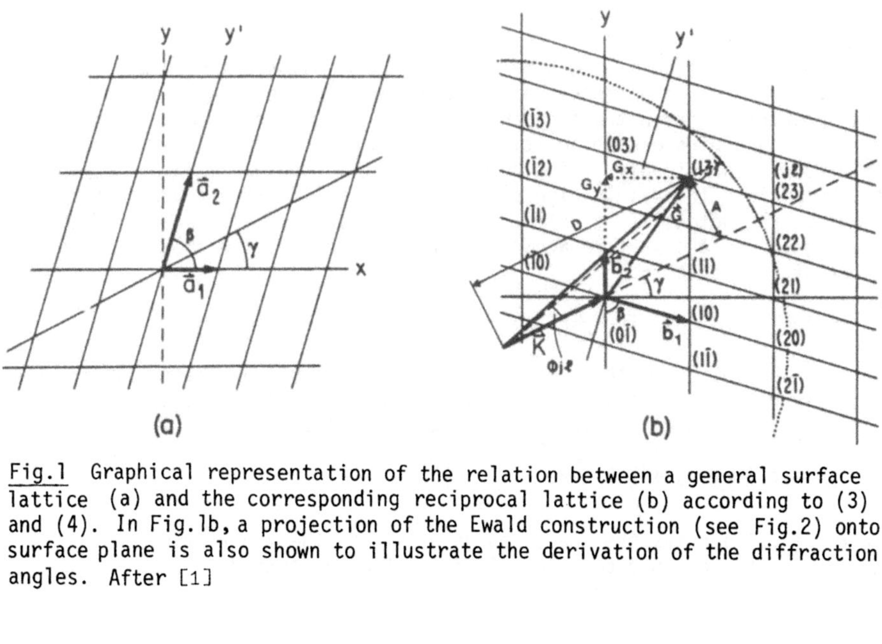

Fig.1 Graphical representation of the relation between a general surface lattice (a) and the corresponding reciprocal lattice (b) according to (3) and (4). In Fig.1b, a projection of the Ewald construction (see Fig.2) onto the surface plane is also shown to illustrate the derivation of the diffraction angles. After [1]

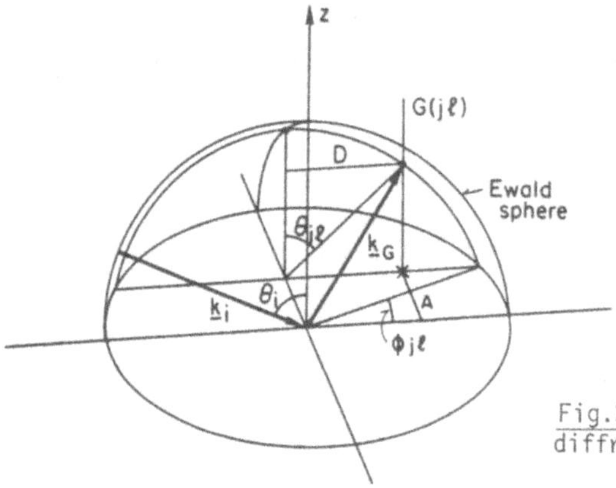

Fig.2 The Ewald construction for diffraction from surfaces. After [1]

from Fig.2, and with $A = -G_x\sin\gamma + G_y\cos\gamma$ and $D = K + G_x\cos\gamma + G_y\sin\gamma$ (see Fig.1b), the following formulae are readily obtained:

$$\sin\phi_{jl} = \left[-\frac{j}{a_1}(\sin\gamma + \cos\gamma\cot\beta) + \frac{l\cos\gamma}{a_2\sin\beta}\right]\lambda_i$$
$$\sin\theta_{jl} = \frac{1}{\cos\phi_{jl}}\left[\sin\theta_i + \lambda_i\frac{j}{a_1}(\cos\beta - \sin\gamma\cot\beta) + \lambda_i\frac{l\sin\gamma}{a_2\sin\beta}\right]. \qquad (5)$$

Sketching the derivation of these formulae seemed worthwhile, as an analysis of the diffraction angles yields the important information on the orientation and the dimensions of the unit cell as well as on the angle between the basic lattice vectors.

The Hard Corrugated Wall Model

For the following discussion of the calculation of diffraction intensities for a given scattering geometry and corrugation function as well as for the inverse problem of reconstructing the corrugation function from a given set of diffraction angles and intensities, we use the so-called hard-wall model. In this model, the steeply rising repulsive part of the potential is approximated by a rigid infinitely high and steep wall [9]. Effects due to the softness of the potential as well as due to its nonrigidness because of the thermal motion of the surface atoms are discussed briefly in the final two sections of this chapter.

Calculations of Intensities for Given $\zeta(\underline{R})$: The Direct Problem

We restrict ourselves to the so-called Rayleigh assumption, which states that the incoming and outgoing beams can be considered as plane waves up to the surface. The total particle wave function is then

$$\psi(\underline{r}) = \exp(i\,\underline{k}_i\,\underline{r}) + \sum_{\underline{G}} A_{\underline{G}} \exp(-i\,\underline{k}_{\underline{G}}\,\underline{r}) \;, \qquad (6)$$

where $\underline{r} = (\underline{R},z)$ denotes the space coordinates with the z direction chosen perpendicular to the surface. In principle, the sum extends over all $\underline{G}$'s, i.e., over propagating ($k_{\underline{G}z}^2 > 0$), as well as over evanescent ($k_{\underline{G}z}^2 < 0$) waves. The intensities of the former, given by

$$P_{\underline{G}} = |A_{\underline{G}}|^2 \frac{k_{\underline{G}z}}{k_{iz}} , \tag{7}$$

must obey the unitarity condition $\sum P_{\underline{G}} = 1$ (particle flux conservation). With the boundary condition $\psi[\underline{R},\zeta(\underline{R})] = 0$, which simply states that no particles can penetrate the hard wall, we arrive at the equation

$$\sum_{\underline{G}} A_{\underline{G}} \exp i[\underline{GR} + k_{\underline{G}z}\, \zeta(\underline{R})] = -\exp i[k_{iz}\, \zeta(\underline{R})] \tag{8}$$

which must be fulfilled for any $\underline{R}$ within the unit cell. Using GARCIA's GR method [9], this equation can be solved numerically by relating a finite number N of $\underline{R}$ vectors uniformly distributed within the surface unit cell with the same number of $\underline{G}$ vectors, and inverting the corresponding matrix equation. Depending on the He wavelength, the unit-cell dimensions and the corrugation amplitude, the dimensions of the matrices required to obtain convergent solutions may be quite large (N = 200).

Multiplying (3) with $\exp[-i(\underline{G}'\underline{R} + k_{\underline{G}'z}\, \zeta(\underline{R})]$ and integrating over the surface unit cell of area S yields the matrix equation

$$\sum_{\underline{G}} M_{\underline{GG}'} A^0_{\underline{G}} = A_{\underline{G}'} \tag{9}$$

with

$$M_{\underline{GG}'} = \frac{1}{S} \int \exp[i(\underline{G} - \underline{G}')\underline{R} + i(k_{\underline{G}z} - k_{\underline{G}'z})\, \zeta(\underline{R})]d\underline{R} \tag{10}$$

and

$$A_{\underline{G}} = \frac{1}{S} \int \exp[i\underline{GR} + i(k_{\underline{G}z} - k_{iz})\, \zeta(\underline{R})]d\underline{R} , \tag{11}$$

which can be solved either by matrix inversion [10] or by iteration [11] for the $A_{\underline{G}}$'s. With the Rayleigh Ansatz, convergent solutions are obtained only for corrugation amplitudes not too large [12]. For example, the maximum corrugation amplitude for a one-dimensional sinusoidal corrugation is 14.3% of the lattice constant; for a two-dimensional quadratic corrugation described by a sum of cosines in both x and y directions, the maximum amplitude is 18.8% [13].

Neglect of off-diagonal terms in (9), which corresponds to neglect of multiple scattering, and implies neglect of evanescent waves, yields $A_{\underline{G}} = A^0_{\underline{G}}$. This so-called eikonal approximation [14] works reliably for small wavelengths, shallow corrugations and small angles of incidence but has a number of serious drawbacks. 1) Neither the threshold behavior of the beams vanishing at the horizon nor the structure occurring in the angular dependence of the intensities of the neighboring observable beams are reproduced correctly in the eikonal approximation. 2) The eikonal formula yields the same intensity distributions for $\zeta(\underline{R})$ and $-\zeta(-\underline{R})$, so that for surfaces with two-dimensional inversion symmetry, for which $\zeta(\underline{R}) = \zeta(-\underline{R})$, no decision can be made whether $\zeta(\underline{R})$ or $-\zeta(\underline{R})$ describes the surface profile [15]. 3) Within the eikonal approximation, the specular intensity as a function of angle of incidence is the same irrespective of the azimuthal orientation of the sample. However, the eikonal formula has the invaluable advantage of being computationally fast.

Methods beyond the Rayleigh limit have also been developed [16,17], but most of the real surfaces investigated up to now have corrugation amplitudes below the Rayleigh limit.

Reconstruction of the Corrugation Function from Measured Intensities: The Inverse Problem

In reality, one faces the inverse problem of that discussed in the last paragraph: the derivation of the corrugation function from a given set of measured intensities. In most cases analyzed up to now, both the shape and the height of the corrugation were changed, until optimum agreement between measured and calculated intensities was obtained. As a measure for the agreement, "reliability factors" like

$$R_0 = \frac{1}{N}\left[\sum_{\underline{G}}(P_{\underline{G}}^{calc} - P_{\underline{G}}^{exp})^2\right]^{1/2} \tag{12}$$

can be used. N denotes the number of measured beams $\underline{G}$. The corrugation function is often written as a Fourier sum

$$\zeta(\underline{R}) = \sum_{\underline{G}} \zeta(\underline{G}) \exp(i\underline{G}\underline{R}) = \sum_{j,l} \zeta(j,l) \exp[2\pi\, i(j\frac{x}{a_1} + l\frac{y}{a_2})] \ , \tag{13}$$

and the best-fit Fourier coefficients $\zeta(\underline{G})$ have to be determined by a systematic variation of their amplitudes. The number of Fourier coefficients necessary may be quite large, and the amplitude steps should not be taken too small, because otherwise one could mistake a local minimum in R_0 as the real one. Therefore, these calculations may be very tedious and time consuming. However, it has to be emphasized that such a procedure is completely free of any model assumptions concerning the structure of the surface.

An investigation of the symmetry properties of $\zeta(\underline{R})$, which can be performed experimentally by observing symmetries in the angular locations and intensities of the beams, is very helpful to restrict the number of possible Fourier coefficients. Let us consider as an example a surface with rectangular unit cell for which mirror planes along the x and y directions could be established in this way. Reality of $\zeta(\underline{R})$ requires that $\zeta(\underline{G}) = \zeta(-\underline{G})^*$ or $\zeta(j,l) = \zeta(-j,-l)^*$. The mirror plane in the x direction states that $\zeta(\underline{R})$ as observed from a high symmetry point at the surface appears the same for +y and -y. This renders $\zeta(j,l) = \zeta(j,-l)$. The mirror plane in the y direction yields $\zeta(j,l) = \zeta(-j,l)$. Therefore, $\zeta(j,l) = \zeta(-j,-l)$ is also valid. The latter condition would also follow from inversion symmetry alone. Thus, for the surface considered here, the indices of the Fourier coefficients may be restricted to values $j,l > 0$, and the sum over exponential functions restricts to a sum over cosine functions. The existence of domain structures can simulate a higher symmetry than actually present on the surface. It is clear that this possibility renders additional complications in analyzing diffraction data [18].

A more direct approach to derive the best-fit corrugation has been developed recently by the author in collaboration with CELLI and GARCIA [19].

Starting point is (14). Assuming that $k_{\underline{G}z}$ can be approximated by $-k_{iz}$, and rewriting the equation as

$$\zeta(\underline{R}) = \frac{1}{2ik_{iz}} \ln \left| - \sum_{\underline{G}} A_{\underline{G}} \exp(i\underline{G}\underline{R}) \right| , \tag{14}$$

one observes that the corrugation function can be expressed in terms of the scattering amplitudes $A_{\underline{G}}$. Application of (14) requires full knowledge of the scattering amplitudes, i.e., absolute values $A_{\underline{G}}$ plus phases $\varphi_{\underline{G}}$. An example for a two-dimensional corrugation should be sufficient to illustrate the range of applicability of (14) [20]. We consider a quadratic lattice of unit-cell length a = 2.84 A, and a corrugation function given by

$$\zeta(\underline{R}) = \frac{1}{2}\,\zeta(10)\,(\cos\frac{2\pi}{a}x + \cos\frac{2\pi}{a}y) + \zeta(11)\,\cos\frac{2\pi}{a}x\,\cos\frac{2\pi}{a}y \,. \tag{15}$$

The diffraction amplitudes for an incoming wavelength of 0.8 A and an angle of incidence of 20° are calculated with the eikonal approximation, and are then used in connection with (14) to derive the complex corrugation function point by point for a quadratic grid. The real parts of these values are then used in a Fourier analysis which comprises four Fourier coefficients $\zeta(10)$, $\zeta(11)$, $\zeta(20)$ and $\zeta(12)$. Everything known about the symmetry of the surface is incorporated into the Fourier representation of the corrugation. For increasing total corrugation amplitudes, the following results are obtained:

Input parameters	Result obtained with (14)
$\zeta(10)$ = 0.12 A, $\zeta(11)$ = 0.08 A	$\zeta(10)$ = 0.118 A, $\zeta(11)$ = 0.080 A $\zeta(20)$ = -0.001 A, $\zeta(12)$ = -0.002 A
$\zeta(10)$ = 0.15 A, $\zeta(11)$ = 0.10 A	$\zeta(10)$ = 0.141 A, $\zeta(11)$ = 0.093 A $\zeta(20)$ = -0.002 A, $\zeta(12)$ = -0.004 A
$\zeta(10)$ = 0.225 A, $\zeta(11)$ = 0.15 A	$\zeta(10)$ = 0.196 A, $\zeta(11)$ = 0.128 A $\zeta(20)$ = -0.009 A, $\zeta(12)$ = -0.007 A

We observe that for small corrugation amplitudes, (14) reproduces the important Fourier coefficients very accurately, and that the extra coefficients come out very small. For larger corrugation amplitudes, the "real" Fourier coefficients come out systematically too small, the deviations being roughly proportional to the corrugation amplitudes; the extra coefficients increase, which indicates a slight deformation of the form of the corrugation. Nevertheless, (14) gives reasonable results for total corrugation amplitudes up to values near to the Rayleigh limit. The explanation for the systematic deviations for large corrugation amplitudes probably lies in the fact that evanescent waves, whose amplitudes have not been taken into account at all, play a more important role the larger the corrugation amplitudes become.

Of course, in real diffraction experiments only beams within the Ewald sphere can be measured, so that no information on the amplitudes of the evanescent beams is available. The situation is actually even worse, as only intensities according to (6) are measured, so that the information on the phases of the diffracted beams is not available either. However, the value of (14) lies in the fact that a rough estimate of the phases suffices to obtain an approximate corrugation function. This function can be uesed to generate a full set of phases, which allows calculation of an improved $\zeta(\underline{R})$. The procedure can be repeated until optimum agreement between the measured and calculated intensities is obtained. To obtain a good estimate for the initial phases, investigation of a coarse mesh of phases (steps $\pi/2$) for three or four of the most intensive beams is almost always sufficient. As the phases exhibit a symmetry similar to the intensities, the phases of symmetry-related beams can be varied in parallel. For clarity, Fig.3 shows the flow diagram of a computer program used by the author. For large corrugation amplitudes, a systematic increase of the amplitude of the corrugations calculated can be performed to improve the calculated $\zeta(\underline{R})$. As an example, we chose the largest of the corrugations cited above. A rather good fit was already obtained after six iterations with all phases set zero initially, and the resulting parameters were: $\zeta(10) = 0.223$ A, $\zeta(11) = 0.167$ A, $\zeta(20) = -0.012$ A and $\zeta(12) = 0.010$ A.

It has to be emphasized that the "inversion procedure" described above yields fits only near the optimum parameters, the reason being that different approximations are used for the direct and the inverse calculations. It is nevertheless very valuable as it not only shows which Fourier coefficients are the important ones, but also yields values of their amplitudes rather near to the optimum values. The final best-fit values can then be easily obtained by direct calculations, as outlined at the beginning of this section,

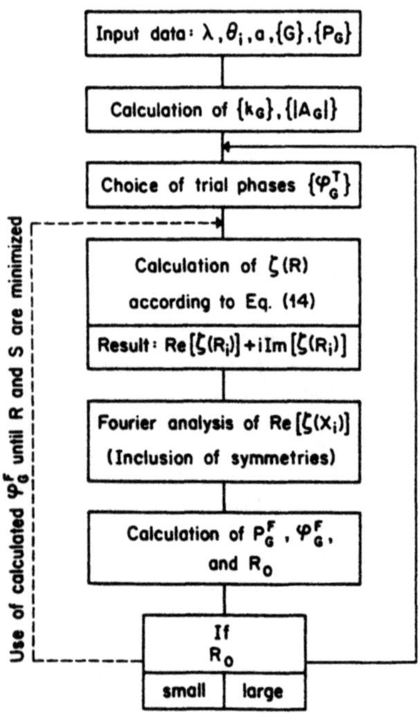

Fig.3 Flow diagram of the computational "inversion procedure" to approach the best-fit corrugation function from a given set of diffraction intensities. After [19]

using small variations around the parameters obtained with the "inverse procedure". An improved method which uses (8) self-consistently for both the direct and the inverse calculations has been developed recently by R.H. SWENDSEN [21]. It has already been successfully applied to real systems exhibiting one-dimensional corrugations.

Influences due to the Softness of the Potential

ARMAND and MANSON [22] were the first to study effects due to softness of the repulsive potential on diffraction intensities. They were able to solve the scattering equation for one- and two-dimensional corrugations for repulsive potentials of the exponential form

$$V(z) = C \exp\left[-\kappa\left(z - \zeta(\underline{R})\right)\right] \qquad (16)$$

which allows wave penetration into the potential region to a degree dependent on the value of κ. For $\kappa = \infty$, the results correspond to that of the hard corrugated wall model. Although the numerical procedures developed so far give convergent results only for very small corrugation amplitudes (ζ_m = 0.03 a), several important conclusions could be drawn from the model calculations: 1) For soft potentials, scattering into the specular beam is enhanced, so that an anlysis of experimental data on the basis of the hard corrugated wall model tends to underestimate the corrugation amplitudes. 2) The discontinuities observed at threshold angles are dependent on κ; ARMAND *et al.* [23] have used the latter effect to estimate the value of κ on the basis of experimental data for the system H2/Cu(100), and arrive at κ = 8 A^{-1}. 3) The diffraction intensities of beams emerging near the surface normal are enhanced relative to those with grazing emergence [24]; this has recently been observed by the author in his studies of the energy dependence of the corrugation function of Ni(110), and gives rise to the hope that, besides the Fourier coefficients describing the corrugation, also the "softness-parameter" κ can be derived from the diffraction data. Thus, a complete mapping of the electron density should become possible, with important consequences for an understanding of bonding at the surface.

Influences due to the Thermal Motion of the Surface Atoms

The thermal motion of the surface atoms gives rise to a decrease in the intensities of the diffracted beams without change in shape in analogy to the diffraction of X rays or neutrons. However, inelastic contributions might contribute appreciably at elevated temperatures under the diffraction peaks as well as at their wings, so that the shape might be distorted. The simple Debye-Waller theory used in X-ray and neutron diffraction for the description of the thermal attenuation of the diffraction beams is not entirely suited in the case of atomic and molecular diffraction [25-27]. It is only reasonable if the interaction time between the particle and the surface is short in comparison with the oscillation period of a surface atom. Due to the long-range attractive part of the interaction potential, this is never the case for heavy particles or for light particles with low velocities. With the depth D of the attractive potential known, one can correct for this effect, provided the particle velocities are sufficiently high, using a formula for the Debye-Waller factor originally proposed by BEEBY [28]. We cite here the form given by HOINKES *et al.* [29],

$$\exp(-2W) = \exp[-(\Delta \underline{k}^2) + (8mD/\hbar^2)] \ \langle u^2 \rangle , \qquad (17)$$

where Δk denotes the change in momentum during diffraction, m the mass of the scattered particle, and $\langle u^2 \rangle$ the mean-square displacement of the surface atoms, represented in terms of the Debye temperature Θ_D by

$$\langle u^2 \rangle = \frac{3\,\hbar^2\,T_S}{M\,k_B\,\Theta_D^2} . \tag{18}$$

These formulae immediately show a number of facts to be borne in mind for an analysis of diffraction intensities: The diffracted beams are more attenuated the larger the surface temperature T_S and the larger the beam energy, and the attenuation is smaller the larger the Debye temperature. The Debye-Waller attenuation becomes smaller with increasing angle of incidence, and increases with increasing potential depth. For any nonzero angle of incidence, diffraction beams scattered towards grazing emergence are enhanced relative to the specular, whereas the reverse is true for beams emerging near the surface normal; this is just opposite to the effect due to the softness of the potential (see preceding section). A serious analysis of diffraction intensities would therefore require measurements at different temperatures and extrapolation to $T_S = 0$. However, in many cases the Debye temperature is sufficiently high, so that the differences in the relative intensities of the beams for different temperatures are smaller than the experimental errors made in the intensity determinations; in such cases, the beams need not be corrected. This is especially the case for small corrugations where only beams very near to the specular are of appreciable intensity. Nevertheless, measurements should always be performed over a temperature range as wide as possible to obtain a feeling for the influence of the Debye-Waller attenuation and also for the contributions of the inelastic background. It is clear that in these respects the possibility of performing measurements at very low surface temperatures is especially valuable.

4. Experimental Aspects

To extract the maximum possible information for a diffraction experiment, both the angular locations and the intensities of the Bragg peaks must be measured. In their historical experiments, STERN and ESTERMANN [30] used low-pressure effusion sources which gave beams with low intensity and a broad Maxwellian velocity distribution, so that only first-order diffraction peaks could be well resolved. It took almost forty years of research in gas dynamics until sources delivering thermal beams with high intensity and narrow velocity distribution became available. In modern high-pressure nozzle sources, a complicated interplay of adiabatic cooling and hydrodynamic flow leads to a narrowing of the velocity distribution and to an enhancement of the mean velocity of the He atoms by a factor of 5/3 relative to the most probable velocity of an effusion source at the same temperature. Such sources are now commonly in use in gas-surface scattering experiments. The gas load due to the high pressure behind the nozzle is handled by several differential pumping stages using high-speed pumps. With such arrangements, pressures of 5.10 Torr can be maintained in the scattering chamber with the beam on. A well-designed nozzle-skimmer geometry is essential for good performance of the beam system. Typical He-beam parameters for a nozzle at room temperature are: $E = 63$ meV or $\lambda = 0.57$ A; with a pressure behind the nozzle of 1 atm, a velocity spread (FWHM) of about 12% of the mean velocity can be readily achieved. Higher pressures behind the nozzle decrease the velocity spread

[31]. The mean energy of the beam can be varied easily by heating and cooling the nozzle.

The particles scattered from the target are favorably analyzed with a quadrupole mass spectrometer mounted on a goniometer, which allows in-plane as well as out-of-plane detection. The possibility of out-of-plane detection also including large scattering angles is not only important for determination of the symmetry properties of the surface under investigation, but also for complete determination of the diffraction intensities, necessary if one wants to apply the "inversion procedure" for determination of the corrugation function as outlined in the previous chapter. The possibility of rotation of the sample around its surface normal is also very helpful in these connections.

Modulation of the beam with a mechanical chopper in the last pumping stage before the scattering chamber, and phase-sensitive detection result in an improved signal-to-noise ratio. Surface cleaning, using ion bombardment and heating cycles, and surface characterization, using standard methods like Auger electron spectroscopy, secondary ion mass spectroscopy and LEED, may be carried out in the scattering chamber.

5. Examples

Metals

Surface-structure investigations on clean surfaces are motivated by two main reasons: First, many processes of technological importance occur at the topmost layers of solids, and are not only sensitive to the chemical composition of the surfaces, but also to their geometry. Second, there is often a subtle balancing of interatomic forces in the bulk, which can be disrupted as soon as such a strong perturbation as the creation of a surface takes place. The system then tries to find a new minimum of potential energy, which is often realized via a rearrangement of the atoms at and near the surface, so that the atomic configuration no longer resembles that of the bulk (surface reconstruction) [32]. Thus, a lot can be learned about atomic interactions through detailed studies of surface structures.

Whereas He diffraction from ionic and covalent crystals yields appreciable diffraction intensities [33-36], diffraction from close-packed metal surfaces was long searched for until BOATO *et al.* [37] and HORNE and MILLER [38] managed to detect extremely weak first-order diffraction peaks (0.2% of the specular) for Ag(111). This indicates a corrugation amplitude as small as $<$ 0.01 A, and clearly demonstrates the smoothing effect of the free electron charge at metal surfaces. More open metal surfaces than the Ni(110) [39] and W(112) [40] show, for small He energies (20-60 meV), larger corrugation amplitudes (0.06 and 0.15 A, respectively) in the direction perpendicular to the close-packed rows, where smoothing is less effective.

He-diffraction investigations on reconstructed surfaces have been performed for Au(110) and Au(100) [41,42]; Au(110) shows a (1 $\times$ 2) reconstruction with LEED. The He-diffraction data indicate a one-dimensional corrugation in the direction perpendicular to the close-packed rows with the double periodicity of the distance between the rows in agreement with the LEED observation. Diffraction data are dominated up to rather high energies by resonant scattering features, which allowed determination of the bound-state energies [41]. Diffraction data at higher beam energies could not be performed with sufficient experimental resolution to resolve the high-order beams. However, a pronounced cutoff in intensities at an angle of $\Delta\Theta = 65^0$ off the specular can be used to estimate the corrugation amplitude by means of the classical rainbow formula

$$\Delta\Theta \doteq 2 \arctan \frac{\pi}{a} \zeta_m , \qquad (19)$$

which is easily derived for a sinusoidal corrugation from purely classical reflection considerations. For the Au(110) (1 × 2) surface, ζ_m was found to lie between 1.2 and 1.5 A, which corresponds to a rather large corrugation amplitude, and strongly supports the missing-row model in which every second of the close-packed rows is missing [41].

LEED studies of the Au(100) surface indicate a (5 × 20) reconstruction with two domains present at the surface [43]. He-diffraction studies show diffraction peaks of 1/5 order in both directions according to the domain structure. No peaks of 1/20 order were observed, although the angular resolution would have been sufficient for a He wavelength of 1.09 A. Therefore, the corrugation function appears one dimensional and has been determined from several sets of experimental data [42]. It is consistent with several of the proposed structures discussed in the literature [43], but no definite decision between them seems to be possible at present. However, the He-scattering result clearly indicates that the topmost layer exhibits a (5 × 1) reconstruction as also observed on the Ir(100) and Pt(100) surfaces with LEED, and that, therefore, the complicated (5 × 20) structure must stem from a rearrangement of the atoms in the next deeper layers, so that it is only observable with LEED.

Stepped metal surfaces, which are of special interest because of their enhanced chemical reactivity [44], have also been investigated with respect to their structure with He diffraction. Especially, a pile-up of valence electron charge at the step sites due to Friedel oscillations was searched for. However, diffraction investigations performed on Cu(117) by LAPUJOULADE *et al.* [45] and on Pt(997) by COMSA *et al.* [46] were not entirely conclusive on this point.

Adsorbate Structures

Structural studies of adsorbate phases are motivated by the fact that in many catalytically activated chemical reactions, surface adsorption of one or more reaction partners is an essential condition for the reaction to occur with high efficiency [47]. A crystallographic study of the adsorption sites and adsorption configurations can yield important information on the adsorbate-substrate as well as on the adsorbate-adsorbate interactions.

The first He-diffraction studies on adsorbate-covered metal surfaces were performed in 1979 by ENGEL and the author [39,48]. Three ordered phases of hydrogen on Ni(110) could be observed and analyzed: A (2 × 6) phase corresponding to a coverage of $\Theta_H = 0.8$ monolayers, a (2 × 1) phase for $\Theta_H = 1.0$ ML, and a (1 × 2) phase for $\Theta_H = 1.6$ ML. Only the (2 × 1) and the (1 × 2) had been previously seen with LEED [49]. The nomenclature used above means that, for a (m × n) structure the surface unit cell has the dimensions ma_1 and na_2, with a_1 being the shorter and a_2 the longer unit-cell vector of the substrate. A coverage $\Theta_H = 0.5$ means that there are half as many adsorbed atoms than atoms in the topmost (unreconstructed) substrate layer.

Typical in-plane ($\Theta = 0^0$) and out-of-plane ($\phi = \pm 7.3$) diffraction spectra for the H(2 × 6) phase are shown in Fig.4. The incident He beam is perpendicular to the close-packed Ni rows. The solid lines correspond to the experimental traces, and the dashed lines to the best-fit eikonal calculations. As obvious from Fig.4, the in-plane spectrum exhibits only even beams of 1/6 order, whereas the out-of-plane spectra shown exhibit only odd beams of 1/6 order. This points to a centered structure, and has the consequence that a description of the surface corrugation within the same coordinate system as that used for the clean surface (but related to the new lattice constants) requires only Fourier coefficients for which the sum j + 1 is even. Besides the data shown in Fig.4, several other spectra obtained for different angles of incidence and different wavelengths corresponding to a total of about 120 intensities were analyzed, so that the best-fit values of the Fourier coefficients describing the corrugation function could bee determined to a high degree of reliability. Its analytical form is given by

$$\zeta(x,y) = -\frac{1}{2}\zeta(02)\cos\frac{2\pi y}{3a_2} - \frac{1}{2}\zeta(04)\cos\frac{4\pi y}{3a_2} - \frac{1}{2}\zeta(06)\cos\frac{2\pi y}{a_2} + \zeta(15)\sin\frac{2\pi x}{2a_1}\sin\frac{10\pi y}{6a_2} + \zeta(13)\sin\frac{2\pi x}{2a_1}\sin\frac{\pi y}{a_2} \tag{20}$$

with the best-fit parameters $\zeta(02) = -0.09$ A, $\zeta(04) = 0.12$ A, $\zeta(06) = 0.09$ A, $\zeta(15) = 0.06$ A and $\zeta(13) = -0.03$ A.

The very structured corrugation of the H(2 × 6) phase is displayed in Fig.5b together with the corrugation of the clean Ni(110) surface (Fig.5b) and that of the H(2 × 1) structure (Fig.5c). The clean surface shows appre-

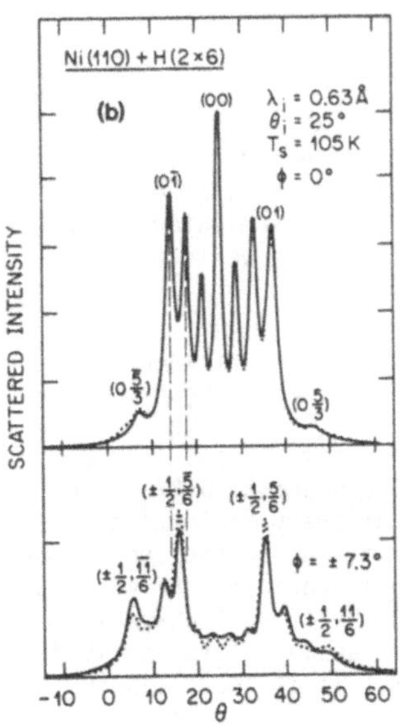

Fig.4 Typical in-plane as well as out-of-plane diffraction traces for the (2 × 6) phase of hydrogen on Ni(110). The full lines correspond to the measured curves, the broken lines to the result of a best-fit calculation taking into account the resolution of the apparatus. After [48]

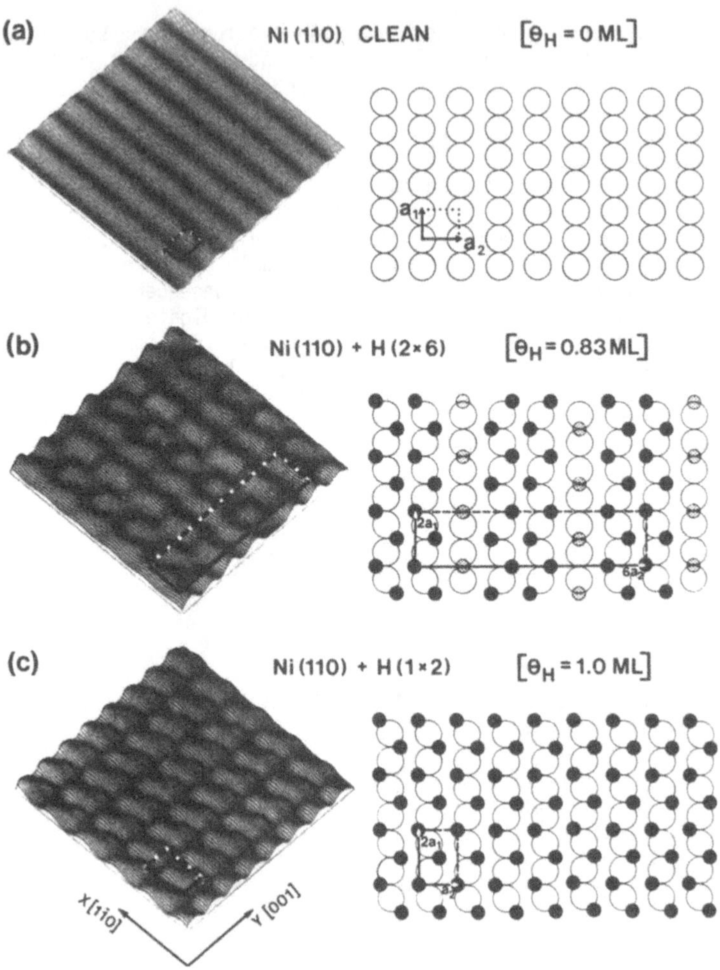

Fig.5 Corrugation functions and corresponding hard-sphere models for (a) the clean Ni(110) surface, (b) the (2 × 6), and (c) the (2 × 1) adsorbate phases of hydrogen on Ni(110). The surface unit cells are indicated. The corrugation functions are expanded by a factor of six in the vertical direction. After [39] and [48]

ciable corrugation only in the direction perpendicular to the close-packed Ni rows (direction y) with an amplitude of only 0.06 A, the corrugation amplitude in the x direction appearing much smaller (≈ 0.01 A) for the small incident He energies (25-60 meV) used. For the H-adsorbate phases (2 × 6) and (2 × 1), the maximum corrugation amplitude has increased to 0.27 A and the corrugation is appreciable in both directions. As every maximum in these corrugations can be attributed to an underlying H atom, the adatom configuration can be inferred direct from these pictures. This conclusion is supported by the fact that the (2 × 6) phase is not at all and the (2 × 1) phase is barely visible with LEED, which means that no Ni displacements are involved in the formation of these structures. Hard-sphere models of the clean surface and the H(2 × 6) and H(2 × 1) structures are shown on the right-hand side of Figs.5a-c. The registry of the H atoms relative to the substrate atoms was chosen such, that the number of equivalent adsorption sites was maximized. It can be seen from Fig.5c, that for $\Theta_H = 1$, all adatoms sit in nearly threefold coordinated sites, but that for $\Theta_H = 0.8$, the lateral repulsion of the hydrogens forces some of them to occupy the energetically less favorable twofold coordinate sites. The coverage of both adsorption phases can be determined simply by counting adatoms per unit cell, and is consistent with flash desorption measurements.

HAMANN [6] has recently succeeded in performing first-principle calculations of surface charge densities of clean and adsorbate-covered systems. His results for the clean Ni(110) and the H(2 × 1) are shown in Fig.6. The experimentally used energy of 60 meV, according to (2), corresponds to a charge density of 8.4×10^{-5} au. The theoretical corrugation amplitude for the clean Ni surface exceeds the experimental value by a factor of three (left side of Fig.6). This might be due to the application of the hard-wall model in analyzing the experimental data, which is known to underestimate the corrugation amplitudes [22]. New data on this surface obtained by the author [5] for a large range of He energies (20-270 meV) do indeed show systematic deviations of beam intensities when fitted with the hard-wall model, and it is presently hoped that these data can be used to find a more realistic value for the corrugation parameter, and also the softness parameter κ, which was also derived in HAMANN's calculations.

Influences due to the softness become less pronounced for larger corrugation amplitudes, and, therefore, the results of HAMANN on the H-covered Ni surface are very important for the future development of He diffraction as a surface crystallographic tool. HAMANN found that with the usual H radius

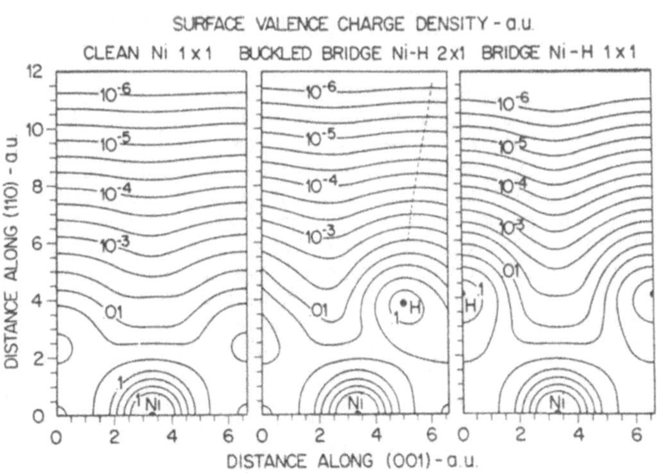

Fig.6 Surface charge-density contours as calculated by HAMANN [6] for clean and hydrogen-covered Ni(110) surfaces. The surface-bonded H and the second Ni layer lie in the plane of the drawings. Calculations were performed for two different adsorption sites. The picture in the middle corresponds to the almost threefold coordinated sites of the H(2 × 1) structure, suggested by the experiment [48].

of 0.6 au (1 au = 0.529 A), the H in the threefold coordinated sites would give a much smaller corrugation than that observed in the experiment. In the twofold coordinated site, on top of the rows, the corrugation would be too large and the (2 × 1) character is not reproduced (right side of Fig.6). A larger H radius of 0.9 au seemed to be more appropriate as indicated by structures of transition-metal cluster molecules. With this value for the H radius, the corrugation amplitudes observed could be reproduced reasonably well establishing a bond length of 3.2 au in the shifted twofold coordination (middle of Fig.6). These results show that He diffraction not only reflects surface atom configurations through the shape of the corrugation function, but that a serious data analysis taking into account all the possible influences discussed in the previous chapter coupled with a theoretical analysis in the sense of HAMANN can also yield information on atomic positions and even bond lengths, although the experiment probes only regions of very low electron density. From the experimental point of view, the theoretical efforts can of course be further supported by performing (and analyzing) measurements over a wide energy range.

The (1 × 2) structure mentioned above corresponds to the saturation phase of hydrogen on Ni(110). Although the amplitude of its corrugation is practically the same as that of the preceding H phases, it appears one dimensional like the corrugation of the clean surface. Therefore, the adatom configuration cannot be determined in the straightforward manner as for the two-dimensional corrugations. However, since a rather dense adatom arrangement is already reached with the (2 × 1) phase, it is conceivable that further H uptake requires drastic changes in the surface geometry. Analyses of the extra LEED spots, which are strong in this case and substantiate the occurrence of substrate reconstruction, were performed by DEMUTH [49] and indicated a pairing of adjacent close-packed Ni rows. Therefore, H can possibly also be adsorbed on the second Ni layer. On the basis of the He-diffraction results, the H-saturation coverage, which was previously unsecure within a factor of five, could be determined to be $\Theta_H = 1.6 + 0.1$.

Oxygen adsorption on Ni(110) leads to a one-dimensional corrugation for the (2 × 1) phase, which occurs at a coverage of about 0.5 ML [50]. The corrugation is orthogonal to that of the clean surface, i.e., it occurs in the x direction (compare Fig.5). This observation rules out the model put forward on the basis of a LEED analysis, in which the oxygens are adsorbed in twofold coordinated sites on top of the Ni rows. It supports a model based on low-energy ion-scattering data by VAN DER BERG *et al.* [51]; their model requires a massive reconstruction of the topmost Ni layer upon adsorption of oxygen in a way that every second row of Ni atoms perpendicular to the close-packed rows is missing. The oxygens are incorporated into the remaining rows, and the dense packing of oxygen and nickel atoms causes the apparent one-dimensional corrugation in the y direction. The corrugation of the O(2 × 1) - Ni(110) system is very similar to that found by LAPUJOULADE *et al.* for O(2 × 1) on Cu(110) [52]. However, these authors propose a different surface structure without substrate reconstruction.

A further He-diffraction study by T. ENGEL and the author was concerned with the (2 × 2) phase of H on Ni(111) [53], which is the only H-adsorption system studied previously with LEED [54]. However, no successful intensity analysis has been possible up to now, so that the proposed open hexagonal structure could not yet be confirmed by He diffraction. A similar difficulty in analyzing experimental data was encountered by the author in his study of the partially disordered c(2 × 4) structure of C_2H_4 on Ni(110) [55]. However, it is hoped that further progress in the development of effective "inversion routines" may help to shed light also onto these interesting structure problems.

References

1. T. Engel, K.H. Rieder: "Structural Studies of Surfaces with Atomic and Molecular Beam Diffraction", Springer Tracts in Modern Physics, Vol. 91 (Springer, Berlin, Heidelberg, New York 1981)
2. H. Hoinkes: Rev. Mod. Phys. 52, 933 (1980)
3. W.E. Carlos, M.W. Cole: Surface Sci. 91, 339 (1980); and E. Zaremba, W. Kohn: Phys. Rev. B 13, 2270 (1976)
4. N. Esbjerg, J.K. Norskov: Phys. Rev. Lett. 45, 807 (1980)
5. K.H. Rieder: to be published
6. D.R. Hamann: Phys. Rev. Lett. 46, 1227 (1981)
7. V. Celli, N. Garcia, J. Hutchison: Surface Sci. 87, 112 (1979)
8. J.M. Soler, V. Celli, N. Garcia, K.H. Rieder, T. Engel: Proc. 4th International Conf. on Solid Surfaces, and 3rd European Conf. on Surf. Sci. (Cannes, France 1980) Suppl. à la Revue "Le Vide, les Couches Minces" No. 201 (1980) p. 815
9. N. Garcia: J. Chem. Phys. 67, 887 (1977)
10. H. Chow, E.D. Thompson: Surface Sci. 82, 1 (1979)
11. R.H. Swendsen, K.H. Rieder: submitted to Surface Sci.
12. P.M. Van den Berg, J.T. Fokkema: J. Opt. Soc. Am. 69, 27 (1979)
13. N.R. Hill, V. Celli: Phys. Rev. B. 17, 2478 (1978)
14. U. Garibaldi, A.C. Levi, R. Spadacini, G.E. Tommei: Surface Sci. 48, 659, (1975)
15. K.H. Rieder, A. Baratoff, U.T. Höchli: Surface Sci. 100, L475 (1980)
16. N. Garcia, N. Cabrera: Phys. Rev. B 18, 576 (1978)
17. R.I. Masel, R.P. Merrill, W.H. Miller: J. Chem. Phys. 65, 2690 (1976)
18. W. Schlup, R.H. Swendsen, K.H. Rieder: to be published
19. K.H. Rieder, N. Garcia, V. Celli: Surface Sci., in press
20. K.H. Rieder: unpublished results
21. R.H. Swendsen: to be published
22. G. Armand, J.R. Manson: Phys. Rev. Lett. 43, 1839 (1979)
23. G. Armand, J. Lapujoulade, J. Lejay: Proc. 4th International Conf. on Solid Surfaces, and 3rd European Conf. on Surf. Sci. (Cannes, France 1980) Suppl à la Revue "Le Vide, les Couches Minces" No. 201 (1980) p. 857
24. G. Armand: private communication
25. A.C. Levi, H. Suhl: Surface Sci. 88, 221 (1979)
26. H. Asada: Surface Sci. 81, 386 (1979).
27. J. Lapujoulade, Y. LeCruer, M. Lefort, Y. Lejay, E. Maurel: Surface Sci. 103, L85 (1981)
28. J.L. Beeby: Proc. 2nd International Conf. on Solid Surfaces (Kyoto 1974) Jpn. J. Appl. Phys., Suppl. 2, Pt. 2, 537 (1974)
29. H. Hoinkes, H. Nahr, H. Wilsch: Surface Sci. 40, 457 (1973)
30. J. Estermann, O. Stern: Z. Phys. 61, 95 (1930)
31. J.P. Toennies, K. Winkelmann: J. Chem. Phys. 66, 3965 (1977)
32. J.A. Appelbaum, D.R. Hamann: Surface Sci. 74, 21 (1978)
33. G. Boato, P. Cantini, L. Mattera: Surface Sci. 55, 191 (1976)
34. P. Cantini, R. Tatarek, G.P. Felcher: Phys. Rev. B 19, 3957 (1979)

35. M.J. Cardillo, G.E. Becker: Phys. Rev. Lett. 42, 508 (1979)
36. M.J. Cardillo, G.E. Becker, S.J. Sibener, D.R. Miller: to be published
37. G. Boato, P. Cantini, R. Tatarek: Proc. 7th International Vac. Congress, and 3rd International Conf. on Solid Surfaces (Vienna 1977) p. 1377
38. J.M. Horne, S.C. Yerkes, D.R. Miller: Surface Sci. 93, 47 (1980)
39. K.H. Rieder, T. Engel: Phys. Rev. Lett. 43, 373 (1979)
40. D.V. Tendulkar, R.E. Stickney: Surface Sci. 27, 516 (1971); and F.O. Goodman: Surface Sci. 70, 578 (1978)
41. K.H. Rieder, T. Engel, N. Garcia: Proc. 4th International Conf. on Solid Surface, and 3rd European Conf. on Surf. Sci. (Cannes, France 1980) Suppl. à la Revue "Le Vide, les Couches Minces" No. 201 (1980) p. 861
42. K.H. Rieder, T. Engel: to be published
43. W. Moritz, D. Wolf: Surface Sci. 88, L29 (1979); and C.M. Chan, M.A. van Hove, W.H. Weinberg, E.D. Williams: Surface Sci. 91, 440 (1980)
44. M. Salmeron, R.J. Gale, G.A. Somorjai: J. Chem. Phys. 70, 2807 (1979)
45. J. Lapujoulade, Y. Lejay: Surface Sci. 69, 354 (1977)
46. G. Comsa, G. Mechtersheimer, B. Poelsema, S. Tomoda: Surface Sci. 89, 123 (1979)
47. T. Engel, H. Kuipers: Surface Sci. 90, 162 (1979)
48. K.H. Rieder, T. Engel: Phys. Rev. Lett. 45, 824 (1980)
49. J. Demuth: J. Colloid. Interface Sci. 58, 184 (1977); and T.N. Taylor, P.J. Estrup: J. Vac. Sci. Technol. 11, 244 (1974)
50. K.H. Rieder: Appl. Surface Sci. 2, 74 (1978)
51. J.A. van den Berg, L.K. Verheij, D.G. Armour: Surface Sci. 91, 218 (1980)
52. J. Lapujoulade, Y. LeCruer, M. Lefort, Y. Lejay, E. Maurel: Phys. Rev. B 22, 5740 (1980)
53. K.H. Rieder, T. Engel: to be published
54. K. Christmann, R.J. Behm, G. Ertl, M.A. van Hove, W.H. Weinberg: J. Chem. Phys. 70, 4168 (1979)
55. K.H. Rieder: unpublished

The Coherence Length in Molecular and Electron Beam Diffraction

G. Comsa

Institut für Grenzflächenforschung und Vakuumphysik, Kernforschungsanlage Jülich, Postfach 19 13
D-5170 Jülich, Fed. Rep. of Germany

1. Abstract

The trouble with the "coherence length" in electron and molecular beam diffraction is that the concept of a transfer width introduced by PARK et al. although essentially correct has often been misinterpreted, while the concept of the so-called "coherence length", found in text books, is in many cases not correct. This lecture will start by recognizing that the diffraction patterns of electrons and molecules are exclusively the result of the interference of each particle with itself. In the case of a perfect surface grating the broadening of the diffractive peaks is the consequence of instrument imperfections. It will be argued that the problem is in fact a visibility problem, i.e., how much information can be extracted from the diffraction pattern in spite of the perturbing effect of the instrument. The main conclusion is that information about periodic structures much larger than the familiar 100 Å length are present in diffraction patterns but that this information can be extracted only by deconvoluting for the instrumental effects.

2. Introduction

The concept of coherence length is used to describe the broadening of diffracted beams due to the nonideality of the instrument. A few years ago a discussion started concerning this concept and its applicability in low-energy electron diffraction (LEED) and molecular beam diffraction (MBD) experiments [1]. In spite of the existence of a correct description of the measured intensities as a function of the "instrument response function" and even of ready to use formulae (for LEED) [2] the discussion seemed to be necessary. Indeed, two other formulae [3,4], leading in general to erroneous results, were still in use at that time, probably because of their intuitiveness. In addition, even the correct description [2] was interpreted in a way which implies the absence of valuable information actually present in diffraction patterns. It seems that the initial discussion [1] and a paper which appeared simultaneously [5] as well as the following controversial discussion [6,7] have contributed to clearing up the main aspects of the matter. Recently, a particularly comprehensive study of "the resolving power of low-energy electron diffractometer and the analysis of surface defects" [8] appeared. The paper contains a very clear discussion of the capabilities of the instrument to detect phase correlations. Moreover it is demonstrated by considering a variety of model surfaces that the size of the ordered domain that can be resolved depends on the type of defects present. Further, "the minimum angle of resolution" is defined which appears to be more appropriate in describing the resolving power of the LEED instrument than the traditional lengths on the surface (coherence length or transfer width).

In view of this situation, there is, at present, no point in writing a new paper on this subject. But, for the benefit of the students of this school, it might be useful to state in a kind of extended abstract the main aspects of the problem by emphasizing the critical points which have been often misunderstood in the past and might, incidentally, also give rise to confusion in the future.

3. The Formation of the Diffraction Pattern

The first point we have to recognize is that in both LEED and MBD experiments we are dealing with perfectly incoherent particles. Electrons and molecules leave their respective source at time intervals distributed completely at random, the same as they arrive at the detector. The contribution of any incidental interference between any two particles having random phases will average out during the measuring time. Therefore the diffraction pattern is exclusively the result of the interference of each particle with itself.

3.1 The Simple-Minded Approach

The formation of this pattern can be described in a simple-minded but intuitive way. Each particle produces its diffraction probability pattern, the shape of which depends on the surface properties. For a perfect periodic surface grating and the usual source-to-sample and sample-to-detector distances the shape of the peaks in this pattern can be considered as a δ function. The probability patterns of the different particles are not identical, the peaks being shifted relative to each other due to the energy and angular spread of the incident beam. The diffraction pattern produced by all the particles results from the incoherent superposition (incoherent because probabilities and not amplitudes add together) of the individual probability patterns weighted according to the energy and angular distributions of the particles. This superposition and the finite acceptance in angle and energy of the detector lead[1] to the shapes of the peaks (angular width $\Delta\theta$) which are actually measured. In the case of a perfect periodic surface grating, these shapes (angular width $\Delta\theta^{\infty}$) represent in fact the "instrument response function", i.e. the broadening effect of the whole instrument.[2]

The real surfaces we are dealing with are not perfect periodic gratings. Even on the best surfaces of monocrystals the well ordered domains have a finite size. This finite size leads also to a broadening $\Delta\theta^{d}$ of the angular width of the diffraction peaks: the smaller the domain size the larger the broadening. Accordingly, the diffraction pattern contains information on the size of the domains. The instrumental width $\Delta\theta^{\infty}$ limits the size of the or-

---

1 From the above it is obvious that the peak shape depends also on the respective diffraction order.

2 Note that the detector is able to do both: to broaden further, but also to narrow the peaks as they result from the angular and energy spread of the incident beam. Indeed, a detector with an energy window narrower than the energy spread of the incident beam will narrow the peaks. Similarly, a well collimated detection may reduce the broadening which results from the finite size of the beam spot on the surface. However, the best collimated detector cannot reduce the broadening due to the finite size of the source, and of course vice versa, a badly collimated detector will broaden the peaks even when using very small sources.

dered domain that can be resolved. This is the reason people have always tried to also express the instrumental effect by a length on the surface. This can be easily done, for instance [1], by considering a finite interval over which a great number of coherent sources are placed and imposing the condition that the beam of wave length λ emitted by these sources at a polar angle θ_f should have an angular width $\Delta\theta^{\infty}$. The length of the interval should be

$$w \cong \frac{\lambda}{\Delta\theta^{\infty} \cos \theta_f} \quad . \tag{1}$$

The artificial character of this length thought to materialize the "instrument response function" is probably somewhat overemphasized in the simplistic argument above. However, from a pedagogical point of view, this emphasis might be more appropriate here.

Similar lengths on the surface were deduced from various approaches and were named "coherence length" or "transfer width". These lengths have been interpreted as the limiting dimension over which phase correlations can be detected. This interpretation was shown to be incorrect [1]. Indeed, as we have just argued, the information is brought in the diffraction pattern by the self-interference of the individual electron wave, which is correlated in phase over distances much greater than these lenghts. As described, the measured diffraction pattern is simply the convolution of the diffraction pattern of the individual electron with the instrument response function. By knowing the latter accurately the information may be, in principle, extracted by deconvolution. This has already been demonstrated in a simple case [5].

Two more or less semantic remarks (semantics often hides or uncovers wrong convictions) concerning the currently used "coherence length of the beam" seem appropriate here:

- not only the beam but the whole instrument determines the shape of the diffraction peaks from a perfect periodic grating and, thus, eventually the "length" on the surface
- this shape is ultimately the result of the superposition of the diffraction pattern of essentially incoherent particles.

If a length is defined at all, to characterize the response of the instrument, then something like "transfer width of the instrument" seems to be a more appropriate expression.

3.2 The Rigorous Approach

The same approach used to obtain the above simple picture of the formation of the diffraction patterns was presented in a rigorous way [2]. The measured intensity $i(\vec{k})_{meas}$ is given by the intensity function $i(\vec{k})$, which is the pattern one would measure with a perfect instrument [3], convoluted with the instrument response function $T(\vec{k})$

$$i(\vec{k})_{meas} = i(\vec{k}) * T(\vec{k}). \tag{2}$$

[3] The intensity function $i(\vec{k})$ results in fact from the incoherent sum of diffraction patterns of individual particles all with the same wave vector $\vec{k}$. For the current source-surface and surface-detector distances the particle waves are in a good approximation plane waves.

The instrument response function $T(\vec{k})$ represents the broadening effect of the instrument and results from the convolution of the different effects leading to this broadening:

$$T(\vec{k}) = T(\vec{k})_S * T(\vec{k})_{\Delta E} * T_d(\vec{k}) * T_D(\vec{k}) \qquad (3)$$

where the terms represent, successively, the effect of finite source size, energy spread, beam width, and finite detector width. The Fourier transform of (2) is

$$F\{i(\vec{k})_{meas}\} = F\{i(\vec{k})\} \cdot F\{T(\vec{k})\} = \phi(\vec{r}) \cdot t(\vec{r}) \qquad (4)$$

where $\phi(\vec{r})$ is the autocorrelation function we are actually looking for, $t(\vec{r})$ the transfer function and $\vec{r}$ the vector connecting grating points. In principle, $\phi(\vec{r})$ can be determined for any $\vec{r}$ unless $t(\vec{r}) = 0$. As $T(\vec{k})$ is, in general, Gaussian, $t(\vec{r})$ is never zero and, thus again, in principle, the limitation can be imposed only by the uncertainty principle. However, in practice the limit is set much earlier by the level of confidence in the measurement of both $i(\vec{k})_{meas}$ and $T(\vec{k})$ from which $F\{i(\vec{k})_{meas}\}$ and $t(\vec{r})$ are obtained. The width of the transfer function $t(\vec{r})$ defined as the transfer width of the instrument [2] is of course only a limited and formal (very much like w in (1)) description of the experimental reality.

The "mininum angle of resolution" of the instrument, defined recently [8] as

$$\theta_{min} = [b_T + x\%b_T)^2 - (b_T - x\%b_T)^2]^{1/2} = 2\, b_T(x\%)^{1/2} \qquad (5)$$

where b_T is the FWHM of $T(\vec{k})$ and x% the error in the determination of the FWHM of both $i(\vec{k})_{meas}$ and $T(\vec{k})$, seems more appropriate. θ_{min} is a measure of the smallest FWHM of $i(\vec{k})$ that can be resolved by the instrument in the particular experiment performed. The advantages of the definition (5) compared to the transfer width are:

- it takes into account the change of the resolving power with the change in accuracy of the determination of the instrument profile. For instance, by repeating the experiment x% can be reduced;
- it is not a function of the type of defect structure under investigation, such as the transfer width was shown actually to be [8].

4. Summary

The measured diffraction pattern results from the coherent interference of each particle with itself and then from the incoherent superposition of the individual patterns. Accordingly, the information contained in the diffraction pattern is partially blurred by the nonideal nature of the instrument, i.e., by the energy and angular spread of the particles at the source and detector. The deconvolution of the information is possible both in principle and, to a significant degree, in practice, if the peak shapes are accurately measured.

References

1. G. Comsa: Surface Sci. 81, 57 (1979)
2. R.L. Park, J.E. Houston, D.G. Schreiner: Rev. Sci. Instrum. 42, 60 (1971)

3. R.D. Heidenreich: Fundamentals of Transmission Electron Microscopy (Interscience, New York, 1964) pp. 101-105
4. J.B. Pendry: Low Energy Electron Diffraction, (Academic Press, London, 1974) pp. 5-6
5. G.-C. Wang and M.G. Lagally: Surface Sci. 81, 69 (1979)
6. D.R. Frankl: Surface Sci. 84, 485 (1974)
7. G. Comsa: Surface Sci. 84, L489 (1979)
8. T.-M. Lu and M.G. Lagally: Surface Sci. 99, 695 (1980)

Charge Density Waves Surface Deformation Studied by Helium Atom Diffraction

P. Cantini

Gruppo Nazionale di Struttura della Materia del CNR and Instituto di Scienze Fisiche dell'Università di Genova, Viale Benedetto XV, 5
I-16132 Genova, Italy

1. Introduction

Helium atom diffraction has been extensively described in this book; the technique appears to be a very effective tool to obtain detailed information on the structural features of complex surfaces such as stepped metal [1], adsorbate-covered [2], and reconstructed [3] surfaces. The angular position of elastic diffraction peaks determines the symmetry and the dimensions of the unit mesh, while the peak intensity is very sensitive to the location of the scattering centers in the mesh. The aim of present paper is to show a typical example in which helium atom diffraction was used to probe surface deformation. I will present data concerning the study of two transition-metal dicalcogenides, the 1T-TaS and the 2H-$TaSe_2$. These layered materials exhibit charge-density wave (CDW) instabilities and superlattice reconstruction at low temperatures, which appear as weak extra spots in X-ray, neutron or electron diffraction. Similar extra spots also appear in the helium diffraction study of the outer layer of these compounds, and the observed superlattice peaks have in some cases the same intensity as that of the main peaks, thus indicating a high sensitivity of atom diffraction to surface deformation. Some of the data presented here have already been published [4]; some other data are still unpublished and their presentation must be considered as the result of a preliminary study [5].

2. Unreconstructed Structure of the Layered Compounds

The transition-element dichalcogenides are layered compounds of the type MX_2, where M = Ta or Nb and X = Se, Te or S. The CDW transition occurs in these materials when a static electronic charge-density wave develops within the conduction electron gas. The crystal lattice is fundamentally involved in the transition since the large Coulomb energy of the CDW must be reduced

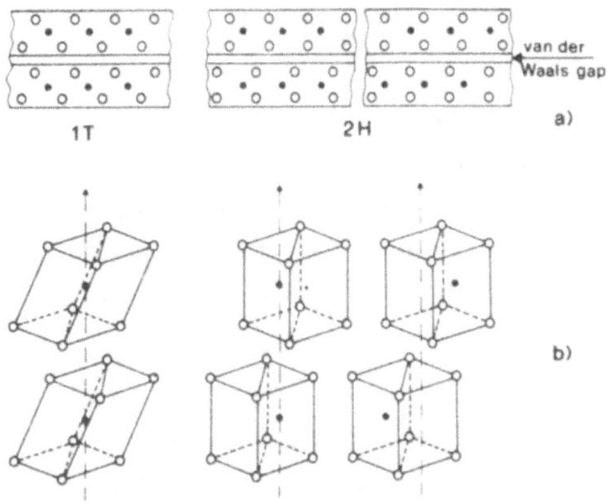

Fig.1a,b. Arrangement of the layers of atoms of the metal (●) and the chalcogens (○) (a) and crystallographic structures (b) of the 1T and of the two possible 2H polytypes

by the accomodating motion of the ions. As a result of the coupling between the lattice and the charge density, a lattice reconstruction develops [6-8].

The unreconstructed compounds contain repeating sandwiches consisting of three layers X-M-X. The bonding of the metal layer to the chalcogen layers in the sandwich is strong and predominantly covalent, while the bonding between adjacent sandwiches is of the Van der Waals type. The structure of the M and X layers is hexagonal; the layers of M and X atoms can be relatively shifted in different ways, so as to generate an octahedral (1T modification) or a trigonal (2H modification) environment of X atoms about M atoms in the sandwich. More generally, adjacent sandwiches can be relatively shifted in different ways, and also sandwiches with different modifications can be interleaved in many ways; this explains the large number of polytypes that can be observed. In Fig.1 the three simpler polytypes with 1T and 2H structures are shown.

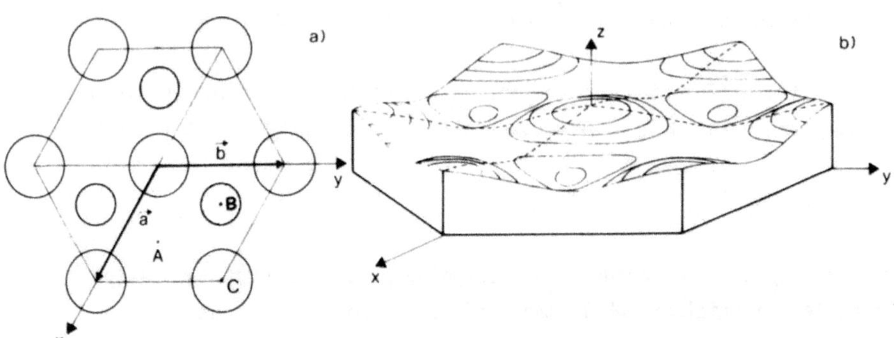

Fig.2a,b. Unit mesh (a) and corrugation profile (b) corresponding to the lower order Fourier component for the unreconstructed surface

Compounds of the same composition in different modifications differ strongly one from another for many properties, like electronic properties and CDW superstructures. As far as the unreconstructed structures are concerned, however, the outer layer of chalcogen atoms, say the surface seen by helium atom probe, appears very similar for all these polytypes; they show a trigonal symmetry with third-order axis perpendicular to the surface. The unit mesh is shown in Fig.2a, where the points labeled A and B are not equivalent, due to the presence of the metal atom in the lower layer.

As is well known, the elastic peak intensity in atom diffraction is explained by assuming a steeply rising gas-surface repulsive potential $V_R[(z - \zeta(\vec{R}))]$ that can be regarded as a hard wall. The shape function $\zeta(\vec{R})$ of the corrugation profile can be expanded in a series

$$\zeta(\vec{R}) = \sum_{G} \zeta_G \exp[i \, \vec{G}\cdot\vec{R}] \quad , \tag{1}$$

and the Fourier components can be guessed from symmetry considerations. In Eq.(1), $\vec{G}$ are reciprocal lattice vectors of the surface. The simpler corrugation profile which describes the unit mesh with the above symmety will have at least two different Fourier components: $\zeta_{10} = \zeta_{\bar{1}1} = \zeta_{0\bar{1}} = \zeta'$ and $\zeta_{01} = \zeta_{1\bar{1}} = \zeta_{\bar{1}0} = \zeta''$. Due to the large radii of the helium atom with respect to the mesh dimensions, we can suppose that, even if $\zeta' \neq \zeta''$, however $\zeta' \simeq \zeta''$. Therefore $\zeta' = \zeta(1-ih)$ and $\zeta'' = \zeta(1+ih)$ with $|h| \ll 1$. Equation (1) becomes

$$\zeta(\vec{R}) = \zeta_1(\vec{R}) + \zeta_2(\vec{R}) = 2\zeta\left[\cos\frac{2\pi}{a}x + \cos\frac{2\pi}{a}y + \cos\frac{2\pi}{a}(y - x)\right] + 2\zeta h\left[\sin\frac{2\pi}{a}x - \sin\frac{2\pi}{a}y + \sin\frac{2\pi}{a}(y - x)\right] \quad . \tag{2}$$

The corrugation function $z = \zeta(\vec{R})$ evaluated with $h = 0.25$ is sketched in Fig.2b.

The elastic diffraction probability P_G can be calculated in the Kirchhoff approximation by assuming a hard corrugated surface (HCS). It is given by

$$P_G = \exp[-2W_G]|S_G|^2 \cos\vartheta_0/\cos\vartheta_G \tag{3}$$

where ϑ and ϑ_G are the incidence and outgoing polar angle and $2W_G$ is the Debye-Waller coefficient, while the diffraction amplitude S_G is

$$S_{m,n} = \frac{1}{a^2}\int_0^a \exp\left\{-i\left[\frac{2\pi}{a}(mx + ny) + |q_{m,n}|\zeta(\vec{R})\right]\right\}d^2\vec{R} \tag{4}$$

where m,n are indices of the $\vec{G}$ vector and $q_{m,n}$ the momentum transfer vertical component. Because $|h| \ll 1$, the deviation from the hexagonal symmetry can be considered as a perturbation,

$$\exp\{-i|q_{m,n}|\zeta(\vec{R})\} = \exp\{-i|q_{m,n}|\zeta_1(\vec{R})\}\exp\{-i|q_{m,n}|\zeta_2(\vec{R})\} \quad , \qquad (5)$$

and if $|q_{m,n}|\zeta_2(\vec{R}) \ll 1$ the following approximation is valid:

$$\exp\{-i|q_{m,n}|\zeta_2(\vec{R})\} \simeq 1 - i|q_{m,n}|\zeta_2(\vec{R}) - \frac{1}{2}|q_{m,n}|^2[\zeta_2(\vec{R})]^2 + \dots \quad . \qquad (6)$$

The scattering amplitude becomes $S_{m,n} = S^0_{m,n} + S^{(1)}_{m,n} + S^{(2)}_{m,n} + \dots$ where $S^0_{m,n}$ is the elastic amplitude for a hexagonal net,

$$S^0_{m,n} = \frac{1}{a^2}\int_0^a \exp\left\{-i\left[\frac{2\pi}{a}(mx+ny) + |q_{m,n}|\zeta_1(\vec{R})\right]\right\}d^2\vec{R} \quad . \qquad (7)$$

The first-order perturbation is

$$S^{(1)}_{m,n} = -h\zeta|q_{m,n}|\left[S^0_{m-1,n} - S^0_{m+1,n} - S^0_{m,n-1} + S^0_{m,n+1} + S^0_{m+1,n-1} - S^0_{m-1,n+1}\right] \quad . \qquad (8)$$

The effect of the deviation from the hexagonal symmetry is clearly different in the two $\phi = 0^o(m \neq 0,\ n = 0)$ and $\phi = 60^o(m = 0,\ n \neq 0)$ azimuthal planes. This effect can be observed in the scattering from a MX_2 layered compound without CDW deformation, or with a weak CDW reconstruction. It was observed in the case of He/2H-$TaSe_2$ [5], a system for which the effect of CDW deformation at the surface is too small to mask the unreconstructed trigonal symmetry. In Figs.3a,b the measured angular distributions are reported; they were taken with a room-temperature beam ($k_o = 11.05$ Å^{-1} or $E_o = 63.8$ meV) and with a crystal at $T_S \simeq 20$ K. The incident angle was $\vartheta_o = 14°$; the position of the diffraction peak corresponds to a surface reciprocal vector $G = 2.11$ Å^{-1} ($a = 3.44$ Å). The more evident difference between the two azimuths is represented by the intensities of the peaks adjacent to the specular peak. Even if less evident, intermediate peaks appear between the main peaks; they are shown in detail in Figs.5a,b and will be discussed in next section. To estimate the corrugation amplitude of the proposed Fourier components, we calculated the diffraction probabilities (3) by numerical integration of (7) for the different values $m' = m,\ m \pm 1$ and $n' = n,\ n \pm 1$ of (7) and (8). The corrugation parameters which are more likely to describe the experimental data are $\zeta = 0.060 \pm 0.005$ Å and $h = 0.07 \pm 0.03$; they give heights $z_A = -0.20$ Å, $z_B = -0.16$ Å and $z_C = 0.36$ Å in the points A, B and C of the unit mesh.

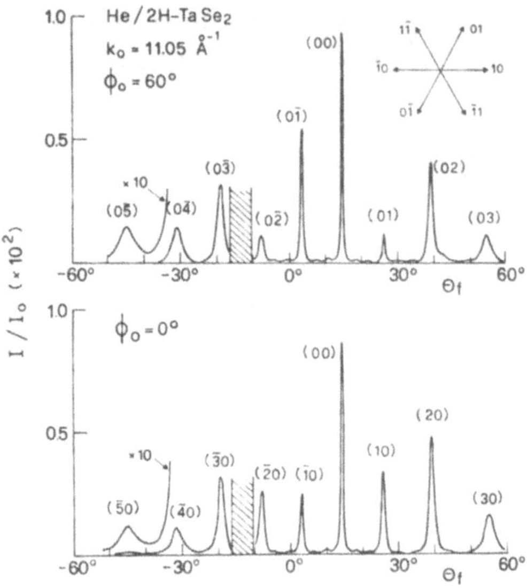

Fig.3a,b. He/2H-$TaSe_2$ diffraction intensity measured in two different azimuths, with the crystal at $T_S = 20$ K. The inset shows the six lower order reciprocal vectors of the unreconstructed structure

3. CDW Deformations

The transition-metal dicalcogenides present CDW superstructures that at lowest temperatures are in general commensurate with the unreconstructed lattice. These structures correspond to metal atom displacements parallel to the layer plane and to calcogen atom displacements both parallel and perpendicular to the layer plane. As has been shown for 1T-TaS_2 [4], such structures can be observed at the surface by helium atom diffraction; they give extra peaks with the same periodicity of the bulk superlattice.

The CDW deformation has in general hexagonal symmetry, with the hexagonal axis normal to the layer. Even in the commensurate phase, the CDW lattice can have different periodicity and orientation with respect to the basic lattice, to generate several different commensurate superstructures. In Figs.4a,b the reciprocal supernets of the 2H-$TaSe_2$ and 1T-TaS_2 compound are reported. As shown in Fig.4, the two supernets show different symmetries for the two systems. The 2H-$TaSe_2$ shows a (3×3) commensurate superstructure; it appears at crystal temperatures lower than 90K and corresponds to a weak surface deformation. It was observed by helium atom scattering, with the crystal at T_S = 20K [5]. The satellite peak intensities are in general proportional to the intensity of the neighboring main peaks, and on the average a factor of 75 weaker, as shown in Fig.5. The 1T-TaS_2 compound shows a quite different structure. The supernet shows a $\sqrt{13} \times \sqrt{13}$ commensurate

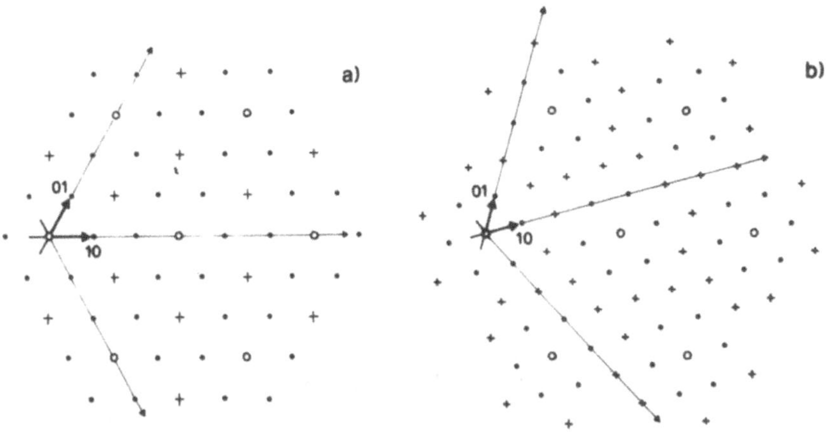

Fig.4a,b. Reciprocal supernets for the commensurate phase of the 2H-$TaSe_2$ (a) and 1T-TaS_2 (b) polytypes. The open circles represents the unreconstructed structure reciprocal vectors

superstructure, rotated by about 14° with respect to the basic structure, as shown in Fig.4b. It appears at a crystal temperature lower than 200K and yields a stronger lattice deformation. The surface reconstruction was observed with helium atom diffraction, with the crystal held at T_S = 80K [4]. The angular distributions plotted in Figs.6a,b show that the superlattice peaks have approximately the same intensity as those corresponding to the

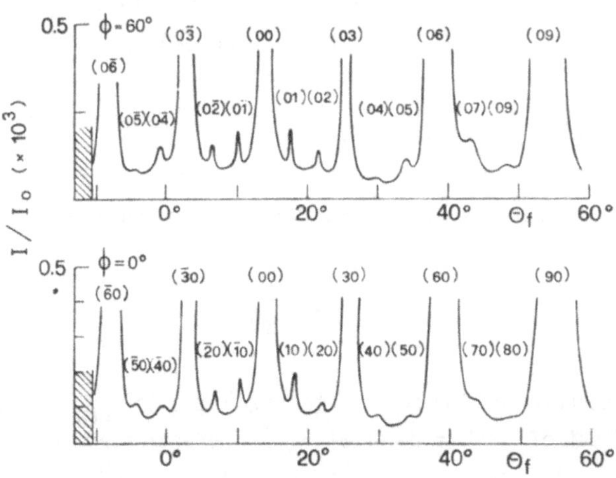

Fig.5. Details of the angular distribution of Fig.3 showing the He/2H-$TaSe_2$ peaks corresponding to the 3×3 commensurate superlattice

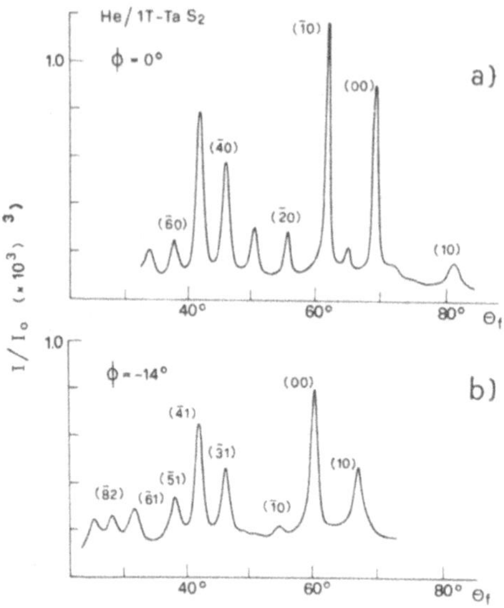

Fig.6a,b. He/1T-TaS_2 diffraction intensity measured in two different azimuths. Curve (a) was measured at $\vartheta_0 = 69°$ in the azimuth containing the CDW peaks. Curve (b) was measured at $\vartheta_0 = 60°$ in the azimuth of the main peaks

basic net. This indicates a strong periodic deformation, the vertical displacement of sulfur atoms being of the same order of magnitude as the surface corrugation.

The amplitude of surface deformation can be obtained by the diffraction peak intensities evaluated in the Kirchhoff approximation and assuming a HCS. The simplest corrugation function which describes the superlattice structure will contain two terms $\zeta(\vec{R}) = \zeta_0(\vec{R}) + \zeta_{CDW}(\vec{R})$. The first term $\zeta_0(\vec{R})$ with the periodicity of the basic net is given by (2); the second term, with hexagonal symmetry, has the periodicity of the CDW's. For the 2H-$TaSe_2$ polytype the simples $\zeta_{CDW}(\vec{R})$ is

$$\zeta_{CDW}(\vec{R}) = 2\zeta_{CDW} \cos\left[\frac{2\pi}{3a} x + \cos \frac{2\pi}{3a} y + \cos \frac{2\pi}{3a} (y - x)\right] \ , \tag{9}$$

and the elastic probability in the Kirchhoff approximation is given by (3), where now the scattering amplitude

$$S_{m',n'} = \frac{1}{9a^2} \int_0^{3a} \exp\left\{-i\left[\frac{2\pi}{3a} (m'x + n'y) + |q_{m',n'}|\zeta(\vec{R})\right]\right\} d^2\vec{R} \tag{10}$$

is obtained as a double integral over the mesh of the supernet. If ζ_{CDW} is sufficiently small, the effect of surface deformation can be estimated as a perturbation. The formalism is completely similar to that developed to evaluate the deviation from the hexagonal symmetry [Eqs.(5) and (6)], where now the unperturbed amplitude

$$S^{0}_{m',n'} = \frac{1}{9a^2}\int_0^{3a} \exp\left\{-i\left[\frac{2\pi}{3a}(m'x+n'y) + |q_{m',n'}|\zeta_0(\vec{R})\right]\right\}d^2\vec{R} \tag{11}$$

gives a nonzero contribution only for diffraction peaks with m',n' = 0; ±3; ±6..., say for the peaks of the main net. The first-order perturbation is

$$S^{(1)}_{m',n'} = -i|q_{m',n'}|\frac{1}{9a^2}\int_0^{3a} \zeta_{CDW}(\vec{R})$$

$$\exp\left\{-i\left[\frac{2\pi}{3a}(m'x + n'y) + |q_{m'n'}|\zeta_0(\vec{R})\right]\right\}d^2\vec{R} \tag{12}$$

and using (9) it becomes

$$S^{(1)}_{m',n'} = -i|q_{m',n'}|\zeta_{CDW}\left[S^0_{m'+1,n'} + S^0_{m'-1,n'} + S^0_{m',n'+1} + S^0_{m',n'-1}\right.$$

$$\left. + S^0_{m'+1,n'-1} + S^0_{m'-1,n'+1}\right] \quad . \tag{13}$$

In (13), $S^{(1)}_{m',n'} \neq 0$ only when (m' ± 1), (n' ± 1) = 0; ±3, ±6;...; say for the diffraction peaks adjacent to the main peaks, and the contribution is proportional to that of the nearest peak

$$S^{(1)}_{m',n'} = -i|q_{m',n'}|\zeta_{CDW}S^0_{m'+1,n'} \tag{14}$$

or similar. We can call these peaks "first-order supernet" peaks.

In the integral the second-order perturbation gives terms like

$$\zeta_{CDW}(\vec{R})^{\;2} = (\zeta_{CDW})^2\left[\exp\left(i\,\frac{2\pi}{3a}\,2x\right) + \ldots + 2\exp\left(i\,\frac{2\pi}{3a}\,x\right) + \ldots + 6\right] \;. \tag{15}$$

Therefore the $S^{(2)}_{m',n'}$ gives a contribution to the main peaks to the first-order and to the second-order supernet peaks. The last one is given by terms like

$$S^{(2)}_{m',n'} = -|q_{m',n'}|^2(\zeta_{CDW})^2 S^0_{m'+1,n'+1} \quad . \tag{16}$$

This contribution is very small when $|q_{m',n'}||\zeta_{CDW}| \ll 1$. Otherwise when $|q_{m',n'}||\zeta_{CDW}| \simeq 1$ all the supernet peaks can be of the same order of magnitude and the perturbation is no longer valid.

From (14) we can estimate the ζ_{CDW} for the 2H-$TaSe_2$ polytype. We measured $\langle P_{m'+1,n'}/P_{m',n'}\rangle \simeq 75$ which corresponds to $(\zeta_{CDW})^2 \; \langle q^2_{m',n'}\rangle \simeq 0.013$; the average value $\langle q_{m,n}\rangle \simeq 20$ Å^{-1} gives $\zeta_{CDW} \simeq 0.006$ Å with vertical displacement of Se atoms of about ±0.025 Å.

The perturbation formalism is not valid in the case of $1T\text{-}TaS_2$; here the diffraction probability can be calculated by direct integration of a $S_{m',n'}$ amplitude like (10), where $\zeta(\vec{R}) = \zeta(\vec{R}) + \zeta_{CDW}(\vec{R})$. The $\zeta_0(\vec{R})$ and the $\zeta_{CDW}(\vec{R})$ must be of the proper periodicity and symmetry. The corrugation parameters which give a rather satisfactory agreement with the experimental results were [4] $\zeta_0 = 0.055 \pm 0.005$ Å and $\zeta_{CDW} = 0.040 \pm 0.005$ Å. The last term corresponds to a surface deformation associated with CDW, with vertical displacements of about ± 0.18 Å, a factor 7 larger than those observed with $2H\text{-}TaSe_2$.

The model potentials proposed to describe the $He/2H\text{-}TaSe_2$ and $He/1T\text{-}TaS_2$ elastic diffraction are surely too simple to describe in detail the unreconstructed structure and the CDW deformation at the surface. They are significant however for showing how sensitive the helium atom probe is both to small deviation from the hexagonal unreconstructed structure and to small CDW surface deformations. We are confident that a detailed study of the complex intensity patter obtained by atom diffraction will give very significant information on the surface structure of these layered compounds.

References

1 J. Lapujoulade, Y. Lejay: Surf. Sci. *69*, 354 (1977)
2 K.H. Rieder, T. Engel: Phys. Rev. Lett. *45*, 824 (1980); To be published
3 M.J. Cardillo, G.E. Becker: Phys. Rev. Lett. *42*, 508 (1979); Phys. Rev. B*21*, 1497 (1980)
4 G. Boato, P. Cantini, R. Colella: Phys. Rev. Lett. *42*, 1635 (1979); Physica B*99*, 59 (1980)
5 P. Cantini, R. Colella: Unpublished results
6 J.A. Wilson, F.J. Di Salvo, S. Mahajan: Adv. Phys. *24*, 117 (1975)
7 C.B. Scruby, P.M. Williams, G.S. Parry: Philos. Mag. *31*, 255 (1975)
8 D.E. Moncton, J.D. Axe, F.J. Di Salvo: Phys. Rev. Lett. *35*, 20 (1975)

Part II

Characterization of Adsorbed Phases

Phase Transitions in Surface Films

M. Bienfait

Département de Physique, Faculté des Sciences de Luminy, Case 901
F-13288 Marseille Cédex 9, France

Abstract

Various two-dimensional phases are stable when adsorbed on crystalline surfaces. It is shown how the reduced dimensionality and the surface symmetry can produce some original properties in two-dimensional matter. The subject is illustrated by a few typical examples taken among the order-disorder, solid-solid and gas-liquid transitions.

1. Introduction

Among the various topics explored very actively during the last past years in surface physics and surface physical chemistry, two-dimensional phase transitions were paid particular attention from numerous groups of experimenters and theoreticians. A lot of new and unexpected results were obtained bringing a comprehensive view on the behavior of two-dimensional adsorbed phases and on their transitions. Several international conferences [1,2], books [3,4] or review papers [5,6] testify to the dynamism of this field. The reader interested in a deep and complete analysis of 2D phase transitions is invited to study the above references because I don't want, in this lecture, to summarize the whole subject. I prefer, by choosing typical examples, to illustrate the richness of surface phase transitions and to show their specificity by pointing out, for instance, the importance of symmetry and dimensionality. I also want to compare, as my lecture progresses, the usual experimental techniques employed in surface phase characterization, and show their main advantages and limitations.

All the examples are chosen, as often as possible, among the latest results obtained on submonolayers of simple molecules adsorbed on well-defined crystalline surfaces.

In Fig.1a is schematically represented the classical phase diagram for bulk matter. The abscissa is the volume V per molecule and the ordinate is the pressure P, the temperature T or the chemical potential μ. We can imagine an adaptation of this bulk phase diagram for surface films. In an adsorbed layer, the volume V is replaced by the surface S occupied by a molecule. So, a phase diagram looking like the one drawn in Fig.Ib is expected in two dimensions. A a matter of fact, the experimenters working on surface phase transition prefer to use a slightly different representation. They usually draw the

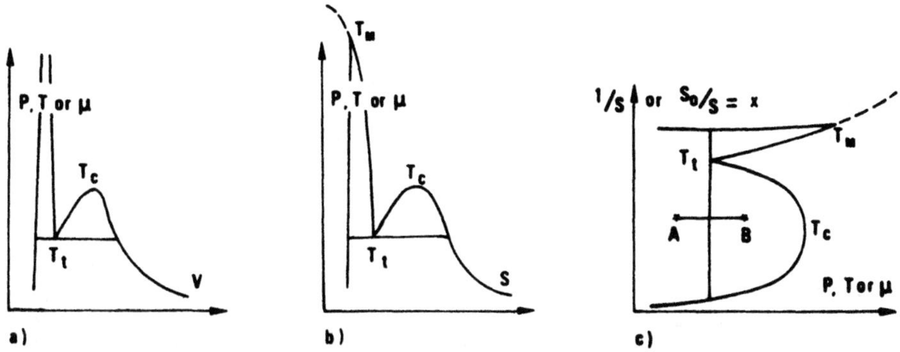

Fig.1 Two-dimensional phase diagrams expected in surface films (Fig.1b,c). They are obtained by analogy with those known in bulk matter (Fig.1a). Note the positions of the triple point T_t and of the critical temperature T_c. In Fig.1b and 1c, a critical line (dotted) is added. It starts from a multicritical point T_M. Diagrams looking like those in Fig.1b and 1c have been obtained by computer simulation [57] and by a group renormalization calculation [58].

surface density 1/S as a function of p, T or μ. Then, it is easy to show that the two-dimensional (2D) phase diagram will exhibit the shape given in Fig.1c. In numerous cases, it is more meaningful to replace 1/S by its relative value with respect to the monolayer completion ($1/S_0$). This relative density $x = S_0/S$ is called coverage. The monolayer is fully completed at x = 1.

The above arguments would have stayed pure intellectual pastimes if experimental results had not supported them. The first decisive step in this direction was cleared by the pioneer work of THOMY and DUVAL in the late sixties [7]. They measured phase diagrams looking like Fig.1c for submonolayers of simple molecules adsorbed on graphite (0001). Coexistence between different 2D phases and their phase boundaries were established by using adsorption isotherm measurements. From the thermodynamic analysis of their results, they were able to infer different surface phases and they called them "2D gas", "2D liquid" and "2D solid". They also demonstrated the existence of a 2D triple point and a 2D critical temperature. However, the nature of the various 2D phases was not clearly established and the major part of the work on 2D phase transitions during the last decade was devoted to the analysis of the properties of the different surface phases and to the comprehension of the mechanism of their transitions. One of the challenges was to show to what extent the so-called 2D solid, 2D liquid and 2D gas resemble their analogues in bulk matter. Various phases and transitions having no equivalent in three dimensions were discovered whereas on the other hand some 2D phases appeared to have the same behavior as in the three-dimensional world. Actually, the known phase diagrams are often more complicated than the one schematically described in Fig.1c. I'd like to show, here, the flavor of the state of the art in this field, approaching successively the order-disorder transitions which include melting and the gas-solid transition, then the solid-solid transformation and the gas-liquid transition. Finally I'd like to point out the influence of surface defects and heterogeneities on 2D phase transitions.

2. Order-Disorder Transitions

The notion of order-disorder transitions covers various physical transformations like crystal-fluid, crystal-glass, crystal-gas and crystal-liquid transitions. Here, three examples are chosen to illustrate the specificity of surface phase transformations. First, it is explained how the conjunction of the variation of symmetry and the reduction of dimensionality can produce quite novel critical behaviors in a 2D solid-fluid transition. Then it is shown that the gas-solid and liquid-solid transitions can be first order.

When a thin film is deposited on a crystalline surface, the competition between adsorbate-adsorbate interactions and the adsorbate-substrate lateral periodic potential can produce a structure which is either commensurate (in registry) with the substrate or incommensurate (out of registry) depending on the chemical potential of the system.

2.1 Critical Exponents and Surface Symmetry

In the last past years, theoreticians succeeded applying the notion of universality class to surface phase transitions [8]. Let us consider a commensurate overlayer (Fig.2) stable over a finite temperature range. As the temperature is raised, the 2D

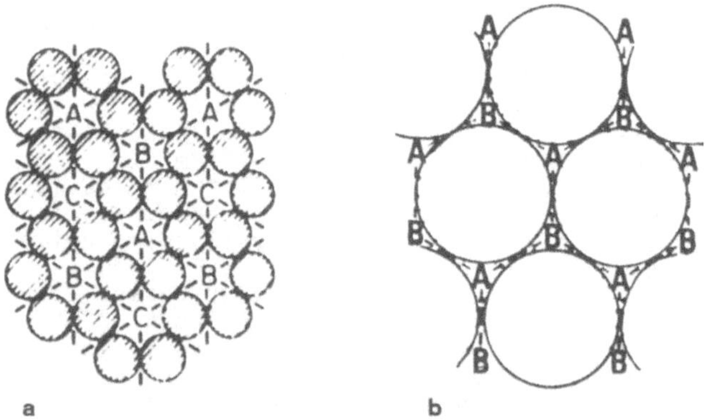

Fig.2(a) The graphite basal plane. A,B and C represent equivalent adsorption sites. The ^{4}He commensurate structure corresponds to all of either A or B or C sites occupied. (b) The Kr plating. A and B represent equivalent adsorption sites. The ^{4}He commensurate structure corresponds to all of either A or B sites occupied. Reprinted from M.J.TEJWANI, O. FERREIRA and O.E. VILCHES [10].

solid can transform into a fluid phase. If the transition is continuous (as for instance along the dotted line in Fig.1c), the specific heat diverges at the transition temperature T_c according to

$$C \propto |T-T_c|^{-\alpha} \qquad (1)$$

where α is the critical exponent. α depends only on the dimensionality of the system, on the number of components of the order parameter and on the substrate symmetry. Theoreticians were able to prepare tables where the different commensurate lattices and superlattices have been classified in a few typical universality classes. Furthermore they were able to calculate the corresponding critical exponents α.

To be more precise, let us consider the two different lattices described in Fig.2. They represent two hexagonal monolayers of helium adsorbed on graphite (0001) or on krypton-plated graphite. In both cases, helium atoms are located at the potential wells of the surface but in Fig.2a three different sites A, B and C are accessible to the adsorbed atoms whereas in Fig.2b only two different sites A and B are available. The first lattice, which has three degenerate ground states, can be mapped to a three-state Potts model. In this case, $\alpha = 1/3$. The second lattice can be represented by a two-state Potts model, better known as the Ising model ; then $\alpha = 0$.

Calorimetry measurements were performed to test those predictions. Two different groups [9|[10] determined α for the He layer adsorbed on bare graphite ; they found 0.36 and 0.28, values not so far from 1/3. One of those results is displayed in Fig.3. As for the experiment with He deposited on Kr-plated graphite, it yielded a logarithmic divergence of specific heat, i.e. $\alpha \simeq 0$ [10]. The results are shown in Fig.4. This verification of the theory convincingly demonstrates the crucial role of symmetry in surface phase transitions.

2.2 2D Gas-Solid Transition

One of the most studied 2D gas-solid transition is that occurring, at low temperature, in a submonolayer of xenon adsorbed on graphite. It has been explored by adsorption volumetry [7],

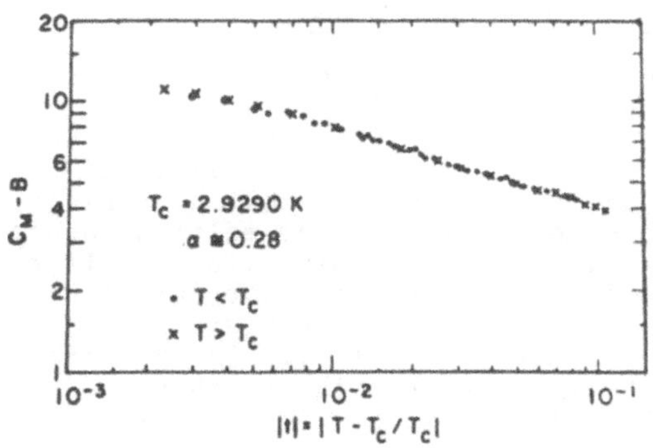

Fig.3 The specific heat for ^{4}He/graphite (0001) at the orderdisorder transition and critical coverage. Reprinted from M.J.TEJWANI, O. FERREIRA and O.E. VILCHES [10].

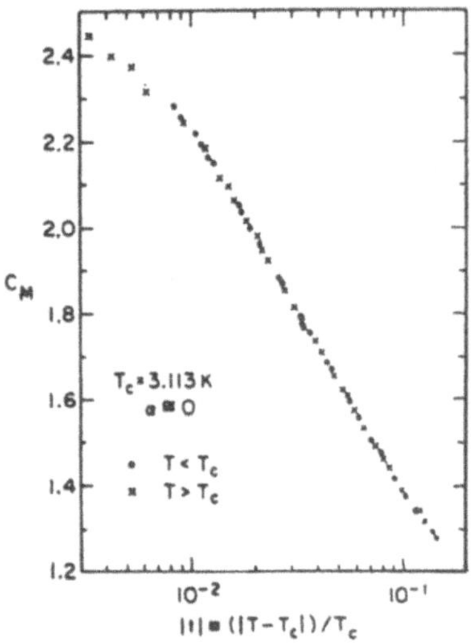

Fig.4 The specific heat for ^4He/Kr-plated graphite at the order-disorder transition. Reprinted from M.J. TEJWANI, O. FERREIRA and O.E. VILCHES [10].

Auger spectroscopy, LEED [11], HEED [12], calorimetry |13|, X-ray diffraction [14] and ellipsometry [15,16]. The main results have already been explained at length [17] and I don't want to repeat them here. In brief, a first order transition between a dilute phase and an incommensurated solid is observed below 100 K. Typical adsorption isotherms are represented in Fig.5. They clearly indicate that the transition is abrupt. To every temperature corresponds a well-defined condensation pressure p. The classical Clausius-Clapeyron equation can be adapted in two-dimensions,

$$\frac{d \ Ln \ p}{d \ 1/T} = - \frac{\Delta H}{R} \quad ; \qquad (2)$$

ΔH is the latent heat of transformation. The value obtained for xenon adsorbed on graphite is 5.5 Kcal mole^{-1}. It is larger than the bulk sublimation latent heat (3.76 Kcal mole^{-1}). This effect is easily explained by the relative strong attraction of xenon atoms by the graphite substrate.

2.3 2D Melting (Existence of a Self-Bound Liquid ?)

Above the 2D triple point, the so-called 2D liquid is stable. If we refer to Fig.1c when the temperature is raised from A to B at constant coverage x, the melting transition occurs at the triple line and should be first order. However this view was questioned by theoreticians [18,19] who thought that a triple point should not exist in two dimensions and that melting should be continuous. According to their views, two-dimensional matter could transform progressively into a disordered isotropic fluid phase, with an intermediate liquid crystal phase having short range orientational order. Theoretical results [20,21,22] obtained by computer simulation are at variance with the above prediction. They show that 2D melting must be first order.

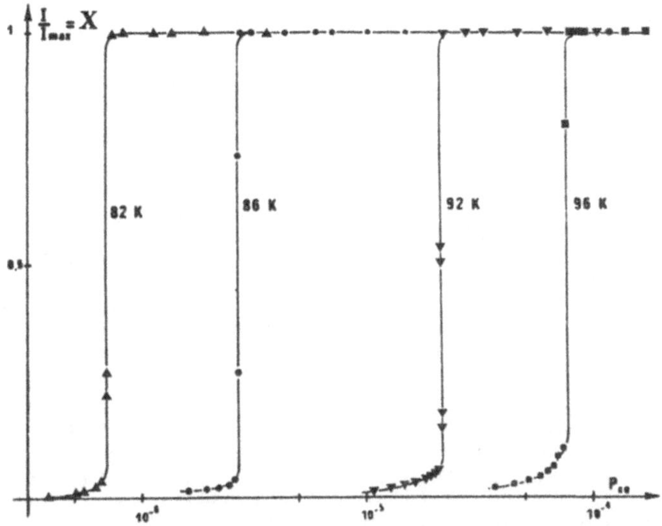

Fig.5 Adsorption isotherms of Xe on graphite (0001). The step is well defined and corresponds to a 2-dimensional gas ⇄ solid phase transition. Reprinted from J. SUZANNE, J.P. COULOMB and M. BIENFAIT [11].

Experimenters also brought their observations in the debate. Two systems have been extensively studied, methane and krypton adsorbed on graphite.

2.3.1 Methane/Graphite

At coverages below 0.8 and at temperatures smaller than ∿ 50 K, the methane submonolayer is commensurate with the graphite basal-plane [23,24]. Above ∿ 50 K, it expands to an incommensurate solid and melts suddenly at 56 K [23,25]. The coherence length

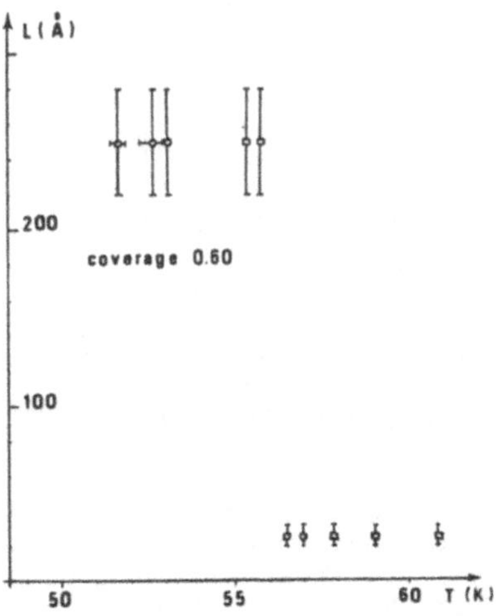

Fig.6 Temperature variation of the spatial correlation range (or 2D cluster size) in adsorbed CD_4 submonolayer. The abrupt change at 56 K within 0.7 K is the signature of a 2D first order melting transition. Reprinted from A.GLACHANT,J.P.COULOMB,M.BIENFAIT and J.G.DASH [25].

of the adsorbed layer measured by neutron diffraction drops abruptly at melting (Fig.6). This tends to show a first order transition. Complementary mobility measurements by quasi-elastic neutron scattering display a sudden enhancement of the diffusion coefficient at 56 K [26].

2.3.2 Krypton/Graphite

The results on the krypton/graphite system are more controversial. Heat capacity measurements show a continuous solid-fluid transition [27]. It has been suggested that an incipient critical temperature could exist. In other words, the liquid phase could be absent if solid stability is so enhanced by registry with the substrate that its regime extends above the critical temperature of the liquid.

However, other experiments carried out by adsorption volumetry strongly suggest the coexistence of 2D solid and 2D liquid in the vicinity of the expected triple point [28]. Their authors propose a classical phase diagram for krypton films adsorbed on graphite, with a first order melting transition and a triple point very close ($\sim$ 0.5 K) and slightly below the critical temperature. They explain the calorimetric measurements by a broadening of the first order transition due to the effect of the non-uniformity of the graphite substrate. This important point, i.e. the influence of heterogeneities on two-dimensional phase transitions will be analyzed in Sect.5. Anyway, the issues formed on this system for temperatures and coverages close and further experiments are needed to resolve them.

However, there is a situation for this system which should not bear to controversy. Above or in the vicinity of the multicritical temperature T_M (Fig.1c), the solid-fluid transition must be continuous and obey a three-state Potts model because the krypton layer is in registry with the substrate (see Fig.2 and Sect. 2.1). X-ray diffraction experiments have been per-performed on this system for temperatures and coverages close to T_M [29]. The intensity I of the 10 diffracted peak must vanish like

$$I \propto |T-T_c|^{2\beta} \quad . \tag{3}$$

Actually, the observed decrease of the intensity at the transition is smooth and can be fitted to the above power law. The obtained critical exponent β is 0.065 $\pm$ 0.015 [29]. From the lattice gas theory, the predicted values of β are $\frac{5}{48}$ or $\frac{1}{9}$ along the critical line [30] and $\frac{6}{96}$ or $\frac{1}{18}$ at the multicritical point T_M [31]. The measured critical index is intermediate between the critical and multicritical values. This result gives strong support to the lattice gas description of the monolayer krypton on graphite behavior in the vicinity of the multicritical temperature.

The different examples analyzed above show that the 2D order-disorder transition can be first order or continuous. The important parameters driving the order of the phase transition seem to be the symmetry of the substrate and of the film, the degree of commensurability of the layer and the location of the experimental conditions with respect to the 2D triple point or 2D multicritical temperature.

3. Solid-Solid Transformation

In this paragraph three kinds of solid-solid transitions will be examined : first, the commensurate-incommensurate transition, then the 2D polymorphism and finally the variation of stoechiometry of surface compounds.

3.1 Commensurate-Incommensurate (C-I) Transition

3.1.1 Kr/Graphite (0001)

At low and medium coverage, the layer of krypton is commensurate with the graphite substrate. Krypton atoms are located at the graphite potential wells (Fig.2). If the coverage is raised, the surface film is forced out of the well and the 2D crystal becomes incommensurate. This transition has been studied at great length by theoreticians and experimenters. Theory [32] takes account of the energy and the entropy of the domain walls and of their intersections occurring when the structure becomes incommensurate (domain walls are "grain boundaries" separating various commensurate regions). It definitely shows that the transition must be first order. However LEED [33][12] and X-ray diffraction experiments [34] display a continuous transition.* It has been suggested that the disagreement could arise from the existence of an important number of dislocations occurring at the fairly high temperature at which the experiments have been carried out. The theory is a low temperature theory and assumes a small density of domain walls. This transition has been re-covered recently by X-ray diffraction but at lower temperature [35]. The results show unambigously a first order transition. Coexistence between the C and the I structures is observed and a hysteresis caused by a nucleation barrier is pointed out. It would be interesting to extend these measurements toward higher temperatures in order to understand the previously observed continuous transition.

3.1.2 Xe/Cu (110)

In that case, the behavior of the layer at the C-I transition is quite different. Instead of having an isotropic compression of the film at the transition, the substrate imposes an uniaxial squeezing. The crystalline Cu (110) surface is made of parallel

(*)Note added in proof : New high resolution X-ray experiments performed above 80 K confirm the value of the critical exponent of the continuous transition (the peak position follows a 1/3 power law versus reduced temperature)[55]. Furthermore a liquid-like disorder phase is observed between the commensurate and the incommensurate solids. This last result gives experimental support to recent theoretical ideas [56].

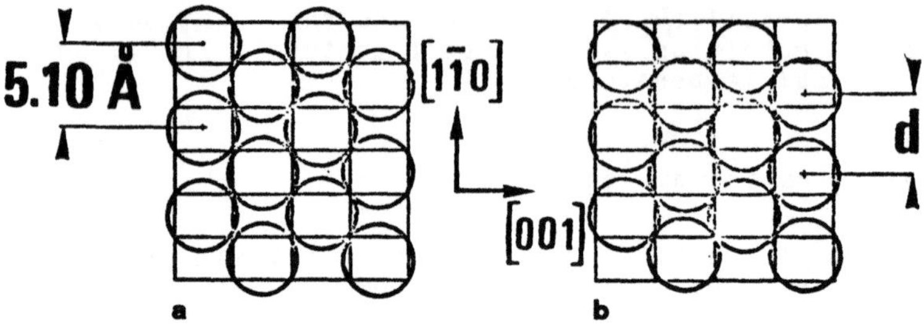

Fig.7 Xe ordered monolayers on Cu (110) surface.
(a) C(2x2) commensurate structure. (b) Pseudo-hexagonal structure obtained by compression along the |1$\bar{1}$0| direction. The 2D solid is incommensurate along this direction only.

atom rows providing well-defined canals (Fig.7). Xenon atoms adsorb first in a commensurate C 2x2 lattice ; eventually, as the chemical potential is increased, they compress themselves uniaxially in the |1$\bar{1}$0| direction. This system seems to be a good realization of models developed by theoreticians interested in 2D C-I transitions occurring on anisotropic substrates [36-41]. Here, the transition is no longer first order and all theoreticians agree in finding the following critical law

$$\Delta d = C(\Delta\mu)^{1/2} \tag{4}$$

where Δd and $\Delta\mu$ are respectively the variation, at the transition, of the overlayer lattice and the chemical potential. C is a proportionality constant.

Δd has been measured by LEED at different temperatures [42]. The results are represented in Fig. 8 in a log-log scale. They are consistent with a critical exponent 1/2. This result shows that the ingredients introduced in the theories, i.e. a strong anisotropy of the substrate potential and short range or long range lateral interactions between adatoms, are sufficient to describe the critical behavior of the uniaxial C-I transition.

3.2 2D Polymorphism

The above solid-solid transitions dealt with surface films made of single and spherical particles (atoms). Other phase transformations can occur when elongated or rectangular molecules are adsorbed on a crystalline substrate.

Neutron diffraction has shown that molecules like N_2O_2 or C_2H_6 can lie flat on the substrate, can be tilted with respect to the surface or can stand up [43,44]. The different structures depend on coverage and temperature. As an illustration, the different low temperature configurations of deuterated ethane are given in Fig.9 [44]. The transformation between the various packing as a function of chemical potential is usually first order. It provides us with good examples of 2D polymorphism. Furthermore, the 2D structures have often no equivalent in bulk matter.

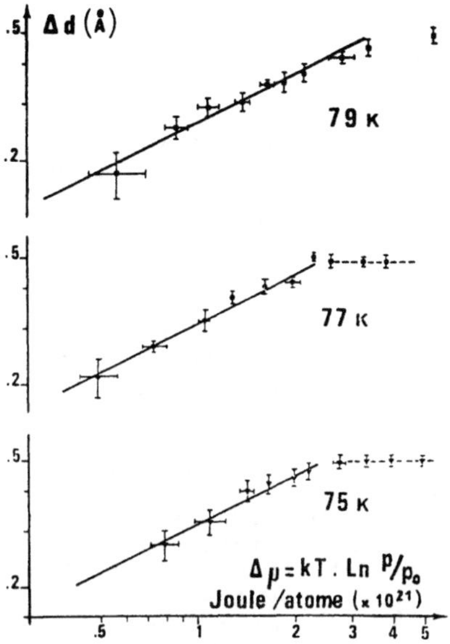

Fig.8 The misfit $\Delta d = d_0 - d$ along the |1$\bar{1}$0| direction as a function of $\Delta\mu = kT\ Ln(p/p_0)$, in a log-log plot. The full lines represent eq. (4). The dashed lines correspond to saturation of the in commensurate structure. The reference pressure p_0 equals $1.7.10^{-7}$, 8.10^{-8}, $3.6.10^{-8}$ for T = 79, 77, 75 K respectively. Reprinted from M. JAUBERT, A. GLACHANT, M. BIENFAIT and G. BOATO [42].

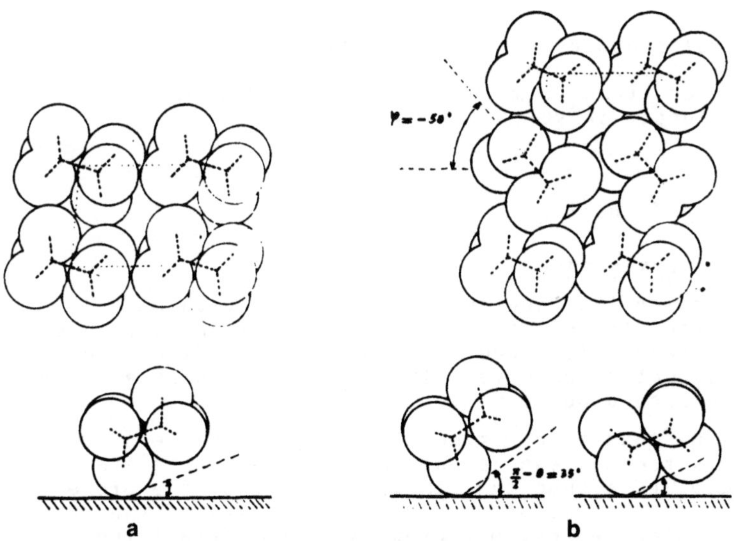

Fig.9 2D polymorphism of a layer of ethane adsorbed on graphite. (a) At low coverage, the molecule nearly lies down on the substrate. (b) As coverage increases, molecules take a tilted configuration. The final orientation of the molecule (upright configuration) is not represented. Reprinted from J.P. COULOMB et al.[44].

3.3 Non-Stoechiometric Surface Compounds

In all the examples quoted above, the adsorbed atoms or molecules did not interact strongly with the substrate and did not form surface compounds. Other classes of surface phases are far from being pure adsorbate layers but have a composition intermediate between that of the substrate and absorbate. For instance, when studying the first stages of oxidation, experimenters discovered new surface compounds having a structure and a stoechiometry quite different from the bulk oxide.

Other techniques can be used to determine the orientation of adsorbed molecules with respect to the substrate. For instance, high-resolution electron energy loss and UV photoemission spectroscopies have been used to determine the various structures of pyridine chemisorbed on Ag (111) [45].

Let us consider an example taken from a review paper [46]. When sulfur is adsorbed on the (100) face of iron, different superstructures are observed in a LEED experiment. They have been interpreted by a steady variation of the surface sulfide composition as a function of the chemical potential. All the 2D sulfides obey the general formula $Fe_{2n-2}S_n$. They are composed of two kinds of chains having a FeS_2 or a FeS composition. The relative number of each type of chain results in a defined non-stoechiometric compound whose unit cell corresponds to the superstructure pattern observed in LEED. When the chemical potential is raised again, the FeS bulk compound is finally obtained. This original 2D chemistry is quite typical of the sulfur-metal interaction.

All the examples quoted in Sect. 3 show that crystalline surfaces impose quite original structures and configurations for adsorbed atoms or molecules and that new solid-solid transformations occur in surface films.

4. 2D Gas-Liquid Transition

4.1 Liquid-Gas Coexistence

In 2.3.1, we showed that neutron diffraction indicated a first order melting transition in a layer of methane adsorbed on graphite. Consequently, according to the phase diagram drawn in Fig.1c, we expect to observe a liquid-gas coexistence above the triple line around point B.

Let us imagine that an experiment, giving the value of the mobility in this region, can be carried out. Let us be even more restrictive and assume that the resolution of the probe will be adapted to the measurement of the mobility of the 2D liquid only. In other words, our probe will not detect the motion of molecules in the 2D gas. Moreover, if the experiment is performed at constant T and varying coverage, we expect to probe an amount of 2D liquid adsorbed on the surface varying relatively with respect to coverage (the 2D liquid islands moving in a sea of 2D gas will occupy a surface roughly proportional to coverage). Then, the measured mobility will be

Table I. Diffusion coefficient (in 10^{-5} cm^2 s^{-1}) of the saturated 2D liquid of methane adsorbed on graphite (0001) as a function of coverage at 61.7 K. Its value is coverage independent within experimental error

Coverage	0.30	0.40	0.55	0.63
Diffusion coefficient	3.7±0.7	4.2±0.5	4.4±0.5	4.6±0.5

coverage independent because we will probe an extensive property of a 2D phase, i.e. its mobility. The experiment has been carried out by quasielastic neutron scattering and reported in regular papers [26,47] and in another summer school [48]. The reader interested in experimental details is invited to refer to them. The major results are given in table I. It can be seen that the mobility of the 2D liquid is constant within the experimental uncertainty ; this behavior confirms the ideas developed above.

To complete this demonstration, it would be interesting to verify if mobility drops abruptly below the triple point and if its behavior changes in a region where coexistence between gas and liquid is no longer observed (i.e. above the critical temperature T_c). The two experiments have been carried out. Below the triple point (56 K), the methane layer freezes. As for the mobility of a 2D fluid in a single phase region (above T_c), the results are summarized in table II. Then the diffusion coefficient is coverage dependent. This behavior was expected because in an hypercritical fluid, molecules tend to occupy all the surface available. The free space between molecules decreases when coverage increases and the molecules are less mobile at higher coverages. This set of experimental results is proof of self-bound liquid existence.

Table II. Diffusion coefficient (in 10^{-5} cm^2 s^{-1}) of the 2D hypercritical fluid of CH_4 adsorbed on graphite (0001) versus coverage at 91.5 K ($T_c \simeq 75$ K). Its value is coverage dependent

Coverage	0.45	0.70	0.90	1
Diffusion coefficient	22±4	12±2	4.0±0.5	∿ 0

4.2 Critical Index

Much effort has been devoted to the determination of the critical exponent of the liquid-gas transition in the vicinity of the critical temperature T_c. The most precise measurement has been published recently [49]. It deals with the determination of the curvature of the liquid-gas boundary near T_c in a second layer of argon adsorbed on the cleavage face of cadmium chloride. The results, i.e. adsorption isotherm measurements,

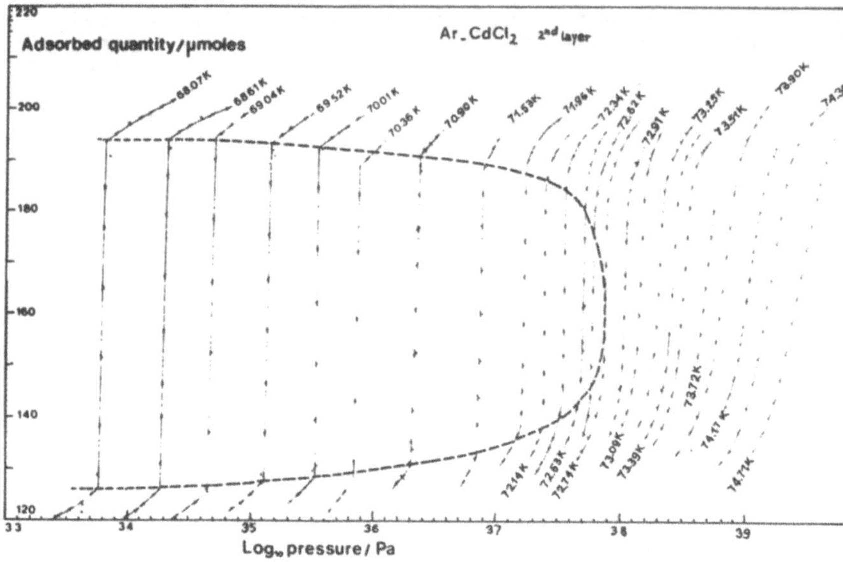

Fig.10 Boundaries of the 2D gas-liquid coexistence domain in the second layer of Ar adsorbed on $CdCl_2$, obtained by adsorption isotherm measurements. Reprinted from Y. LARHER [49].

are represented in Fig.10. The dotted line defines the limit of the liquid-gas coexistence region. The variation of the concentration ΔN between the 2D gas and 2D liquid must diverge in the vicinity of T_c as

$$\Delta N \sim \left(\frac{T_c - T}{T_c}\right)^{\beta}$$

where β is the compressibility critical exponent (it can also be determined by diffraction measurements - see Sect.2.3.2). A value of β equal 0.16 fits very well the experimental data. Let us recall that the critical gas-liquid behavior can be mapped to an Ising model. The bulk value of β (0.31) is definitely larger than the experimental determination whereas the 2D Ising β (1/8) is not so far from the measured one. Accordingly the 2D character of the gas-liquid transition is demonstrated.

5. Influence of Heterogeneities on Surface Phase Transitions

When theoreticians determine the mechanism of a 2D phase transition, they can imagine either molecules confined in a plane and experiencing lateral interactions only or, in addition, interacting with the periodic potential of a substrate. They usually assume a homogeneous and uniform two-dimensional world. Unfortunately, real surfaces have imperfections and their behavior upon adsorption can be far from the one expected. Although experimenters will prepare and characterize carefully their substrates, a few heterogeneities like steps, adatoms different to the ones being studied, vacancies and impurities are always present on the crystalline surfaces. They can conceal the nature of a 2D transition. For instance, it has been pointed out [50] that surface imperfections can broaden an otherwise first order transition.

This effect is beginning to be studied theoretically [51-53]. The calculations try to estimate the relative weight of size effects and interaction energy distributions on experimental singularities in the vicinity of 2D phase transitions. Experimental data quoted in Sect.2.1 [10] have been analyzed by exmploying renormalization-group approximations. It has been shown [53] that in the system of helium adsorbed on krypton-plated graphite, 50 % of the condensed helium does not participate in the order-disorder transition. However, it seems that only the intensity of the heat capacity signal is reduced and that the critical exponent is not modified.

Experiments dealing with the influence of heterogeneities on 2D phase transitions are very rare. This is related to the difficulty of fully characterizing the surface defects. One attempt in this direction has been published recently [54]. Surfaces of single crystals of copper and sodium chloride have been perturbed on purpose by producing periodic monoatomic steps or by ion bombardment. The gas-solid transition in a layer of xenon has been studied when the layer was adsorbed on the uniform and smooth (100) faces or on the imperfect surfaces. The effect of heterogeneities is far from being negligible as seen in Fig. 11.

The vertical step observed in the adsorption isotherm measurement of Xe on the uniform (100) surface of copper indicates a first order gas-solid transition. When edges, steps, disorder structures are produced on this surface, the transition

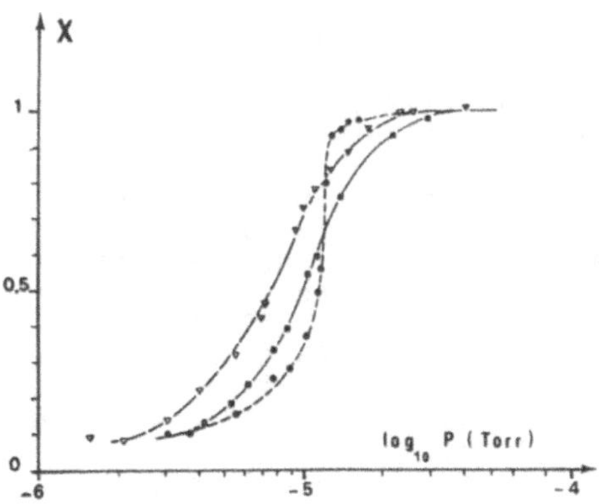

Fig.11 Comparison between xenon adsorption isotherms at 99 K obtained on (610)Cu (▽), disordered (100)Cu (■), and flat (100) Cu (•) surfaces. θ is the coverage of xenon on each face (θ = 1 for 5.7 x 10^{14} xenon atoms cm^{-2} on the (100)Cu flat face); P is the absolute pressure of xenon in torr. On the disordered (100) and stepped (610)Cu faces, transition becomes continuous. Reprinted from U. BARDI,A. GLACHANT and M. BIENFAIT [54].

becomes continuous. In the same paper, it is also shown that the epitaxial relation between the overlayer and the substrate can change dramatically when parallel monoatomic steps are present on the surface.

Although the above efforts point out the importance of defects or size effects on 2D phase transition, one must be aware that our knowledge in this field is still in a primitive state. The next years should provide more information about this important point.

6. Conclusion

The last few years have seen more and more research groups working in surface science and particularly in the field of two-dimensional phase transitions. The resulting acceleration in the published works leads me to expect in the near future decisive steps in the comprehension of the questions still open. One can imagine that the relations between symmetry and critical exponents will be deepened, that 2D melting will be finally understood, that the questions of commensurate-incommensurate transitions will be clarified and that the field of 2D polymorphism will be enlarged.

References

1 Two-Dimensional Adsorbed Phases, J. Physique C4 (1977).
2 Ordering in Two-Dimensions, ed. S.K. Sinha, North-Holland (1980).
3 J.G. Dash, Films on Solid Surfaces, Acad. Press N.Y. (1975).
4 Phase Transitions in Surface Films, ed. J.G. Dash and J. Ruvalds, Plenum N.Y. (1980).
5 M. Bienfait, Surf. Science 89, 13 (1979).
6 O.E. Vilches, Ann. Rev. Phys. Chem. 31, 463 (1980).
7 A. Thomy and X. Duval, J. Chim. Phys. 67, 1101 (1970).
8 M. Schick, in Ref.4 p. 65.
9 M. Bretz, Phys. Rev. Lett. 38, 501 (1977).
10 M.J. Tejwani, O. Ferreira and O.E. Vilches, Phys. Rev. Lett. 44, 152 (1980).
11 J. Suzanne, J.P. Coulomb and M. Bienfait, Surface Science 40, 414 (1973) - 44, 141 (1974) - 47, 204 (1975).
12 J.A. Venables, H.M. Kramer and G.L. Price, Surface Science 55, 373 (1976) - 57, 782 (1976). P.S. Schabes-Retchkiman and J.A. Venables, Surface Science 105, 536 (1981).
13 J. Régnier, J. Rouquerol and A. Thomy, J. Chim. Phys. 3, 327 (1975).
14 E.M. Hammonds, P. Heiney, P.W. Stephens, R.J. Birgeneau and P. Horn, J. Phys. C 13, L 301 (1980).
15 G. Quentel, J.M. Rickard and R. Kern, Surface Science 50, 343 (1975).
16 G. Quentel and R. Kern, Surface Science 55, 545 (1976).
17 M. Bienfait in Ref.4, p.29.
18 B.I. Halperin and D.R. Nelson, Phys. Rev. Lett. 41, 121 (1978), 519(E) - D.R. Nelson and B.I. Halperin, Phys. Rev. B 19, 2457 (1979).
19 D. Frenkel and J.P. Mc Tague, Phys. Rev. Lett. 42, 1632 (1979).

20 F. Tsien and J.P. Valleau, Mol. Phys. 27, 177 (1974).
21 S. Toxvaerd, Phys. Rev. Lett. 44, 1002 (1980).
22 F.F. Abraham, Phys. Rev. Lett. 44, 463 (1980).
23 P. Vora, S.K. Sinha and R.K. Crawford, Phys. Rev. Lett. 43, 704 (1979) and in Ref.2, p. 169.
24 A.Glachant, J.P. Coulomb, M. Bienfait, P. Thorel, C. Marti and J.G. Dash in Ref.2, p. 203.
25 A. Glachant, J.P. Coulomb, M. Bienfait and J.G. Dash, J. Phys. Lett. 40, L 543 (1979).
26 J.P. Coulomb, M. Bienfait and P. Thorel, J. Physique 42, 293 (1981).
27 D.M. Butler, J.A. Litzinger, G.A. Stewart and R.B. Griffiths, Phys. Rev. Lett. 42, 1289 (1979) - D.M. Butler, J.A. Litzinger and G.A. Stewart, Phys. Rev. Lett. 44, 466 (1980).
28 Y. Larher and A. Terlain, J. Chem. Phys. 72, 1052 (1980).
29 R.J. Birgeneau, G.S. Brown, P.M. Horn, D.E. Moncton and P.W. Stephens, J. Phys. C 14, L 49 (1981).
30 A.N. Berker in Ref.2, p.9 - A.N. Berker and S. Ostlund, J. Phys. C 12, 4961 (1979) - A.N. Berker, S. Ostlund and F.A. Putnam, Phys. Rev. B 17, 3650 (1978).
31 B. Nienhuis, E.K. Riedel and M. Schick, to be published.
32 J. Villain in : Ordering in Strongly Fluctuating Systems, ed. T. Riste, Plenum N.Y. (1980) p. 214 - J. Villain, Surface Science 97, 219 (1980).
33 M.D. Chinn and S.C. Fain, Phys. Rev. Lett. 39, 146 (1977) - S.C. Fain, M.D. Chinn and R.D. Diehl, Phys. Rev. B 21, 4170 (1980).
34 P.W. Stephens, P. Heiney, R.J. Birgeneau and P.M. Horn, Phys. Rev. Lett. 43, 47 (1979) - R. Birgeneau, E.M. Hammons, P. Heiney, P.W. Stephens and P.M. Horn in Ref.2, p.29.
35 J. Bohr, M. Nielsen and J. Als-Nielsen, Proceedings of 3rd European Conference on Surface Science, Supplt. Le Vide, Vol.1, 112 (1980) - M. Nielsen, J. Als-Nielsen, J. Bohr and J.P. Mc Tague, to be published.
36 V.L. Pokrovsky and A.L. Talapov, Phys. Rev. Lett. 42, 65 (1979).
37 H.J. Schulz, Phys. Rev. B 22, 1 (1980).
38 T. Nattermann, J. Physique 41, 1251 (1980).
39 Y. Okwamoto, J. Phys. Soc. Jap. 49, 8 (1980).
40 M.B. Gordon and J. Villain, J. Phys. C 12, L 151 (1979).
41 F.D.M. Haldane and J. Villain, to be published.
42 M. Jaubert, A. Glachant, M. Bienfait and G. Boato, Phys. Rev. Lett. 46 (1981) 1679.
43 J. Suzanne, J.P. Coulomb, M. Bienfait, M. Matecki, A. Thomy, B. Croset and C. Marti, Phys. Rev. Lett. 41, 760 (1978) and J. Physique 41, 1155 (1980).
44 J.P. Coulomb, J.P. Biberian, J. Suzanne, A. Thomy, G.J. Trott, H. Taub, H.R. Danner and F.Y. Hansen, Phys. Rev. Lett. 43, 1878 (1979).
45 J.E. Demuth, K. Christmann and P.N. Sanda, Chem. Phys. Lett. 76, 201 (1980).
46 J. Oudar, in Ref.1, p. 141.
47 J.P. Coulomb, M. Bienfait and P. Thorel, Phys. Rev. Lett. 42, 733 (1979).
48 M. Bienfait in Proceedings of NATO Summer School "Surface Mobilities on Solid Materials", Les Arcs 1981, to be published.

49 Y. Larher, Molecular Physics 38, 789 (1979).
50 T.T. Chung and J.G. Dash, Surface Science 66, 559 (1977).
51 Y. Imry and M. Wortis, Phys. Rev. B 19 (1979) 3580 ; 21 (1980) 2042.
52 J.G. Dash and R.D. Puff, to be published.
53 D.J.E. Callaway and M. Schick, Phys. Rev. B 23, 3494 (1981).
54 U. Bardi, A. Glachant and M. Bienfait, Surface Science 97, 137 (1980).
55 D.E. Moncton, P.W. Stephens, R.J. Birgeneau, P.M. Horn and G.S. Brown, Phys. Rev. Lett. 46, 1533 (1981).
56 S.N. Coppersmith, D.S. Fisher, B.I. Halperin, P.A. Lee and W.F. Brinkman, Phys. Rev. Lett. 46 (1981) 549, 869(E).
57 J.M. Phillips, L.W. Bruch and R.D. Murphy, to be published.
58 S. Ostlund and A.N. Berker, Phys. Rev. B 21 (1980) 5410.

Universal Laws of Physical Adsorption

M.W. Cole and G. Vidali
Department of Physics, The Pennsylvania State University
University Park, PA 16802, USA

1. Introduction

Physics aims at understanding a diversity of phenomena in terms of a small number of "laws". The search for such unifying simplicity can take on either theoretical (i.e. ab initio) or empirical character. In the case of physisorption, we are only beginning to make progress [1-4] in first principles calculations of atom-surface interaction $V(\vec{r})$. A technical but fundamental difficulty is our lack of knowledge of electronic properties in the region where the electronic density falls to a value of order 10^{-4} Å^{-3}. This is the region where overlap of surface and adatom charges produces the repulsive part of $V(\vec{r})$.

This paper describes recent developments in a semi-empirical description of the interaction. As the title suggests, we are searching for universal behavior. One specific proposal we consider is that the lateral average of $V(\vec{r})$ satisfies a scaling relation

$$V_0(z) = Df(z/z_0) \tag{1}$$

where D is the well depth, z_0 is a characteristic length, and f is a universal function which we call the reduced potential.

Equation (1) means that two parameters suffice to characterize V (z) for a particular combination of atom and surface. If (1) is true, properties of adsorbed phases will obey laws of corresponding states. For example, at low vapor pressure P, the Henry's law relation [5] between the surface density n_s and the vapor density n_v becomes

$$\frac{n_s}{n_v} = A^{-1} \int d^3r \, \{\exp[-\beta V(\vec{r})] - 1\} \simeq z_0 \int dx \, \{\exp[-f(x)/T^*] - 1\} \quad ,$$

where $\beta^{-1} = kT \equiv DT^*$ and the approximation neglects terms of second order in the Fourier components $V_G(z)$. Thus $n_s/n_v z_0$ will be a universal function of the reduced temperature T^*.

On physical grounds one might expect (1) to be invalid because the potential energy contains such diverse ingredients. The repulsive part of the interaction arises from charge overlap. This alone must be characterized by at least two parameters -- a strength (V_r) and a length. There is no a priori relation to the attractive part of the interaction, which is due to dispersion and varies at large distances as

$$V_0(z) \sim -C \, z^{-3} \; . \tag{2}$$

Thus one might expect at least three parameters to be relevant for each system, in disagreement with (1). Nevertheless, two observations motivated us to explore (1): (a) the interaction between identical noble gas atoms satisfies [6] a relation similar to (1), and (b) as noted by HOINKES [7], C_3 is roughly proportional to D, suggesting that one of the three parameters is redundant. This is not particularly surprising because the minimum of the potential is determined by the equality of the attractive and repulsive parts of the force; thus D is implicitly related to both C_3 and V_r. A simple analogy is the Lennard-Jones 6-12 potential. In this case the well depth is one-half the magnitude of the attractive part at the potential minimum; the latter is proportional to the coefficient C_6 of the r^{-6} term.

2. Evidence for Universality

The most complete information available to date about $V_0(z)$ comes from bound state resonance measurements in atomic and molecular beam scattering experiments [7]. These are restricted, thus far, to the light gases because the resonance is a diffractive scattering into a bound state. Information about more classical gases is discussed below.

The resonance positions yield the eigenvalues E_n of $V_0(z)$. To a good approximation [8] these satisfy the Bohr-Summerfeld condition

$$\frac{(n + 1/2)\hbar\pi}{(2m)^{1/2}} = \int [E_n - V_0(z)]^{1/2} \, dz$$

where the integration domain lies between the classical turning points. Assuming (1) to be true and defining $\varepsilon_n = E_n/D$, this yields

$$y_n \equiv (n + 1/2)a = J(\varepsilon_n) \quad , \tag{3}$$

where

$$a = \hbar\pi/[z_0(2mD)^{1/2}] \tag{4}$$

and

$$J(\varepsilon) = \int [\varepsilon - f(x)]^{1/2} \, dx \tag{5}$$

If (1) is true, $J(\varepsilon)$ is a universal function. In order to test (3), we need to know z_0. Since D can be estimated well if E_0 (the lowest eigenvalue) is known and since C_3 is known for many systems [9], a convenient possibility is

$$z_0 = (C_3/D)^{1/3} \quad . \tag{6}$$

This choice is in fact <u>required</u> by (2); asymptotically

$$f(x) \sim -x^{-3} \quad . \tag{7}$$

This formulation thus assures universality at large x; the only other requirement of the reduced potential f(x) is that it equal (-1) at its minimum.

We may test (3), and hence (1), by collecting bound-state resonance data for a variety of systems. Each value gives a point y_n for every ε_n. Note that (6) means that $y_n \propto D^{-1/6}$, which is fairly insensitive to any uncer-

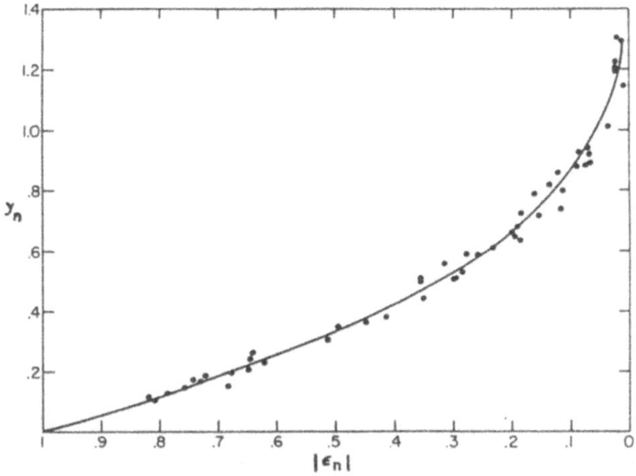

Fig. 1 y_n vs ε_n from data as reported in [7]. Neither experimental uncertainty nor the 10% uncertainty in D [7] is indicated. The solid line corresponds to Eq. (5) using the Exp.-3 potential (9) with u = 5.52.

tainty in D; ε_n is more sensitive to errors in D since $\varepsilon \propto 1/D$. We used estimates of D given by HOINKES to obtain Fig. 1. One sees there that the existing data do correspond closely to a universal curve; the deviation in ε_n from the smooth curve is less than 0.05. Note that revisions of D could be made to yield closer agreement with the curve; lacking further information about the potential, this may be appropriate.

Obvious applications of the curve include (a) predicting E_n if D is known, and (b) using it as an easy consistency test of eigenvalues. The obvious question to ask is what reduced potentials are consistent with it. First, we note that Eq. (5) yields the width of the potential as a function of

$$x_2(\varepsilon) - x_1(\varepsilon) = \frac{2}{\pi} \int_{-1}^{\varepsilon} \frac{J'(\varepsilon')d\varepsilon'}{(\varepsilon - \varepsilon')^{1/2}} \quad . \tag{8}$$

A common curve $J(\varepsilon)$ means therefore that the width of the reduced potential is a universal function of ε. Neither this nor (7) precludes the possibility of reduced potentials differing by an arbitrary translation outward from the surface; this corresponds to ambiguity in the x=0 reference plane. Atomic scattering resonances are insensitive to this. In contrast, diffraction intensities yield information about the fully three-dimensional potential which is particularly useful if one has a model of the interaction (e.g. pair-wise summation [10]). Supplementing these data are results of LEED and neutron scattering [11] which yield the absolute position of the potential minimum, z_m. If the hypothesis (1) is valid, z_m/z_0 should be universal. Probably the most complete information available is for He/graphite: C_3 = 185 meV-Å^3, [12] D = 16.4 meV, and z_m = 2.73 Å, [10] differing from the mean distance ($\langle z \rangle$ = 2.85 Å [11], for ^{4}He) because of zero-point motion in the anharmonic potential. Thus z_m/z_0 = 1.22 in this case. Results for other systems are presented in Table 1.

The potentials for rare gases on graphite have been obtained [13] using a summation of pair potentials with carbon parameters obtained from the

Table 1

	C_3 [a] [meV Å³]	D [meV]	z_0 [Å]	z_m [Å]	z_m/z_0
He-graphite	185	16.4	2.24	2.73	1.22
Ne-graphite	346	32.9	2.19	2.82	1.29
Ar-graphite	1208	77.1	2.50	3.14	1.26
Kr-graphite	1730	98.9	2.60	3.24	1.25
Xe-graphite	2460	132.9	2.65	3.49	1.32
He-Ag(111)	249	4.95	3.69	4.48[d]	1.21
Ar-Ag(111)	1620	74[b] 70[c]	2.80 2.85	3.69[b]	1.32 1.29
Kr-Ag(111)	2260	112[e]	2.72	3.30[c]	1.21
Xe-Ag(111)	3380	172[e]	2.70	3.55[c]	1.32

Data from Ref. [13] except as otherwise noted.

[a]From table in Ref. [9].

[b]Ref. [3].

analysis [10] of He/graphite scattering data. The predictions of z_m are in good agreement with LEED results [14] and neutron scattering results [11].

Note that z_m/z_0 varies by about 10 percent for the quite diverse systems of Table 1; this indicates that a revised reduced variable is appropriate,

$$z^* = (z - z_n)/z_0 = x - z_m/z_0 \quad .$$

Recall that the previous discussion concerning the bound state resonances and (3) was insensitive to such a translation of origin. Thus the universal curve of Fig. 1 is consistent with this formulation. Fig. 2 shows that the reduced potential $f(z^*)$ is indeed quite universal, with the exception

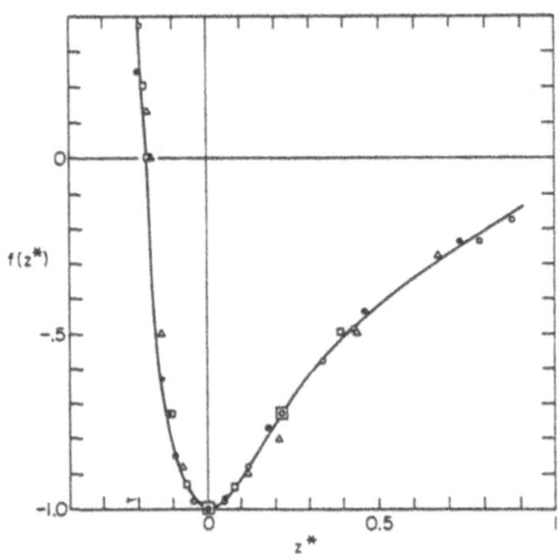

Fig. 2 Reduced potential versus distance z* for He on graphite (circles), He on Cu (squares), He on Ag (dots), He on Au (triangles), from Ref. [13].

of the theoretical result of Lang for Ar on $r_s=3$ jellium. The latter calculation underestimates the attraction because it violates (2). Somewhat puzzling, however, is the relatively softer repulsion than the other reduced curves in Fig. 2.

3. Analytical Forms of the Potential

Many model potentials have been proposed to fit resonance data [7]. A suitable candidate should also satisfy (2), which excludes the otherwise convenient Morse and "variable exponent" potentials [15]. The shifted Morse hybrid potential [16] achieves this at the cost of simplicity. In contrast, the 3-9 reduced potential

$$f_{3-9}(x) = \frac{4}{27x^9} - \frac{1}{x^3}$$

(with a minimum at $x_m = (2/3)^{1/3}$) is quite simple and has the correct asymptotic form. Its spectrum is also simple [8] and yields

$$J_{3-9}(\varepsilon) = 2.233\ (1 - \varepsilon^{1/6}) \quad .$$

We have not shown this function in order to avoid confusing Fig. 1. The result falls somewhat below the empirical universe curve. Note that there is no free parameter to adjust.

The exponential -3 potential [17]

$$f_{exp-3} = \frac{3}{u - 3}\, e^{-u(\frac{x}{x_m} - 1)} - \frac{1}{x^3} \qquad (9)$$

with $x_m = (1 - 3/u)^{1/3}$ has one free parameter, either u or x_m. The curve in Fig. 1 has been drawn with this potential. It clearly fits the data rather well.

4. Conclusion

Figs. 1 and 2 provide substantial evidence for the existence of a universal potential of the form $V_0 = Df(z^*)$. Thus, three parameters (D, z_m, and z_0) are sufficient to characterize a given system. Of these, z_m is irrelevant to both the bound state resonances and Henry's law coefficient (relating vapor density to adsorbate coverage). We hope that this investigation will stimulate efforts to understand the existence and nature of the functional form f.

Research supported in part by Department of Energy Contract DE-AS02-79ER10454 and NSF grant 77-22961.

References

1. E. Zaremba, W. Kohn, Phys. Rev. B13, 2270 (1976).

2. D. R. Hamann, Phys. Rev. Lett., to be published.

3. N. D. Lang, Phys. Rev. Lett. 46, 842 (1981).

4. N. Esbjerg, J. K. Nørskov, Phys. Rev. Lett. 45, 807 (1980).

5. W. A. Steele, The Interaction of Gases with Solid Surfaces (Pergamon, Oxford) Chapter 3.

6. G. Scoles, Ann. Rev. Phys. Chem. 31, 8 (1980).

7. H. Hoinkes, Rev. Mod. Phys. 52, 933 (1980).

8. M. W. Cole, T. T. Tsong, Surface Science 69, 325 (1977).

9. G. Vidali, M. W. Cole, Surface Science, in press.

10. W. E. Carlos, M. W. Cole, Surface Science 91, 339 (1980).

11. K. Carneiro, L. Passell, W. Thomlinson, H. Taub, Phys. Rev. B, to be published.

12. G. Vidali, M. W. Cole, C. Schwartz, Surface Science 87, L273 (1979).

13. S. Rauber, J. R. Klein, M. W. Cole, to be published.

14. C. G. Shaw, S. C. Fain, M. D. Chinn, M. F. Toney, Surface Science, in press.

15. L. Mattera, F. Rosatelli, C. Salvo, F. Tommasini, V. Valbusa, G. Vidali, Surface Science 93, 515 (1980).

16. C. Schwartz, M. W. Cole, J. Pliva, Surface Science 75, 1 (1978).

17. H. Chow, Surface Science 66, 221 (1977). The parameter u is written as αz_m in this paper.

18. J. Unguris, Thesis; J. Unguris, L. W. Bruch, E. R. Moog, M. B. Webb, Surface Science 87, 415 (1979).

19. L. W. Bruch, Mein Sieng Wei, Surface Science 100, 481 (1980).

The Dynamical Parameters of Desorbing Molecules

G. Comsa

Institut für Grenzflächenforschung und Vakuumphysik
Kernforschungsanlage Jülich, Postfach 19 13
D-5170 Jülich, Fed. Rep. of Germany

1. Abstract

Experiments performed in recent years have confirmed the theoretical expectations that, unless the sticking probability is unity, the dynamical parameters of desorbing atoms and molecules need not follow the hitherto accepted general laws. These laws are: the Knudsen cosine law for the angular distribution of the flux, the Maxwell law for the velocity distribution and the Boltzmann law for the internal energy distribution. Deviations from these laws are sources of information about the dynamics of the desorption process. This will be illustrated in the case of the associative desorption of hydrogen after permeation through metals. It will be shown that, starting from the trivial statement that for the systems considered desorption from the chemisorbed state is an endothermic process and comparing the experimental desorption data with bulk permeation data, a new desorption mechanism emerges. In conjunction with the classical desorption mechanism a new picture is obtained, which explains the main features of the associative desorption data.

2. Introduction

The current experimental procedures used for the study of the desorption of atoms and molecules from surfaces average out the basic parameters which actually characterize the desorbing atoms and molecules. Indeed, the only quantity currently measured is the desorption rate. This is the total flux of molecules ϕ desorbing in the unit time from a given surface as a function of time, surface temperature and coverage, and of other parameters. This total flux is, by definition, a multiple integral over the various differential distributions of the desorbing molecules:

$$\phi = \int \int \int \int \frac{\partial^4 \phi(\theta,\psi,v,\varepsilon)}{\partial\theta \partial\psi \partial v \partial\varepsilon} \, d\theta d\psi dv d\varepsilon \qquad (1)$$

where θ and ψ are the polar and azimuthal angles, v the velocity and ε the internal energy of the desorbing molecules. It was believed for a long time that the shape of each of the differential distributions $\frac{\partial\phi}{\partial v}$ is identical for all desorbing systems. It was thought that these differential distributions follow well known laws:

$$\frac{\partial\phi}{\partial\theta} \propto \cos\theta \quad \text{(Knudsen law)} \ , \qquad (2)$$

$$\frac{\partial\phi}{\partial\psi} \propto \text{const} \ , \qquad (3)$$

$$\frac{\partial\phi}{\partial v} \propto v^3 \exp\left(-\frac{mv^2}{2kT_s}\right) \quad \text{(Maxwell law)} \ , \qquad (4)$$

$$\frac{\partial\phi}{\partial\varepsilon} \propto \exp\left(-\frac{\varepsilon}{kT_s}\right) \qquad \text{(Boltzmann law)} \quad , \qquad (5)$$

where T_s is the surface temperature, m the mass of the desorbing molecule and k the Boltzmann constant[1]. If this were true, no information whatsoever would be lost by measuring only the total flux, i.e. by performing the multiple integral (1) by the measuring process itself. However, since 1968 it was successively shown (see below) that none of the four "general" laws (2)-(5) are generally followed, and that the shape of the actual distributions strongly depends on the adsorbate/substrate system. Accordingly, the study of the various differential distributions of the desorbing molecules is able to supply new information about the details of the desorption process. This information might lead eventually to an understanding of the microscopic mechanism of the different desorption processes.

3. The Failure of the General Desorption Laws

Based upon detailed balance considerations, it was first explicitely stated in 1968 [1] that, unless the sticking probability does not depend on the incident polar angle θ_i, the desorbing molecules will not follow the Knudsen law, even under equilibrium conditions. A few months later [2] the experimental evidence for strong deviations from the Knudsen law was already presented: the angular distributions of the flux of H_2 molecules desorbing from Fe, Ni and Pd appear to be, as predicted, strongly peaked in the normal direction[2]. The distributions could be formally described by $\cos^n\theta$ with $2<n<9$.

In 1971, it was shown by time-of-flight (TOF) measurements that the mean energy of H_2 molecules desorbing in the normal direction from a Ni surface is about $3kT_s$ [3], i.e. 50 % larger than predicted by the Maxwell law (4). Improvements in the signal to noise ratio enabled the measurements of the mean energy of the desorbing molecules as a function of the polar angle of desorption [5,6]. The mean energy of D_2 molecules desorbing from Ni(111) varies from $3.4kT_s$ for $\theta=0°$ to $1.3kT_s$ for $\theta=80°$. This is of course also at variance with the Maxwell law (4) which predicts that the mean energy should be $2kT_s$ independent of the desorption angle θ. An important result was obtained by averaging the measured energies of the desorbing molecules weighted by the angular dependence of the flux over the whole half space: the "overall" mean energy is $(3\pm0.2)kT_s$ [6], i.e. again much higher than expected from the Maxwell law in spite of the very cold (but relatively few) molecules desorbing in glancing directions.

The recent implementation of the cross-correlation TOF technique for the study of desorbing molecules [7] led to further progress: TOF distributions almost negligibly convoluted by the gate function can now be obtained in reasonable measuring times. These TOF distributions give a much more detailed characterization of the desorbing molecules. They show even more strikingly the failure of eq. (4) to describe the velocity distribution of these molecules. In Fig. 1 a set of such TOF distributions is presented. The curves

---

[1] Note that none of these laws contain any parameter characterizing the surface except the surface temperature.

[2] As shown more recently [4], the angular distribution of the flux of D_2 desorbing from the stepped Ni(997) surface depends also on the azimuth. This invalidates also eq.(3). The dependence on ψ will not be considered in the following, its consequences seeming to be less general.

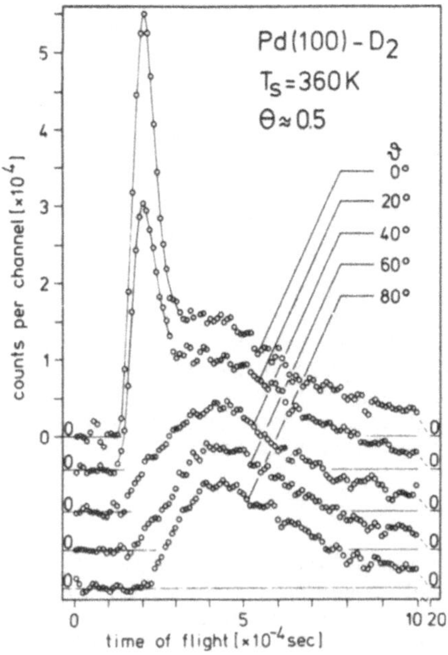

Fig.1 Sequence of TOF curves with desorption angle θ as parameter. Measuring time for one curve: 120 min.

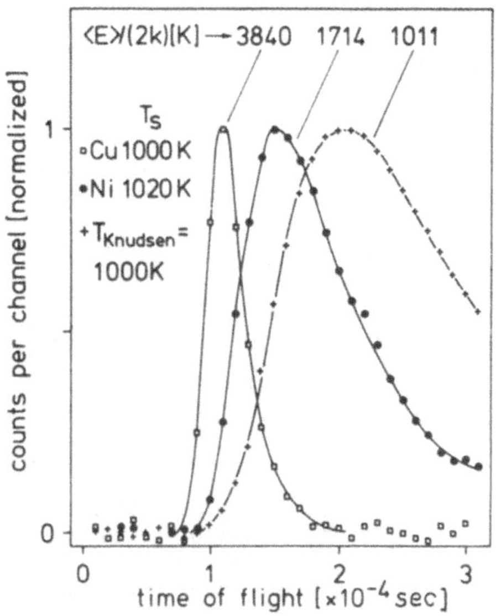

Fig.2 TOF curves of D_2 desorbing at θ = 0° from Cu and Ni and of D_2 effusing from a Knudsen cell

are the TOF distributions of D_2 molecules desorbing at five polar angles θ from a sulphur covered (Θ_S = 0.5) Pd(100) surface at T_S = 360 K. The curves clearly consist of two distinct distributions: one is "Maxwellian"-like and being also cosθ-like appears in all five curves, while the other consists of much faster molecules and is clearly evident only at θ=0° and 20° because of its strong angular dependence (~cos $^{10}\theta$). The comparison of the TOF curves of D_2 desorbing from Cu and Ni at about T_S = 1000 K with the Maxwellian TOF curve expected from eq.(4) at the same surface temperature (Fig.2) is also self explanatory. Fig.2 also shows the strong dependence of the distributions on the nature of the surface.

Very recently it was shown experimentally that also the Boltzmann law eq. (5) is not generally applicable for characterizing the desorbing molecules [8]. Indeed, the vibrational temperature of N_2 desorbing from Fe was found to be significantly greater than T_S.

The experimental proofs for the failure of eqs.(2)-(5) to characterize the desorbing molecules have been confined so far to the associative desorption of H_2, D_2, and N_2 and, in particular, to the case where the atoms are supplied to the surface by bulk permeation. This might give the impression that the failure of these general laws is restricted to this special case. A number of experiments show that the failure is much more general. Specifically, CO_2 molecules desorbing as a product of the catalytic oxidation of CO on Pt have a strongly peaked angular distribution of the flux [9,10] and mean energies corresponding to temperatures up to ~3650 K at T_S = 880 K [11]; N_2

desorbing from W(110) have a peaked angular distribution [12] and even Ar desorbes from Pt with energies corresponding to a lower temperature than T_s [13], etc.

The deviations of the experimental distributions from all the general laws eqs.(2)-(5) are not unexpected. Indeed, as in the case of the angular distribution of the flux [1] it was also demonstrated [14] that there is no reason why, even at equilibrium, the velocity distribution of desorbing atoms and molecules should be described by the Maxwellian distribution eq.(4). On the contrary, by using detailed balance arguments it was concluded that if the sticking probability of the atoms or of the molecule depends on their individual incident velocity, the velocity distribution of the desorbing atoms and molecules must perforce deviate at equilibrium from eq.(4). Since it seems likely that there will be, in general, some dependence of the sticking probability on the incident velocity, eq.(4) should be, in principle, the exception and not the rule[3]. The question of whether or not the deviations of the actual velocity distributions are experimentally measurable in specific cases is, of course, a different problem.

Some remarks now seem appropriate:

- The parameters of the desorbing molecules can be determined experimentally only under non-equilibrium conditions. Indeed, under equilibrium conditions one cannot separate experimentally the desorbing molecules from those which leave the surface after having undergone other kinds of surface scattering like diffraction, specular reflection or inelastic scattering. Accordingly, while detailed balance arguments demonstrate that eqs. (2)-(5) need not and, in general, do not describe properly the desorbing molecules even under equilibrium conditions, experiments can prove the failure of these general laws to describe the desorption only under non-equilibrium conditions.

- The cognitive power of the application of the detailed balance to this kind of problem is remarkable but of course rather limited. It enables the prediction of the shape of the distribution of desorbing molecules as a function of a given parameter (θ, ψ, v or ε) when the dependence of the sticking probability on the same parameter is known and vice-versa. But the detailed balance gives no answer to the question "why has the distribution of desorbing molecules (or the sticking probability dependence) a specific shape?". The microscopic mechanism leading to a given shape of the distributions can be found only by analyzing the experimental distributions, by proposing appropriate models and, of course, by aiming for a first principle description of the desorption (adsorption) process. The existence of a close connection between the distributions characterizing the desorbing molecules and the shape of the sticking probability dependences suggests that a unified theory of adsorption and desorption might be set up. Again the detailed balance gives no hint on what this theory should look like, but only that adsorption and desorption have to be properly related.

- The last remark is of a rather philosophical nature. Up to this point we have busied ourselves with demonstrating the failure of general laws to describe the various distributions of the desorbing molecules. At the beginning we have stated that this failure enables us to obtain additional information about the desorption process, which may eventually lead to an understanding of the microscopic mechanism of the desorption process. The idea

---

[3] All these arguments can, of course, be applied without any change to the problem of the use of eq.(5) for the description of the distribution of the internal energies of desorbing molecules.

that the failure of a general law opens the way for the understanding of the physics of a process should not be surprising. The more general a law is, the less it contains specific material parameters. But the lack of specific material parameters hinders the investigation - at least experimentally - of the features of the physics present behind the general law. It is as if the general laws would hide the actual details of the process. Famous examples of this can, in fact, be given: the very general laws of attraction between masses or of attraction and repulsion between electrical charges have been known for centuries and we still do not know very much about the actual nature of gravitation or electricity. Let us thus concentrate on deviations from general laws with the conviction that these are fruitful sources of knowledge.

4. The Associative Desorption of Permeating Atoms

In the following we will exemplify the way in which deviations from eqs.(2)-(5) have been used up to now to obtain insight in the mechanism of the desorption process. Most of the data obtained so far are confined to the associative desorption of H_2, D_2, and N_2 and for the particular case in which the respective atoms are supplied to the surface by bulk permeation. This is, in fact, the easiest way to study the different distributions characterizing the desorbing molecules. Indeed, by supplying the atoms to the surface by bulk permeation under UHV conditions, essentially all molecules exiting the surface are desorbed molecules; i.e. no molecules that exit the surface after undergoing other kinds of scattering processes are present. On the other hand, we should consider carefully the possible influence of the kind of supply used on the distribution of the atoms over the surface states before generalizing the conclusions.

In figures 1 and 2 one sees that, at least for not too large polar angles θ, the mean kinetic energy of the desorbing molecules is larger than expected from a Maxwellian equilibrium distribution, i.e. larger than $2kT_s$. As already mentioned, even in the case of Ni where the mean energy of molecules desorbing at $\theta > 65°$ is sensibly less than $2kT_s$, the energy averaged over all molecules desorbed in the half space is still $3kT_s$ [6]. Also, the vibrational temperature of N_2 molecules appears to be up to 2.5 times larger than the surface temperature [8]. The first question we are confronted with is: where is this excess energy coming from? Probably the most widespread opinion is that the excess energy is a fraction of the atom recombination energy which remains with the desorbing molecules as internal energy [8] or is transformed in kinetic energy [15]. This assumption has received apparent support by the recent observation of the excess of vibrational energy [8]. It should be obvious, however, that the recombination energy cannot be the ultimate origin of the excess energy (kinetic or internal). Indeed, the associative desorption process of atoms originating in the chemisorption well is an endothermic process in all cases considered here. The atoms in the chemisorption well need the entire recombination energy and in addition, some further energy from the crystal phonons in order to be able to break their bonds to the crystal. Clearly then we cannot speak about some fraction of the recombination energy remaining with the desorbing molecule. In Fig. 3 this argument is exemplified for the systems D_2/Pd and D_2/Cu. For instance, each D atom is bound to the Pd surface (Fig. 3a) with E_n = 2.74 eV (63 kcal/mole). In order to desorb two atoms need 5.48 eV (126 kcal/mole) which is considerably more than the energy they gain by recombination, 4.6 eV (106 kcal/mole). We are thus again back at our initial question concerning the ultimate origin of the excess energy.

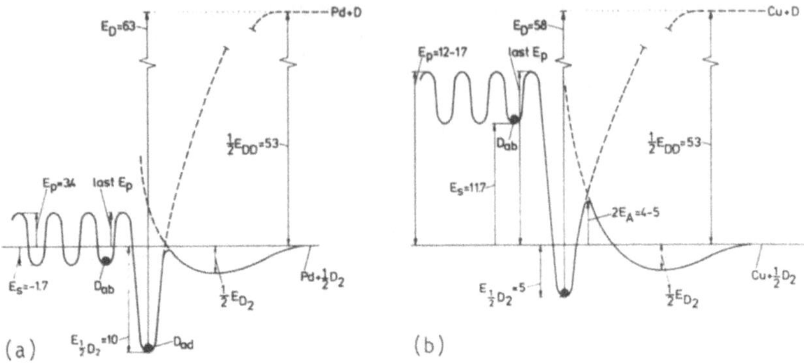

Fig.3 Potential energy diagrams for the systems (a) Pd/D_2 and Pd/D as well as (b) Cu/D_2 and Cu/D. All energies are in kcal/mole

It appears that the only reasonable explanation given to date that accounts for the origin of the excess energy involves, in one way or another, the existence of a region of elevated potential energy. This can be, for instance, the barrier for activated adsorption (E_A in Fig. 3b) ([2] and further variations in Ref. [16 and 17])or the barrier for absorption (last E_p in Fig. 3a and b) [18]. The latter was proposed recently in the frame of a model which takes into account the supply of atoms by permeation, and assumes that the atoms arriving at the surface from the bulk can recombine without first thermalizing in the chemisorption well. In both of the above models the role of the elevated potential is to provide for the excess energy, i.e., to transform the desorption into an exothermic process[4]. The transformation takes place in the first model in the last stage of the desorption process, while in the second model the process is exothermic from the beginning, the desorption of atoms permeating the bulk is for most systems[5] an exothermic process if the atoms do not thermalize in the chemisorption well before recombining.

If the desorption process is transformed at some stage of its evolution by means of a potential into an exothermic process, we may now address again the question as to whether the excess energy is a part of the recombination energy which remains with the molecule as vibrational energy. If some fraction of this internal energy is then transformed into translational energy, then this should show up as a significant isotopic effect when measuring the excess kinetic energy of desorbing H_2 and D_2 molecules.

---

[4] One may imagine of course also other models in which the desorption process changes at some stage of its evolution to an exothermic process. For instance, let us assume that the atoms adsorbed in the chemisorption well repel each other. If the energy necessary to bring them close enough together for their recombination to start is high enough, the desorption is transformed at this stage into an exothermic process. Here again the presence of an elevated potential in the desorption path is the crucial factor for the appearance of excess energy.

[5] There are of course exceptions: for instance according to hydrogen bulk permeation data for Pd - Fig. 3a - the desorption should become exothermic only in the moment when the atoms have overcome the last hump of the bulk potential.

However, none of the desorption distributions of H_2 and D_2 measured up to now (N_i [5], Pd [18], Cu [19]) show any isotopic effect with respect to the kinetic energy. The desorption from Cu seems to be particularly appropriate for the investigation of the features of the distributions of molecules desorbing with excess kinetic energy and in particular, the isotopic effect. Indeed, recent measurements [19] have shown that the desorption from Cu is the first known case in which almost all desorbing molecules have an excess of kinetic energy: the velocity distribution is a "pure" distribution of fast molecules[6]. In Fig. 4 the TOF spectra of H_2 and D_2 desorbing at $\theta = 0°$ from a clean Cu(100) surface is presented. The time scales on the abscissae are plotted in the ratio of the square root of the masses. It is obvious that within the limits of experimental error both the mean kinetic energies ($\langle E \rangle/2k = (4050 \pm 100)$K) and the shape of the properly scaled TOF curves for H_2 and D_2 are identical. We may thus conclude that the excess energy of the exothermic process is directly transformed into kinetic energy and not via internal energy. The experimental data do not yet provide pertinent answers to questions like: may the excess energy be also transformed directly in internal energy (e.g. as a remainder of the recombination energy) or is an intermediate step necessary?

Let us consider again in more detail the two models which tend to materialize the region of elevated potential energy, which we know now to be at the origin of the excess energy of the desorbing molecules.

The first model, in which the activation barrier for adsorption (E_A in Fig. 3b) plays the role of the region of elevated potential energy, is reas-

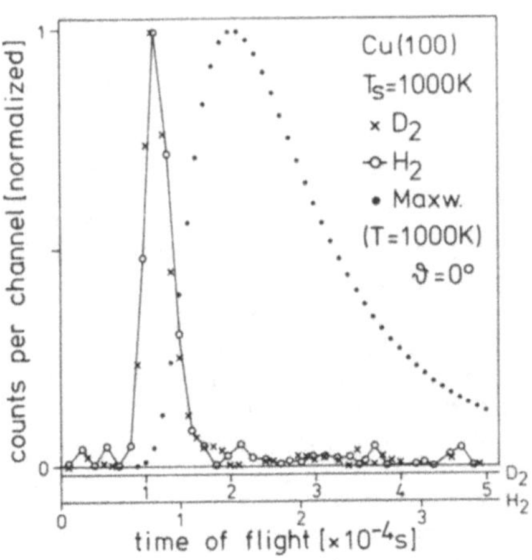

Fig.4 TOF-curves of D_2(x) and H_2(-o-) desorbing from Cu(100). The time scales on the abscissae are plotted in the ratio of the square root of the masses

[6] This is at variance with Pd where in general two distinct distributions of fast and slow molecules are present (Fig. 1 and [18]) and with Ni where the two distributions are strongly mixed (Fig. 2 and [17]).

[7] The height of a potential E_b leading to a velocity distribution at $\theta = 0°$ of mean energy $\langle E \rangle$ is obtained from

$$E_b = \frac{1}{2}\left[\langle E \rangle - 2kT_s + \left(\langle E \rangle^2 - (2kT_s)^2\right)^{1/2}\right]$$

derived from eq.(2) in ref. [17] for $r = 0$. The equation was obtained by calculating the mean energy of a truncated ($v \geq (2E_b/m)^{1/2}$) Maxwellian distribution.

suring because no new assumption has to be made. The trouble with the model is that for clean Ni the hydrogen adsorption is known to be not activated ($E_A = 0$) and still excess energy shows up. In the case of Cu the adsorption is activated, the activation energy measured in a direct molecular beam experiment being E_A = 4-5 kcal/mole [16]. However, the height of the potential necessary for explaining the very fast molecules desorbing from Cu in permeation experiments (Fig. 4) should be about 13.5 kcal/mole, i.e. three times larger[7].

The second model, in which it is assumed that atoms permeating through the bulk may recombine at the surface without thermalizing in the chemisorption well, is a proposal for an additional desorption mechanism. In this case the last potential-energy hump of atoms in the crystal (last E_p in Fig. 3a) plays the role of the elevated potential energy. The mean energies obtained from the fast velocity distributions measured so far scale reasonably well with the activation energies for bulk permeation [20].

We may, therefore, describe the associative desorption of molecules supplied to the surface by permeation as follows: The atoms arriving at the surface can either recombine directly (second model) and desorb with an excess energy associated with the bulk permeation potential, or first thermalize in the chemisorption well and then recombine. In the latter case, according to the first model, the distribution of the desorbing molecules will have an excess energy if $E_A > 0$ or will be Maxwellian-like if $E_A = 0$. The probability of the atoms to take the one or the other way appears to depend strongly on the nature of the metal (compare e.g. Fig. 1 and 4). In Fig. 1, Pd covered with sulphur ($\theta_S \cong 0.5$), the two distinct distributions have comparable fluxes, i.e., the probabilities of taking one way or the other are comparable. In accordance with our opinion the excess energy of the "fast" distribution corresponds to the activation energy for permeation in Pd while the "slow" distribution is Maxwellian-like (corresponding to $E_A = 0$, i.e., non-activated adsorption on Pd). The ratio of the fluxes of the two distributions appears to be almost independent of T_S in the range 360 K < T_S < 800 K. This is also in accordance with the overall picture; i.e., there is no direct competition between desorption barriers of different heights but between different processes. The TOF distribution from Cu shown in Fig. 4 is very different. It seems that the probability of the atoms to recombine directly is in the case of Cu almost unity. Indeed, if an important fraction of them thermalize first, we should expect - according to the first model- to see in the TOF spectra also molecules with an excess energy corresponding to E_A = 4-5 kcal/mole and not only the much faster ones (corresponding to 13.5 kcal/mole) as actually seen.

The ratio of the fluxes of the "fast" and "slow" distributions is in the case of Pd (and also of V) a strong function of surface coverage. In Fig. 5a the TOF distributions of D_2 from Pd(100) at 500 K are presented as a function of sulphur coverage Θ_2. The flux of the "fast" molecules appears to increase with Θ_S at the expense of the flux of the "slow" ones. This can be, of course, interpreted in two ways: either the sulphur atoms open the direct recombination channel or they are simply blocking the adsorption sites in the chemisorption well; once the sites are blocked, the D atoms are no longer able to thermalize in the well and must recombine directly. The second variant seems to be, at present, the more probable. Indeed, recent experiments show that adsorbed CO has an influence similar to that of sulphur atoms. It is known that adsorbed S interacts much stronger with D atoms than adsorbed CO. The effect of S and CO on the distribution being similar suggests that their influence proceeds in a more neutral way than for instance by the blocking of

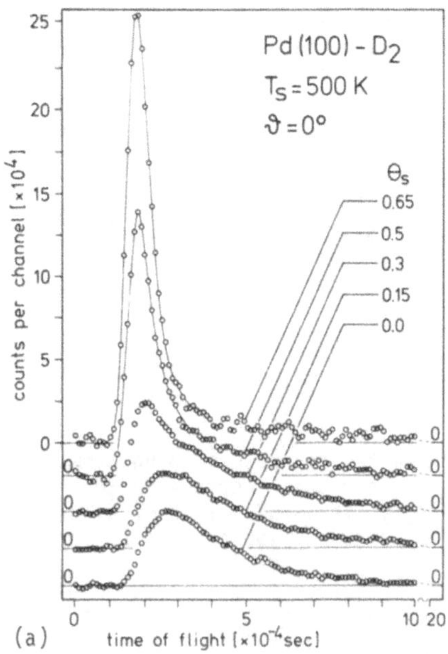

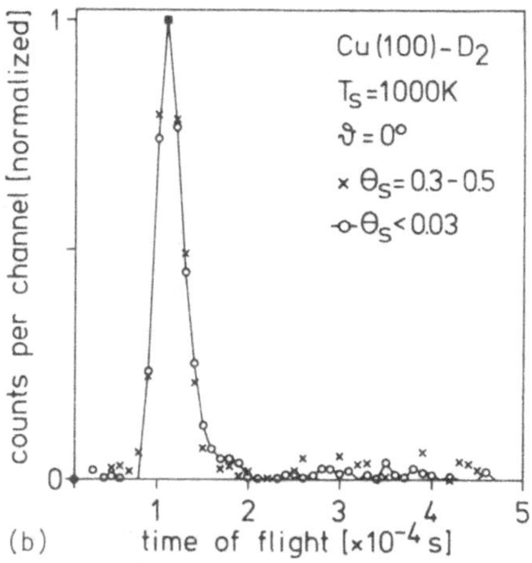

Fig.5 TOF curves of D_2 desorbing from (a) Pd and (b) Cu at different S coverages (Θ_s).

sites [8]. At variance with the strong influence of adsorbates in the case of Pd (Fig. 5a), the presence of sulphur has no influence at all on the shape of the distributions in the case of Ni [6] and of Cu (Fig. 5b). We presently have no explanation for this behavior of the Ni surface. In the case of Cu one may argue that, the probability of thermalizing being very low, even for a clean surface the blocking of the chemisorption well sites is not expected to have any further influence.

In summary: the data obtained so far show that in the case of metals where in the bulk the H (D) atoms have a low potential energy, as in Pd and V (overall exothermic absorption), the ratio of the "fast" to "slow" fluxes is strongly increasing with the sulphur coverage (the clean surface distribution consists only of "slow" molecules). In contrast, in the case of metals with endothermic absorption like Ni and Cu the distributions are independent of sulphur coverage and may consist only of "fast" molecules (Cu).

We have gained so far a qualitative picture of the associative desorption process in the particular case when the atoms are supplied to the surface by bulk permeation. We are now, of course, tempted to verify and complete this picture by experiments in which the supply is originating from the gas phase. As stated at the beginning, there are unfortunately only few results which may be used to this end. The dissociative adsorption of hydrogen on Cu was studied in great detail in a molecular beam experiment [16] leading to the already quoted value E_A = 4-5 kcal/mole for the activation energy of adsorption. Unfortunately, the highest energy of the molecules in the beam was less than 10 kcal/mole and thus too low for the molecules to take the direct ab-

[8] The fact that the excess energy with adsorbed CO is about 20 % less than with S can be explained by the influence of adsorbates on the height of the "last E_p" potential hump.

sorption channel, i.e. to absorb without first thermalizing in the chemisorption well. About 10% of the incident molecules with energies larger than E_A were adsorbed. The angular distribution of their desorbing flux correlated reasonably with the E_A value and thus confirmed the existence of the first model. On the other hand recent measurements with the system H_2/Pd [21] seem to confirm also the second model and, thus, the overall desorption picture: indeed, the experimental data show that the absorption rate at low H_2 pressures is much higher in the case of a CO covered Pd surface than for a clean one. If the surface is clean, the hydrogen is first chemisorbed, and only after the sites are filled with hydrogen can absorption begins. In the case of the CO covered surface, the chemisorption sites being blocked, the H atoms cannot thermalize in the well and the absorption process starts from the beginning of the exposure.

5. Conclusions

The distributions of the parameters which characterize the desorbing atoms and molecules do not follow general laws, but their shape depends on the specific adsorbate/substrate system. Accordingly, the study of these distributions provides information about the different desorption processes.

The molecules desorbed from Cu, Ni and Pd, as a result of the recombination of atoms permeated to the surface, are characterized by overall mean energies sensibly in excess of $2kT_S$. The feature responsible for this excess energy is considered to be a region of elevated potential energy which transforms at least one branch of the desorption process into an exothermic one.

6. Acknowledgement

Many of the conclusions in this lecture are based on the experimental data obtained by Rudolf David with the technical aid of Karl Veltmann. During some limited periods of time Klaus Rendulic and B.-J. Schumacher have also contributed significantly to the progress of the work.

I would like to express to them as well as to many friends and colleagues my gratitude for their patience in discussing my obsessive "cosine" syndrom.

References

1. G. Comsa: J. Chem. Phys. 48,3235 (1968)
2. W. Van Willigen: Phys. Letters, 28A, 80 (1968)
3. A.E. Dabiri, T.J. Lee and R.E. Stickney: Surface Sci. 26, 522 (1971)
4. B.-J. Schumacher: "Pseudostatistische Flugzeitbestimmung des von Nickeloberflächen desorbierenden Wasserstoffes" Ph D. Thesis, Jül-Report Nr. 1628 (1979)
5. G. Comsa, R. David and K.D. Rendulic: Phys. Rev. Letters 38, 775 (1977)
6. G. Comsa, R. David and B.-J. Schumacher: Surface Sci. 85, 45 (1979)
7. G. Comsa, R. David and B.-J. Schumacher: Rev. Sci. Instrum. 52, 789 (1981)
8. R.P. Thorman, D. Anderson and S.L. Bernasek: Phys. Rev. Letters 44, 743 (1980)
9. R.L. Palmer and J.N. Smith: J. Chem. Phys. 60, 1453 (1974)
10. C.T. Campbell, G. Ertl, H. Kuipers, and J. Segner: J. Chem. Phys. 73, 5862 (1980)

11. C.A. Becker, J.P. Cowin, L. Wharton, and D.J. Auerbach: J. Chem. Phys. 67, 3394 (1977)
12. R.C. Cosser, S.R. Bare, S.M. Francis, and D.A. King (to be published)
13. K.C. Janda: 181 ACS-Meeting, Atlanta, Ga. 29. March - 3. April 1981.
14. G. Comsa, in: Proc. 7th Intern. Vacuum Congr. and 3rd Int. Conf. on Solid Surfaces, Vienna, p. 1317 (1977)
15. R.F. Willis and B. Fitton: Astrophys. Space Sci. 34, 57 (1975)
16. M. Balooch, M.J. Cardillo, D.R. Miller and R.E. Stickney: Surface Sci. 46, 358 (1974)
17. G. Comsa and R. David: Chem. Phys. Letters 49, 512 (1977)
18. G. Comsa, R. David, and B.-J. Schumacher: Surface Sci. 95, L210 (1980)
19. G. Comsa and R. David (to be published)
20. G. Comsa, R. David, and B.-J. Schumacher: Proc. IVth Int. Conf. on Solid Surfaces & IIIrd ECOSS (D.A. Degras & M. Costa Eds.) p.252(1980)
21. I. Ratajczykowa: Proc. IVth Int. Conf. on Solid Surfaces & IIIrd ECOSS (D.A. Degras & M. Costa Eds.) p.513 (1980) and private communication

Atomic-Beam Diffraction as a Method for Studying Two-Dimensional Solids

S. Iannotta

Dipartimento di Fisica Universitá di Trento
I-POVO (Tn), Italy

1. Introduction

Two-dimensional adsorbed phases have been the object of very active investigation since their first experimental observation [1]. Several different theories have been developed which are capable of quantitative predictions; meanwhile new experimental techniques have been produced to check them. This probably has been made possible by two favorable conditions. On one hand, two-dimensional systems (or, more correctly, quasi-two dimensional) are somewhat easier to visualize and to treat theoretically than the three-dimensional ones: the geometry is simpler, the numerical computations are easier and fewer particles are needed in computer simulations. On the other hand, many different experimental methods have been successfully used to study the thermodynamics of these systems [2]. The fact that the different techniques complement each other has been particularly determinant in achieving the present stage of research.

Whereas the ólder techniques where based on thermal and vapour pressure measurements, the ones developed in more recent years are all based on the analysis of emission products following the bombardement of the systems with particles of various kind (electrons, ions, neutrons, photons etc.). In 1974 HOBSON [3] estimated the number of recognized echniques to be about thirty but, in the meantime, their number has certainly increased.

Because of their ability to measure accurately both the orientational order (epitaxial orientation) and the lattice parameter of an adsorbed layer, diffraction techniques have attained rapidly growing importance. With methods such as LEED [4],[5], THEED [6] and neutron diffraction [7],[8] it is now possible to measure the lattice parameter of an adsorbed layer with an accuracy of 0.3% (at least in the most favorable cases) and therefore to get very accurate data on the phase diagrams of solid layers.

Molecular beam diffraction, just as in the case of LEED, has its origins in the pioneering experiments aimed at testing the general validity of the de BROGLIE's relationship [9]. Nevertheless its development as a method for studying surface properties has been quite slow compared to LEED. This is mostly due to the experimental difficulties in producing monoenergetic beams of neutral particles, in detecting them and in analyzing them in velocity or in energy, after the scattering. Only in the last ten years has molecular beam diffraction grown to the point of being fully recognized as a standard method in surface physics. The list of the surfaces which up to now have been investigated by this technique includes not only alkali halides but also substances which present more difficult experimental problems such as tungsten [10], silver [11], nickel [12], copper [13], silicon [14] and graphite [15].

Atomic diffraction is discussed in some detail in sec. 2 and it is shown shown to be a particularly suitable technique to study adsorbates on single crystal surfaces, as is also pointed out, by other authors in this book. However, wever surprisingly little work has been done in this direction and sometimes under poorly characterized experimental conditions; a brief review of the work done done up to now is given in sec. 3. The study of rare gases adsorbed on graphite deserves particular consideration. Indeed these systems are interesting because the physisorption of noble gases on graphite has long been studied, and it is now possible to model with some precision both the interaction of rare gas atoms with the graphite substrate and among the rare gas atoms themselves. Knowledge of these interactions, coupled with a realistic description of the motion of the adatoms and of the substrate, is the key for the construction of a viable statistical mechanical model of the physisorbed phases [1]. The first data obtained on one of these systems are those reported a few months ago on diffraction of H atoms from a Xe monolayer adsorbed on the (0001) surface of graphite [16]. In sec. 4. this experiment is fully described and an outlook is given at the possible future developments of this field which seems to be very promising for a better understanding of two-dimensional solids.

2. Atomic diffraction

The growing interest in surface diffractive experiments with atoms and molecules derives from the variety of information which may be obtained from them. Quite a number of reviews have been recently published on this topic [17-20] and here it is therefore worthwhile to summarize them briefly, outlining their relevance to the studies of physisorbed phases on single crystal surfaces.

2.1 Surface Crystallography

Atomic diffraction is in principle an ideal probe for studying directly the structure of a cristalline surface. In fact the *complete lack of penetration of the thermal atoms into the surface* and the *very small probability of multiscattering effects* (at least for sufficiently flat surfaces and at normal incidence) makes the structure analysis with atomic diffraction a purely kinematic problem. This is not the case for electron and neutron diffraction. In LEED the multiscattering effects are in most cases not negligeable and the penetration is at least three atomic layers.
Therefore, the interpretation of electron diffraction data of the majority of the systems depends on dynamical theories which are approximate and of different degree of accuracy, reliability and applicability [2]. Neutrons, on the other hand, are highly penetrating particles and are not a real surface probe. They can be used in some favorable cases of adsorbed layers on highly dispersed solids (for example graphoil) because of large variations of interaction strength [8]. Indeed, *atomic diffraction and LEED are the only available*

methods for studies of monolayers adsorbed on single crystal surfaces. All this can be summarized by saying that the atomic beam is more surface-sensitive than other diffraction probes. Moreover the coherence length (or transfer width) in an atomic diffraction apparatus (a maximum of about 600 Å for a well collimated He nozzle beam at normal incidence) is of the same order of magnitude of that reported for LEED (ranging between 100 to 500 Å).

Atomic diffraction is observable whenever a quasi-monocromatic beam of atoms of mass m has an average wave vector $k_i = (2mE)^{\frac{1}{2}}/\hbar$ such that $\lambda = 2\pi/k_i \cong D$ where D is the surface lattice parameter and λ is the de BROGLIE'S wavelength of the beam. The positions of the diffracted peacks for a purely monochromatic beam are given by the energy and momentum conservation laws:

$$|\underline{k}_i|^2 = |\underline{k}_{\underline{G}}|^2 \quad ; \quad \underline{K}_{\underline{G}} = \underline{K}_i + \underline{G} \tag{1}$$

where $\underline{G}$ is a surface reciprocal lattice vector; $\underline{k}_i \equiv (\underline{K}_i, k_{iz})$ and $k_{\underline{G}} \equiv (\underline{K}_{\underline{G}}, k_{\underline{G}z})$ the wave vectors of the beam before and after the scattering process respectively; $\underline{K}_i \equiv (k_{ix}, k_{iy})$ and $\underline{K}_{\underline{G}} \equiv (k_{Gx}, k_{Gy})$ are the components of $\underline{K}_i$ and $\underline{K}_{\underline{G}}$ parallel to the surface while k_{iz} and k_{Gz} the ones normal to the surface. In the case of in plane scattering from a two-dimensional lattice having equal unit vectors (1) can be rewritten as follow:

$$\sin\Theta_{\underline{G}} = \sin\Theta_i \pm \frac{2\pi}{D|\underline{k}_i|\sin\phi} (m^2 + n^2 - 2mn\cos\phi)^{\frac{1}{2}} \tag{2}$$

where $G = (m, n)$ and ϕ is the angle between the lattice unit vectors. The angle of incidence Θ_i and the angle of scattering $\Theta_{\underline{G}}$ are measured from the normal to the surface.

In a real experiment however the beam has a finite angular spread and a wave-vector distribution. Therefore, eq. (1) (or (2)) have to be convoluted with these characteristics of the beam and the geometry of the apparatus (finite size of the source and of the detector, the size of the beam spot on the sample etc.). The surface structure and the lattice parameter is then obtained by a fitting procedure to the measured positions of the diffracted peaks.

Additional structural information may be obtained from the specular scattering which is phase sensitive to vertical displacements within the unit mesh and over the coherence area of the beam. This has been shown by CARDILLO and BECKER in two experiments where atomic diffraction has been successfully used as a crystallographic probe. An He beam is diffracted from a Si(111) 7 x 7 surface [21] and from a reconstructed surface of Si (100) [22]. The diffraction data show how this technique can give information complementary to LEED and other surface probes. Another very interesting experiment has been carried out by the Genoa group [23] on the 1T-TaS_2 basal plane. They were able to demonstrate for the first time the existence of strong modification of the surface of this layered compound due to the presence of charge density waves (CDW).

In fact for this system LEED, X-rays and neutron diffraction cannot give a final answer to the problem of the real structure of the outmost layer. On the other hand on the basis of the atomic diffraction data it has been possible to model the structural modification of the TaS_2 surface due to the presence of CDW. These are two examples of the very few atomic beam experiments aimed at studying surface crystallography which have been carried out up to date. Further studies are needed to calibrate the technique for surfaces with large unit mesh and to improve the experimental resolution. The accuracy in determining the lattice parameter ($\gtrsim$1%) is still not comparable to the resolution achieved by LEED or neutron diffraction ($\simeq$ 0.3%) and is limited by the velocity spread of the beam ($\frac{\Delta V}{V} \simeq 2\%$ FWHM) and its angular spread (0.06°). Furthermore there are several different surfaces from which has not been observed any atomic diffraction while the number of surfaces studied is still too small. Nonetheless atomic diffraction is a quite promising technique in surface crystallography because, when coupled with techniques such as LEED and low energy ion scattering, it can give a complete instrument for surface analysis.

2.2 Atom-surface interaction potential

The interpretation of the intensity distribution of the atomic diffraction patterns as a function of the scattering parameters (incident and azimuthal angle, incident energy etc.) yields detailed information on the atom-surface interaction potential. This can be considered the most highly developed aspect of atomic diffraction.

A detailed survey on this topic is given in the review paper that has been recently written by HOINKES [20] and in the several reviews which can be found in literature [17-19].

Among the different theoretical approaches [19] the hard corrugated wall (HCW) model is the one most commonly used. It is in fact a computationally simple model which reproduces the majority of the experimental data quite satisfactorily. Furthermore, LEVI et al. [24] have generalized the HCW model in the eikonal approximation to the case of adsorbates commensurate to the substrate surface. For these reasons the HCW model is the only one which is discussed in the following.

The HCW model was proposed in 1975 by LEVI and coworkers [25]. The basic assumption is that the gas-surface potential may be approximated by

$$V(\underline{r}) = 0 \text{ for } z > \xi(\underline{R}) \quad \text{and} \quad V(\underline{r}) = \infty \quad \text{for} \quad z \leq \xi(\underline{R}) \qquad (3)$$

where $\xi(\underline{R})$ defines the mathematical shape of the surface that can be written as a Fourier expansion

$$z = \xi(\underline{R}) = \sum_{\underline{G}} \xi_{\underline{G}} \, e^{i(\underline{G}\cdot\underline{R})} \quad ; \qquad (4)$$

$\underline{r} \equiv (x,y,z)$ is the three-dimensional vector in the space and $\underline{R} = (x,y)$ is the vector in the plane parallel to the surface. $\xi_{\underline{G}}$'s are the Fourier coefficients of the potential usually called corrugation parameters. With the interaction potential (3) only perfectly elastic events are considered and all effects caused by resonance channels (see below) are neglected.

In the HCW model the Schrödinger equation can be solved approximately [25] or exactly [26] for obtaining the probabilities of the different diffracted peaks. A fitting procedure to the experimental data is then used to determine the corrugation parameters. Only the first few $\xi_{\underline{G}}$ are usually different from zero. The shape function $\xi(\underline{R})$ may give some information about the structure of the surface [23] [24] as has been already mentioned but some care has to be taken in interpreting the corrugation parameters. It has been shown that different atoms "see" the same surface differently corrugated. This is, for example, the case of the corrugations of LiF (001) [28] and of C(0001) [29] which are respectively 2 or 3 times larger when an H or an He beam is used. This effect proves that the repulsive part of the interaction potential for H atoms is softer than that for He atoms. This would lead in fact to a higher specular intensity which corresponds to a smaller corrugation as has been explained by Rieder in a chapter of this book.

The HCW and the other available models are purely elastic theories which do not take into account the inelastic and incoherent effects due to the thermal vibration of the surface atoms. These effects are usually schematized in a Debye-Waller factor. There is still discussion in literature [30] on the applicability of a Debye-Waller-like theory to the atomic diffraction. Nonetheless there is some experimental evidence that this approach can give a satisfactory description of the temperature dependence of the diffracted beam intensities at least in some cases [31]. In the frame work of the Debye-Waller theory the temperature dependence of the elastically scattered intensities gives information about the average termal displacement of the atoms of the surface

Since the first experiments [32] on atomic diffraction from surfaces, anomalies were observed in monitoring the specular beam intensity as a function of the scattering parameters (incident or azimuthal angle and incident energy). These structures can be easily explained when associated with the bound states of the laterally averaged gas-surface interaction potential. In fact, whenever $k^2_{\underline{G}z}$, the perpendicular energy of the atom leaving the surface, is negative and equals the energy E_n of the n-th bound state of the laterally averaged potential, the (1) can be rewritten as

$$k^2_{\underline{G}z} = (2m/\hbar)^2 E_n = k_i^2 - (\underline{K}_i + \underline{G})^2 \tag{5}$$

where m is the mass of the gas atom. Whenever there is a $\underline{G}$ vector satisfying equation (5), the impinging atom may be transferred via diffraction to a state where it is bound with energy E_n normal to the surface but is free to move move parallel to the surface with the energy $(\underline{K} + \underline{G})^2 \hbar^2/2m$. Such a state causes resonances in the diffracted beams and therefore may give rise to extrema in the elastic intensities. A mechanism in which inelastic events are also involved may give the same effect as well.

The energies of the bound states as well as the $\underline{G}$ vector involved in the process may be obtained by fitting with eq. (5) the positions of the extrema as a function of the incident angle, of the azimuthal angle and of the incident energy. Beams of different isotopes are often used and studied for a more accurate determination of the well of the interaction potential. This is possible because isotopes of the same species have the same interaction potential but different bound states.

In the last few years dynamical scattering theories have been also developed to predict the shape and intensities of bound state resonances.

These models have been quite remarkable in fitting the experimental data and in predicting new features such as the splitting of resonances [33]. The study of these splittings gives detailed information about the strength of the periodic Fourier components of the interaction potential (the (0,0) component being the laterally averaged potential).

A very good example of the accuracy that can be reached with the bound state resonance experiments is given by the quite complete work that has been carried out on the scattering of ^3He and ^4He from the (0001) surface of graphite [34]. For this system, the complete band structure has been studied and the ground state of the interaction potential has been obtained in perfect agreement with the thermodynamical data on adsorbed He films.

2.3 Inelastic scattering

Time resolved diffractive scattering is a probe of phonon structure of the surface and can bring some light on the theory of the Debye-Waller factor. Unfortunately this aspect of the atomic diffraction is the least studied and very few systems have been investigated up to date. This is mostly due to the experimental difficulties involved in the energy analysis of the diffracted atoms. Only very recently it has been possible to get high resolution data on the surface phonons of LiF, NaF, KCl and metals [35]. From the point of view of the studies of adsorbed layers on crystal surfaces, it seems likely that the inelastic scattering may be easier for these systems because of the possible excitation of discrete or quasi-discrete vibrational modes of the adsorbed monolayer.

3. Atomic diffraction from adsorbates

The early atomic diffraction experiments on gases adsorbed on single crystal surfaces can be considered much more a study of the sensitivity of this technique to surface contamination than a real surface analysis. The first observation of this kind for a layer of C atoms adsorbed on the (110) surface of W was published in 1970 [36], but an accurate analysis of the data [37] showed that if the experimental conditions were the ones reported, it was impossible to observe diffraction.

MASON and WILLIAMS in 1972 [38] studied the diffraction of He from the (001) surface of LiF contaminated with molecules of H_2O, C_2H_5OH, CH_3OH CCl_4 , C_6H_6 introduced in the He beam. The LiF diffraction peaks were monitored as a function of the contamination. The main features observed are:

a) the sensitivity of the specular peak to the contamination is higher than that of the higher order diffraction peaks (the ratio of the attenuations observed is roughly 5);

b) lobes were found in the vicinity of the specular peak when also a small contamination was present (attenuation of the (0,0) peak ≈ 1%). These features were present only if the adsorbed molecules were deposited via the nozzle beam system;

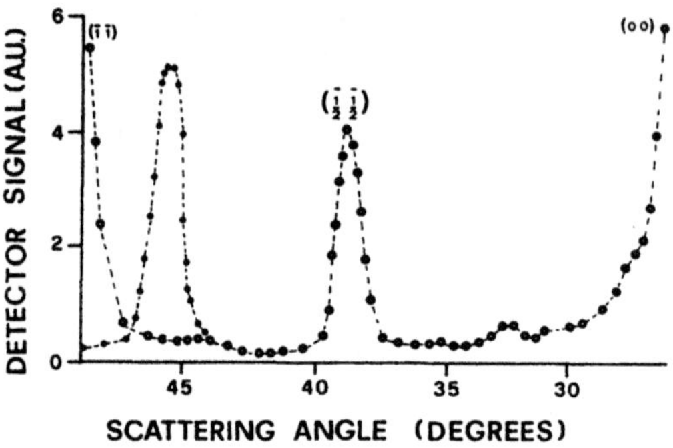

Fig. 1 Half order diffractive peak in the in plane scattering of He from a layer of H_2O adsorbed on the (001) surface of NaF. The peak position changes when the energy of the beam is changed (peak on the left with closed circles). Small quarter order peaks are present at about 33° and 44°[39].

c) the lobes increase their separation and decrease their intensity when contamination increases. Furthermore they do not depend on the ordering of the adsorbed molecules;

d) ethanol is the only molecule showing additional diffraction peaks. A oblique cell was proposed for the explanation of the diffraction patterns.

The authors explained the presence of the lobes with the mobility of the adsorbed molecules on the partially covered surface but there are still some doubts about the ultimate explanation.

The same authors carried out similar experiments on the adsorption of CCl_4, E_tOH and H_2O on the (001) surface of NaF [39]. Ordered structures were observed only for H_2O (fig. 1) and E_tOH (fig. 2). While H_2O shows a structure commensurate to the substrate, E_tOH shows a diffraction pattern related to its molecular dimension. The scattering from the H_2O was studied in greather detail as a function of degree of adsorption, temperature of deposition and annealing behavior. Also in the case of NaF the lobular structures on the shoulders of the specular peak appear as a function of contamination and are not sensitive to changes of the energy of the incident beam. Therefore their presence cannot be due to some structural effect of the surface. A possible interpretation is that they are related to the diffusion of the adsorbed molecules. In fact they are not present if the coverage of the surface is high and uniform. Furthermore the measurements of the diffractive intensity coming from a region of the surface different from that where the molecular layer is adsorbed seem to confirm this interpretation. MASON and WILLIAMS carried out this experiment as a function of temperature and time and were able to estimate the diffusion coefficient for the water molecules adsorbed on NaF.

The interpretation of the experiments of Mason and Williams in terms of surface structure is quite difficult because the adsorption of rather complex molecules (such as water and ethanol) produces rather complicated diffraction patterns. Furthermore the theoretical understanding of the adsorption mechanism for such complicated systems is still poor.

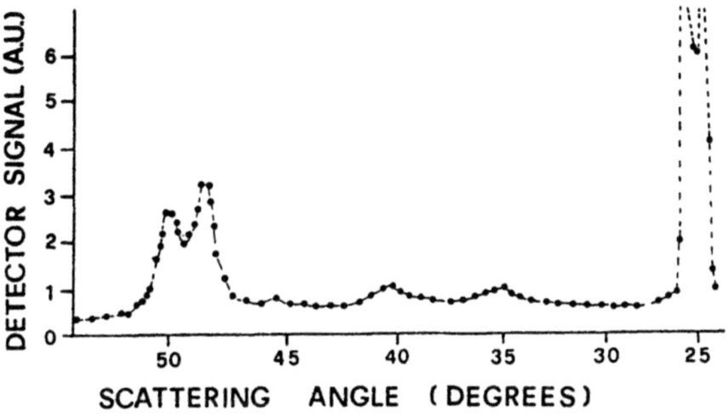

Fig. 2 In plane diffractive intensity of He from a layer of E_tOH adsorbed on (110) NaF. The (0,0) peak is at ≈ 25° and the $(\bar{1}\bar{1})$ is at ≈ 49°. Both peaks presents large dips at their centers. Small fractional order diffraction peaks appear at 35.7°, 40.7° and 45.4°. These peaks are not related to the dimensions of the substrate. The peak at 40.7° could be associated with the packing distance of the methyl group of E_tOH [39].

In 1977 FRANK et al. [39] carried out H-atom diffraction experiments on water covered KCl(001). From the diffractive intensities interpreted with the HCW wall model they obtained a corrugation parameter ξ_0 = 0.76 Å which shows a highly corrugated surface. The overall interpretation of their data seems to suggest that water adsorbes in a commensurate phase with only one molecule per unit cell of KCl (001). The attractive part of the interaction potential H-H_2O/(001)KCl has been studied via bound state resonances with both hydrogen and deuterium beams. They were able to identifie a total of 10 levels and to determine the best Morse potential fitting them. The resulting well depth is 33.4 meV and the range parameter is 0.54 $Å^{-1}$. The authors claim that the attractive part of the interaction potential of water covered KCl(001) and of the clean substrate are practically the same. Unfortunately they were not able to carry out measurements on the clean surface and their claim seems not sufficiently grounded.

In the same paper they report the temperature dependence of the elastic intensities. They have measured to the Debye-Waller factor for both the water covered KCl(001) and the clean substrate from which they have estimated the surface Debye temperatures.

Better characterized experiments have been carried out on atomic species chemisorbed on metal surfaces. RIEDER and ENGEL [40] have studied the clean and hydrogen saturated (110) surface of Ni with the diffraction of an He nozzle beam. These experiments are the first example of a successful application of atomic diffraction to the study of the structure of an atomic layer chemisorbed on a metal surface and show how valuable atom diffractive scattering is for characterizing these systems. This is particularly true in the case of chemisorption of hydrogen. LEED is in fact insensitive to it and therefore can give information only about the changes induced on the substrate by the adsorbate if they are present. The details on these measurements are discussed by Rieder in his lectures in this same book.

LAPUJOULADE et al. [41] carried out a similar experiment diffracting an He nozzle beam from oxygen adsorbed on the (110) surface of copper. The oxygen covered $C_u(110)$ shows all the features of a high corrugated surface. Using the HCW model the authors fitted the experimental diffraction intensities using a shape function of the surface constructed on the basis of realistic radii of atomic oxygen (1.40 Å) and of He (1.44 Å). The resulting corrugation is 0.7 Å which is more than twice the corrugation of the hydrogen covered (110) Ni [40].

They carried out also selective adsorption and Debye-Waller factor measurements.

From the bound state resonances they identified five energy levels of the laterally averaged potential corresponding to a well depth of 24.6 meV. This value has been used to calculate the mean square displacement $\langle u^2 \rangle$ of the oxygen atoms from the measured Debye-Waller factor. A comparison with the analogous measurements for the clean (110) Cu shows that the mean square displacements for atoms of both the oxygen monolayer and the copper surface are practically the same. This effect is due to the strong chemical bonds involved in the adsorption of gases such as H and O on metal surfaces. In the case of physisorption of gases the expected behavior is, in fact, quite different. An example is given by the xenon covered (0001) surface of graphite. For this system LEED measurements [42] have given an estimate for the ratio between the mean square displacements of the atoms of the xenon commensurate overlayer and of the graphite surface. The $\langle u^2 \rangle$ of the Xe atoms in the adsorbed phase is smaller than in solid Xe and is a factor 10 higher for the carbon atoms in the graphite surface. This shows that the xenon atoms lie on a vibrationally frozen graphite substrate and that the contribution of inelastic scattering to the atomic beam diffraction from an overlayer could be rather large compared to the clean surface.

4. Diffraction of H atoms from a Xe overlayer adsorbed on the (0001) surface of graphite.

The main interest in the experiment of diffraction of H atoms from a Xe monolayer adsorbed on the basal plane of graphite [16] resides in the following two reasons:

a - this is the first well-characterized experiment in which atomic diffraction is shown to be a viable technique for studying the properties and structure of weakly bound two-dimensional solids adsorbed on a single crystal surface;

b - the interaction potentials involved in the system studied (more generally rare gases adsorbed on graphite) are quite well known and therefore it is now possible for the first time to test directly the different theoretical model developed for interpreting the diffraction intensities.

The phase diagram of the adsorption of xenon on graphite has been widely studied by means of several different techniques [43] and shows a variety of phases and phase transitions. In particular, both a commensurate and incommensurate two-dimensional solid of xenon are present even though the commensurate structure is confined in a narrow region of pressures and temperatures.

The atomic diffraction experiment has been carried out in non equilibrium conditions. Since an highly efficient cryopumping is used to obtain clean vacuum conditions in the scattering chamber, it is practically impossible to produce a finite xenon vapour pressure around the surface sample. The overlayer is therefore produced bombarding the surface with a Xe beam and lowering the surface temperature while the graphite specular peak is monitored to check the formation of the solid phase.

As has been shown recently in a THEED experiment [44], the adoption of a calibrated nozzle source would also make it possible to use the same apparatus for studying phases in equilibrium.

The basal plane of graphite has been studied with the diffraction of H atoms in previous experiments [29]. It shows a high percentage of coherent elastic scattering confined in the (0,0) (1,0) and ($\bar{1}$,0) peaks. The other kinematically allowed peaks are not present in the diffraction patterns; in particular there is no evidence of the (2,1) and (-2, -1) peaks along the azimuth $\phi = 30°$. The top of fig. 3 shows the reciprocal lattice of the basal plane of natural graphite (open circles) and the convention adopted for the azimuths. The interpretation of the intensity distribution in terms of the HCW model gives a rather small corrugation of $\simeq 3.6\%$.

Before adsorbing the xenon overlayer the diffraction from the clean substrate has been observed as a check of the surface conditions. The in-plane scan

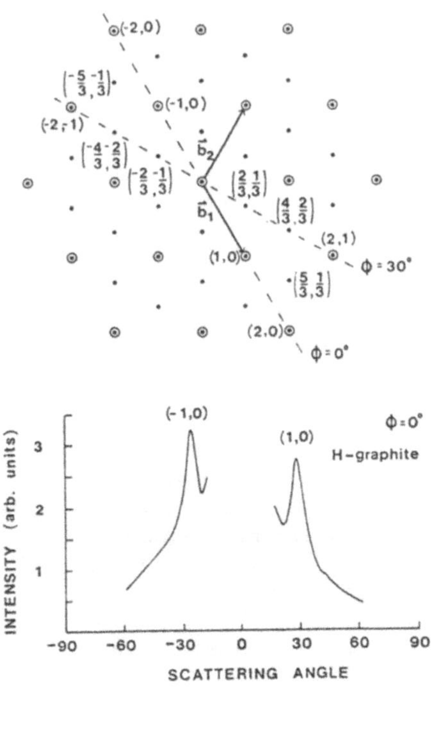

Fig. 3 Top: reciprocal lattice of the (0001) C substrate (open circles) and the in-registry xenon overlayer (black dots). The lattice constants are 2.465 Å and 4.26 Å respectively. Bottom: H atom scattering from the (0001) C substrate $\Theta_i = 1.7° \pm 0.3°$; $\phi = 0°$; surface temperature = 67°K.

of the intensity distribution along the azimuth $\phi = 0°$ is shown at the bottom of fig. 3. When the xenon overlayer is adsorbed and the temperature of the surface is lowered down to 26°K, the scattered intensity distribution results drastically changed: fractional order peaks appear along the $\phi = 30°$ and the $\phi = 10.9°$ azimuths while the peaks (2, 1) and (-2, -1) become evident. The in-plane scans along the three azimuthal direction $\phi = 0°$, $\Phi = 10.9°$ and $\phi = 30°$ are shown in fig. 4 while the reciprocal lattice of a commensurate (3 x 3) structure of a two-dimensional solid of xenon is shown on the top of fig. 3 (black dots).

The analysis of the positions of the peaks (see sect. 2) shows that the data are consistent with the commensurate structure within the experimental accuracy ($\approx$ 2-3%). Nevertheless it is difficult to compare this result with the data obtained with other methods such as LEED and THEED since the latter experiments have been carried out on the overlayer in equilibrium with a vapour pressure of xenon. Further atomic diffraction measurements are needed, aimed at studying the formation of the bidimensional solid phase, its melting and the existence of an incommensurate phase at higher temperatures.

In the interpretation of the intensities notice must be taken of the incoherent background which is due to irregularities of the substrate, defects in the xenon layer and inelastic scattering. The lack of symmetry in the diffraction patterns from both the clean graphite (fig. 3) and the adsorbate (fig. 4) can be ascribed to the irregularities of the substrate. This effect has been observed in other beam experiments carried out by different groups

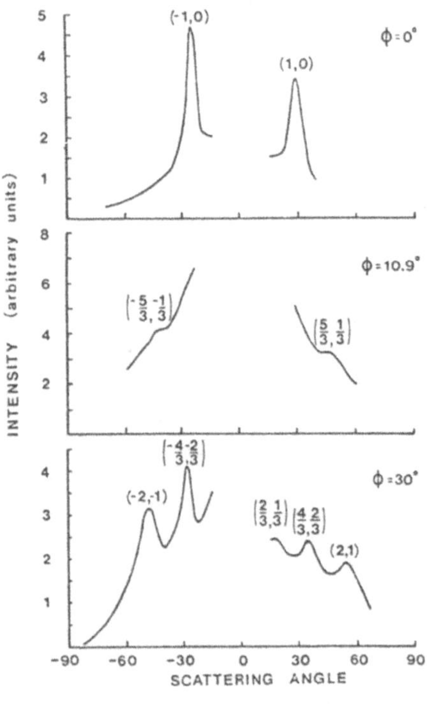

Fig. 4 In plane diffractive patterns of H atoms from the xenon covered graphite surface. The angle of incidence is $\Theta_i = 1.7° \pm 0.3°$ and the surface temperature is 26° ± 3°K.

on the (0001) surface of graphite [15]. However a complete analysis of the different contributions to the background can be carried out only by means of temperature dependence and time resolved measurements. The interpretation of the elastically diffracted intensity is however possible, taking into account only the relative diffraction probabilities. The analysis was carried out with the HCW in the eikonal approximation [25]. The best agreement between the calculated relative probabilities and the experimental ones is obtained with a corrugation of the xenon overlayer of $\approx$27%, that is 7.5 times larger than for the clean substrate. This is just a preliminary analysis which gives only an indication of the corrugation of the system. In fact the eikonal approximation is not correct for highly corrugated surfaces.

In the same paper [29] a comparison is also given of the results of the HCW model with the isopotential surfaces of the H/xenon-(0001) C interaction. This interaction can be modeled as the sum of two terms: one giving the proper long range interaction ($\frac{1}{z^3}$ behavior) as calculated by VIDALI et al. [44] and the other describing the H-xenon pairwise potential. This last term has been obtained through a combination of short repulsion and semiempirical long range calculations [45]. The maximum difference of the classical turning points of this interaction potential computed above the surface unit cell gives 1.00 ± 0.05 Å which corresponds to a corrugation of $\approx$ 23% in good agreement with the HCW model. As already mentioned, knowledge of a realistic gas-surface potential can be also used for calculating directly the diffraction probabilities. In this way it would be possible for the first time to directly compare the validity and accuracy of the several different methods developed for calculating diffraction probabilities.

5. Conclusions

Atomic diffraction is a valuable technique for studying two-dimensional solids physisorbed on single-crystal surfaces and can give information complementary to the other diffractive methods commonly used. The best example in this direction is given by the recent experiment on diffraction of H atoms from a commensurate $(\sqrt{3} \times \sqrt{3})$ xenon layer adsorbed on (0001)C. The analysis of the experiment shows also that the study of rare gases on graphite could be a good test for the different theoretical models available for calculating the diffraction probabilities. An extensive study of these systems using different atomic beams is therefore needed.

In particular the use of He-nozzle beams would give very high resolution data which are indispensable for studying in detail the different phase transitions and the phonon structure of the adsorbate.

References

1. A. Thomy, X. Duval: in Adsorption et Croissance cristalline (ed. CNRS) 81 (1965);
A. Thomy, X. Duval: J. Chim. Phys. 66, 1966 (1969)
2. a) G.A. Somorjai, M.A. Van Hove: Adsorbed Monolayers on Solid Surfaces, Structure and Bonding, Vol. 38 (Springer, Berlin, Heidelberg, New York 1979)

b) J.G. Dash, J. Ruvalds: Phase transitions in surface films (Plenum Press - New York) (1980)
c) J.G. Dash: Films on solid surfaces (Academic Press - New York) (1975)
3. J.P. Hobson: Proc. Int. Vacuum Congr. 6th, Kyoto (1974)
4. M.D. Chinn, S.C.jr. Fain: Vac. Sci. Technol. 14, 314 (1977)
5. M.D. Chinn, S.C.jr. Fain: Phys. Rev. Lett. 39, 146 (1977)
6. J.A. Venables, H.M. Kramer, G.L. Price: Surf. Sci. 55, 373 (1976); 57 782 (1976)
7. J.K. et al.: Phys. Rev. B13, 1446 (1976)
8. M. Nielsen, J.P. McTague, L. Passel: Neutron scattering studies of physisorbed monolayers on graphite in Phase transitions in Surface Films, editors J.G. Dash and J. Ruvalds (Plenum Press - New York) (1980)
9. I. Estermann, O. Stern: Z. Physik 61, 95 (1930)
10. D.V. Tendulkar, R.E. Stickney: Surface Sci. 27, 516 (1971)
A.G. Stoll, R.P. Merrill: Surface Sci. 40, 405 (1973)
11. G. Boato, P. Cantini, R. Tatarek: J. Phys. F6, L327 (1976)
12. K.H. Rieder, T. Engel: Phys. Rev. Letters 43, 373 (1979)
13. J. Lapoujoulade, Y. Lejay, M. Papanicolaou: Surface Sci. 90, 133 (1979)
14. M.J. Cardillo, G.E. Becker: Phys. Rev. B21, 1497 (1980)
15. G. Derry et al.:Surface Sci. 94, 221 (1980)
G. Boato et al.: Phys. Rev. B20, 3957 (1979)
E. Ghio et al.: J. Chem. Phys. 73, 556 (1980)
L. Mattera et al.: Surface Sci. 93, 515 (1980)
16. T.H. Ellis, S. Iannotta, G. Scoles, U. Valbusa: Phys.Rev.B 24, 2307 (1981)
17. D.R. Frankl: ISISS (1979) Surface Science: "Recent Progress and perspective", CRC Press, Cleveland, OHIO
18. M.W. Cole, D.R. Frankl: Surface Sci. 70, 585 (1978)
19. F.O. Goodman: CRC Crit. Revs. in Solid State and Math. Sci. 7, 33 (1977)
20. H. Hoinkes: Rev. Mod. Phys. 52, 933 (1980)
21. M.J. Cardillo, G.E. Becker: Phys. Rev. Lett. 40, 1149 (1978)
22. M.J. Cardillo, G.E. Becker: Phys. Rev. B21, 1497 (1980)
23. G. Boato, P. Cantini, R. Colella: Phys. Rev. Lett. 42, 518 (1979)
P. Cantini, G. Boato, R. Colella: Physica 99B, 59 (1980)
24. A.C. Levi, R. Spadacini, G.E. Tommei: Surf. Sci. 108, 181 (1981).
A.C. Levi: this book
25. U. Garibaldi, A.C. Levi, R. Spadacini, G. Tommei: Surface Sci. 48, 649 (1975)
26. N. Garcia, N. Cabrera: Phys. Rev. B18, 576 (1978)
27. G. Armand, J.R. Manson: Phys. Rev. B18, 6510 (1978)
28. G. Caracciolo, S. Iannotta, G. Scoles, U. Valbusa: J. Chem. Phys. 72, 4491 (1980)
29. T.H. Ellis, S. Iannotta, G. Scoles, U. Valbusa: to appear in J. Vac. Sci. Technol.
30. A.C. Levi, H. Suhl: Surf. Sci. 88, 221 (1979)
31. H. Hoinkes, H. Nahr, H. Wilsch: Surf. Sci. 33, 516 (1972); 40, 457 (1973)
F.O. Goodman: Surf. Sci. 46, 118 (1974)
H. Wilsch, H.U. Finzel, H. Frank, H. Hoinkes, H. Nahr: Proc. Int. Conf. Solid Surfaces, IInd, Kyoto (1974) (published in Jpn. J. App. Phys. Suppl.)
J. Lapoujoulade, Y. Lejay: J. Chem. Phys. 63, 1389 (1975)
H. Frank, H. Hoinkes, H. Wilsch, H. Nahr: Surf. Sci. 63, 121 (1977)

32. R. Frish, O. Stern: Z. Physik 84, 430 (1933)
33. V. Celli, N. Garcia, J. Hutchinson: Surf. Sci. 87, 112 (1979)
H. Chow, E.D. Thompson: Surf. Sci. 71, 731 (1978)
H. Chow, E.D. Thompson: Surf. Sci. 82, 1 (1979)
34. M.W. Cole, D.R. Frankl, D.L. Goodstein: Rev. Mod. Phys. 53, 198 (1981) and references therein
35. G. Brusdeylines, R. Bruce Doak, J.P. Toennies: Phys. Rev. Lett. 46, 437 (1981)
L. Mattera, C. Salvo, S. Terreni, F. Tommasini, U. Valbusa (to be published). See also paper of the same authors in this book
36. W.H. Weinberg, R.P. Merril: Phys. Rev. Lett. 25, 1198 (1970)
W.H. Weinberg, R.P. Merril: J. Chem. Phys. 56, 2893 (1972)
37. F.O. Goodman: J. Chem. Phys. 64, 1051 (1976)
38. B.F. Mason, B.R. Williams: J. Chem. Phys. 56, 1895 (1972)
39. B.F. Mason, B.R. Williams: Surf. Sci. 45, 141 (1974)
40. K.H. Rieder, T. Engel: Phys. Rev. Lett. 43, 373 (1979)
41. J. Lapoujoulade, Y. Le Creüer, M. Leport, Y. Lejay, E. Maurel (to be published)
42. J.P. Coulomb, J. Suzanne, M. Bienfait, P. Masri: Solid State Comm. 15, 1585 (1974)
43. P.S. Shabes-Retchkman, J.A. Venables: Surface Sci. 105, 536 (1981)
44. G. Vidali, M.W. Cole, G. Schwartz: Surface Sci. 87, 1273 (1979)
45. C. Douketis, G. Scoles, S. Marchetti, M. Zen, A.J. Takkar: J. Chem. Phys. in press

Atom Scattering from Overlayers

A.C. Levi

Instituto di Fisica Teorica dell'Università and
Gruppo Nazionale Struttura della Materia
I-Napoli, Italy

1. Introduction

Overlayers attract a great deal of attention - probably as much attention as clean surfaces - ranging from statistical mechanics (in particular, phase transitions in 2D) to lattice dynamics, to epitaxy etc. In fact, this is a special chapter of the subject of dirty surfaces, an optimistic chapter where the dirt is rearranged into a more or less ordered phase (or perhaps several such phases). The existence of overlayers requires the temperature to be high enough that the barriers hindering the approach to equilibrium can be overcome.

Overlayers may form different kinds of phases:

a) *commensurate phases* where an adsorbed atom must necessarily sit at (or near) a site of the *site lattice*, a lattice whose symmetry group is a subgroup of that of the substrate lattice; commensurate phases in general contain vacancies and the vacancy density distinguishes rarefied from dense phases;

b) *incommensurate* (solid) *phases* where the adsorbate forms a lattice bearing no relationship (except perhaps in orientation) to the substrate lattice;

c) *fluid phases* where the adsorbed atoms are distributed at random on the surface.

Also according to the nature of the forces binding the adsorbed atoms to the substrate, adsorbates may be distinguished in chemisorbates and physisorbates; generally chemisorbates form commensurate phases while physisorbates may form any possible kind of phase.

At low temperatures (but, of course, still high enough for the overlayer to form under the effect of diffusion) *phase separation* is expected with the accompanying effects of phase transitions, critical points, etc.

Overlayers have been studied extensively with a variety of techniques, from thermodynamics (equation of state, calorimetry) to scattering and related effects (X-ray, electron, neutron or atom scattering, ellipsometry, Mössbauer effect). It may be useful to distinguish three cases:

a) physisorption on exfoliated graphite,

b) physisorption on other substrates,

c) chemisorption.

Case a) has been studied with a variety of techniques including neutron scattering, since the very large area of exfoliated graphite permits one to ignore the (otherwise dominant) scattering from the bulk.

Case b) is more interesting (because it may involve single crystals) but more difficult (because the small surface area rules out a number of techniques). Case c) is similar but even more difficult.

In cases b) and c) the only techniques available (apart from infrared absorption that gives energetical information only) seem to be electron and atom scattering; the former is much better known, and more extensively used, but I am going to talk about the latter, which is promising. As usual, electron scattering is much simpler experimentally (electrons can easily be deflected and detected) but pays for this by being sensitive to several substrate layers; moreover hydrogen can hardly be detected with electrons due to the small e-H cross section; also, multiple Coulomb scattering is a highly complicated problem that can only be handled with fast computers.

But the true advantage of atom scattering will probably lie in inelastic scattering where low frequencies (atom-surface oscillations are mostly low-frequency) cannot be studied with infrared spectroscopy and can be studied with high resolution EELS only with some effort.

2. Experiments of atom scattering from adsorbates

The experiments are few but rapidly increasing in number. A list of selected references is given in Table I.

a) Since Refs. |1,3,6| mostly deal with attenuation of substrate diffraction due to adsorption and bear only indirect relation to scattering from adsorbates, I will ignore these for the time being.

b) Chemisorption - The systems studied are H on Ni (110) (by Rieder and Engel) and O on Cu (110) (by Lapujoulade et al.) (Hayward has studied several other systems but his work is unpublished and I cannot review it). The two systems are remarkably similar since the (110) face of fcc metals is describable in terms of parallel ridges and is essentially flat along the ridges and strongly corrugated across. However H on Ni shows a number of

Table I: Classification of experiments on atom scattering from adsorbates.

a) Adsorption-induced attenuation of substrate diffraction:	- Smith and Merrill, 1970 |1|
	- Engel, 1978 |3|
	- Poelsema, Mechtersheimer and Comsa, 1980 |6|
b) Chemisorption:	- Rieder and Engel, 1979 |4|
	- Hayward, 1978 |5|
	- Lapujoulade, Le Cruer, Lefort, Lejay and Maurel, 1981 |7|
c) Physisorption:	- Mason and Williams, 1972-1974 |2|
	- Ellis, Iannotta, Scoles and Valbusa, 1981 |8|

different phases (some of which very complicated) and Rieder and Engel have been able to determine the positions of hydrogen atoms (with some uncertainty), LEED being unable to accomplish this feat. The beautiful analysis of Rieder and Engel has been made with the help of a very elementary theory (see below).

Lapajoulade et al., whose work includes very complete study of the scattering of noble gas atoms from copper surfaces, have studied the (110) face of copper covered with an adlayer of oxygen whose atoms are (presumably) in the position shown. The situation is very different here since the oxygen atoms (whose nuclei seem to be *below* the copper nuclei in the ridges) produce a strong corrugation along the ridges while H on Ni only amplified very much (and distorted due to Ni reconstruction) the corrugation across.

Bound state resonances have been studied by Rieder and Engel and by the French group and allow a better determination of the potential: not only the well can be determined, but in one instance even a sign ambiguity in the shape function can be resolved.

c) Physisorption - Mason and Williams pioneering work involved physisorption of water, alcohol and carbon tetrachloride on NaF. More recent measurements refer to a noble gas (xenon). Most interesting is the experiment of the Waterloo group where diffraction of H atoms is observed from an ordered overlayer of Xe atoms on the basal plane of graphite. The adsorbate seems to be commensurate, which would be somehow in contrast to other evidence from LEED. Anyhow, the H-Xe two-body potential is quite well known, which means that a gas-surface potential can be estimated, in particular a corrugation amplitude which can be compared to that obtained from the experiment; and the agreement is satisfactory. Very recently, Mason and Williams performed a scattering experiment of He atoms from an ordered overlayer of Xe atoms on Cu(001) with special attention on inelastic scattering.

3. Theory of atom scattering from adsorbates

Scattering from adsorbates resembles scattering from clean surfaces, but for one point: there is an intrinsic randomness of the adsorbate, which (apart from extreme cases) can hardly be described as a perfect two-dimensional crystal. There may be random vacancies in the adsorbate; or it may happen that the adsorbate virtually only exists in certain restricted areas, the *adsorbate islands* (there will also be random vacancies in an island and random atoms outside); or finally, the adsorbate may be completely randomized.

Such situations are difficult to treat all together. However, it would be nice to have rather general formulae which then should be specialized to get different cases.

I will employ the eikonal (or Kirchhoff) approximation. This is by far the simplest and shows most of the relevant effects, although multiple scattering is neglected.

If the adsorbed atoms have a distribution $A \equiv \{\vec{R}_{Aj}\}$, let dP_A be the differential probability for scattering of beam atoms. This, in the eikonal approximation, is given by |10-12|

$$dP_A = \frac{k|k_{iz}|}{8\pi^3\hbar L^2}\, d\Omega\, d\Delta \int \exp\{i\vec{Q}\cdot(\vec{R}'-\vec{R}) + i\left[\eta_2(\vec{R}';A)-\eta_2(\vec{R};A)\right] - \\ - W(\vec{R};A) - W(\vec{R}';A)\}\ \Phi\ d^2R d^2R', \tag{1}$$

where η_2 is a phase shift which for a hard wall potential reduces to $q_z\zeta(R;A)$ and Φ equals simply $2\pi\hbar\delta(\Delta)$ for elastic scattering but for inelastic scattering

$$\Phi = \int e^{-it\Delta/\hbar}\,\{e^{2\mathcal{W}(\vec{R},\vec{R}',t;A)} - 1\}\,dt \tag{2}$$

where $\mathcal{W}(\vec{R},\vec{R}',t;A) = \frac{1}{2}<\eta_3(\vec{R},0;A)\eta_3(\vec{R}',t;A)>$, reducing for $\vec{R}' = \vec{R}$ and $t = 0$ to $W(\vec{R};A) = \mathcal{W}(\vec{R},\vec{R},0;A)$. For a hard wall potential

$$\eta_3(\vec{R},t;A) = q_z\,\{u_z(\vec{R},t;A) - \vec{U}(\vec{R},t;A)\cdot\nabla\zeta(\vec{R};A)\} \tag{3}$$

and the assumption is made here that the adsorbed atoms do not diffuse away from their initial positions in time on the order of $\hbar/\Delta$. Under reasonable statistical assumptions, then

$$dP = \frac{k|k_{iz}|}{8\pi^3\hbar L^2}\,d\Omega d\Delta \int e^{i\vec{Q}\cdot(\vec{R}'-\vec{R})} \ll \exp\{i[\eta_2(\vec{R}')-\eta_2(\vec{R})] - W(\vec{R})-W(\vec{R}')\}\Phi\gg d^2R\,d^2R', \tag{4}$$

and for elastic scattering

$$dP_{el} = \frac{k|k_{iz}|}{4\pi^2L^2}\,\delta(\Delta)\,d\Omega d\Delta \int e^{i\vec{Q}\cdot(\vec{R}'-\vec{R})} \ll \exp\{i[\eta_2(\vec{R}')-\eta_2(\vec{R})] - W(R)-W(R')\}\gg d^2R\,d^2R'. \tag{5}$$

Now commensurate and incommensurate adsorbates differ in the way the average over configurations of adsorbed atoms

$$\ll \exp\{i[\eta_2(\vec{R}')-\eta_2(\vec{R})] - W(\vec{R})-W(\vec{R}')\}\gg \tag{6}$$

is to be evaluated.

In the case of a commensurate adsorbate, adsorption can only take place in sites of a *site lattice* S and A may be identified as the set $A \subset S$ of the occupied sites. Then the integrals over $\vec{R}$ and $\vec{R}'$ may be split into integrals over unit cells of the site lattice to obtain

$$dP_{el} = \frac{k|k_{iz}|}{4\pi^2N}\,A\,\delta(\Delta)\,d\Omega d\Delta\,\sum_{ij} e^{i\vec{Q}\cdot(\vec{R}_j-\vec{R}_i)}\,\overline{H}_{ij} \tag{7}$$

where A is the area of the unit cell, and where

$$\overline{H}_{ij}(A) = A^{-2}\,K^*(A_i)\,K(A_j). \tag{8}$$

Here A_j is the distribution that obtains from A by displacing the whole adsorbate by $-\vec{R}_j$, and

$$K(A) = \int_{\text{unit cell}} \exp\{i\vec{Q}\cdot\vec{R} + i\eta_2(\vec{R};A) - W(\vec{R})\}\,d^2R. \tag{9}$$

The adsorbate produces incoherent scattering when not all adsorption sites are occupied. If elastic coherent scattering is identified with diffraction, then it is connected to the asymptotic value of $\overline{H}_{ij}$ (for j very far from i). Call this asymptotic value $\overline{H}_{far}$; then the elastic coherent scattering probability is given by a set of diffraction peaks (corresponding to the reciprocal lattice vectors $\vec{G}$ of the site lattice) with scattering probabilities

$$P_{\vec{G}} = \frac{|k_{iz}|}{k_{\vec{G}}} \overline{H}_{far,\vec{G}} , \tag{10}$$

where $\overline{H}_{far,\vec{G}}$ is given by

$$\overline{H}_{far,\vec{G}} = A^{-2} | \langle\langle K_{\vec{G}} \rangle\rangle |^2 . \tag{11}$$

The elastic incoherent scattering probability, on the other hand, is given by

$$dP_{el,inc} = \frac{k|k_{iz}|}{4\pi^2 N} A\, \delta(\Delta) d\Omega d\Delta \sum_{ij} e^{i\vec{Q}\cdot(\vec{R}_j-\vec{R}_i)} (\overline{H}_{ij} - \overline{H}_{far}) . \tag{12}$$

In the case of an incommensurate adsorbate one can write (for a hard wall)

$$\begin{aligned} &\langle\langle \exp\{i[\eta_2(\vec{R}')-\eta_2(\vec{R})] -W(\vec{R})-W(\vec{R}')\} \rangle\rangle = \\ &= \int e^{iq_z(\zeta_2-\zeta_1)-W_1-W_2}\, p(\zeta_1,W_1,\zeta_2,W_2;\vec{R}'-\vec{R})\, d\zeta_2 d\zeta_1 dW_2 dW_1 \end{aligned} \tag{13}$$

which allows the application of the stochastic methods of Beckmann and Spizzichino (BS). |13|

For example, assuming that W may assume only one value, the average reduces to

$$e^{-2W} \int e^{iq_z(\zeta_2-\zeta_1)}\, p(\zeta_1,\zeta_2;\vec{R}'-\vec{R})\, d\zeta_1 d\zeta_2 , \tag{14}$$

where the integral is a quantity extensively used by BS. Assuming ζ to be a function of r, the distance from the nearest adsorbate atom, $p(\zeta_1, \zeta_2; \vec{R})$ may be written in terms of the joint probability of distance :

$$p(\zeta_1,\zeta_2;\vec{R}) = p(r_1,r_2;\vec{R}) \Big/ \left| \frac{d\zeta_1}{dr_1} \frac{d\zeta_2}{dr_2} \right| . \tag{15}$$

The latter in turn may be expressed in terms of the radial distribution function $g(\vec{R}_{12})$ of adsorbed atoms. For example for an isotropic phase (two-dimensional fluid)

$$p(r_1,r_2;\vec{R}) = \frac{\rho^2 r_1 r_2}{4\pi^2} \int g(R_{12})\, d\theta_1\, d\theta_2 \tag{16}$$

where $\theta_{1,2}$ are the angles between $\vec{r}_{1,2}$ and $\vec{R}$, and where $\vec{R}_{12}=\vec{R}+\vec{r}_2-\vec{r}_1$.

An exact scattering theory for random rough surfaces (as opposed to the eikonal approximation used by Beckmann and Spizzichino) has been presented by García, Celli and Nieto-Vesperinas |14| allowing them to treat highly random surfaces (where the correlation length is of the same order of magnitude as the wavelength or even shorter) which are beyond the scope of the eikonal approximation.

More recently, García and Cabrera |15| have studied the problem how the diffraction intensities from the substrate decrease when the surface is covered with random impurities. Letting I to be the intensity, Θ the coverage, and

$$I = I_0 (1 - \alpha\Theta) \tag{17}$$

and assuming a normally distributed surface, hence an exponential, Debye-Waller-like behaviour of the intensities, they determine α to be proportional to the square of the perpendicular momentum transfer and to depend also on the height distribution of the impurities.

Coming back to commensurate adsorbates, we have performed specific calculations for two examples: Kr atoms on the basal plane of graphite and O atoms on W (110) |11|.

Krypton on graphite - Krypton has been chosen because of its ability to form commensurate adsorbates due to its size almost matching 3 times the C-C bond length (xenon is too large and LEED data show incommensurate behaviour, although the recent, above-mentioned experiment by Ellis et al. seems to correspond to a commensurate phase).

Commensurate adsorption only occurs at coverage $\Theta = 1$, so that no incoherent scattering is predicted. The sinusoidal function

$$\zeta(\vec{R}) = 2\zeta_{10}\left(\cos\frac{2\pi X}{a} + \cos\frac{2\pi Y}{a} + \cos\frac{2\pi(X+Y)}{a} \right) \tag{18}$$

corresponding to the " 3 " triangular lattice is assumed, with ζ_{10}=0.126 Å (this value being obtained from steric arguments, using the known He-Kr potential and deriving $\zeta(\vec{R})$ from the equation $V(\vec{R}, \zeta(\vec{R})) = E = 600K$). From the diffraction intensities so obtained it has been determined that corrugation is so strong that the rainbow lies below the horizon.

Oxygen on Tungsten - Oxygen on W(110) forms several different structures, but our attention will be focused here on the p(2x1) structure since Lu, Wang and Lagally have measured for the p(2x1)-vapour coexistence a beautiful phase diagram resembling that of the 2-D Ising model. The p(2x1) structure consists of parallelograms of sides a,2a (where a=2.74 A is the nearest neighbour distance in W) and angles arccos(± 1/3). Island formation takes place in this system, which has interesting consequences on incoherent scattering.

Indeed, coherent scattering (i.e. diffraction) has nothing special about it, except for the fact that the strong anisotropy of the adsorbate shows up in the distribution of diffraction intensities and, in turn, can give information about the electron densities, hence about the chemical bonds.

Incoherent scattering is more interesting. To calculate the incoherent line shape, a single cell approximation may be used (where the corrugation is assumed to depend only on the adsorption in the same cell). Then

$$\bar{H}_{ij} - \bar{H}_{far} = \frac{\Theta^2}{A^2} | K^+ - K^- |^2 \tilde{p}_{ij}, \tag{19}$$

where K^+ and K^- are diffraction integrals for occupied and empty cells respectively and $\tilde{p}_{ij}$ is a correlation coefficient given by

$$\tilde{p}_{ij} = \frac{1}{\Theta} P(j|i) - 1 , \tag{20}$$

where $P(j|i)$ is the conditional probability for cell j to be occupied if cell i is. In an island model $\tilde{p}_{ij}$ includes two effects: the correlation of sites inside an island, and the conditional probability $p^{island}(\vec{R})$ that, if a point A lies inside an island, another point B at a distance $\vec{R}$ from A lies inside the same island too. If the Ising model is adopted, then

$$\tilde{p}_{ij} = \left(\frac{1+p_{ij}}{2\Theta} - 1 \right) p^{island}(\vec{R}_j-\vec{R}_i) \tag{21}$$

where p_{ij} is the correlation coefficient in the Ising model, tending at long distances to

$$\lambda = \{ 1 - 16 [(\sqrt{2}+1)^{T_c/T} - (\sqrt{2}+1)^{-T_c/T}]^{-4} \}^{1/4} \tag{22}$$

and behaving as

$$p_{ij} - \lambda \simeq A a^{1/4} K^{-7/4} R_{ij}^2 e^{-KR_{ij}}, \tag{23}$$

where

$$K = \frac{F^-}{a} \frac{T_c-T}{T_c} \tag{24}$$

(for a square lattice the constants are F^-=3.525, A=.17356), and where $p^{island}(\vec{R})$, for circular islands of diameter L_0, is given by

$$p^{island}(\vec{R}) = 1 - \pi^{-1}(\psi + \sin\psi) \qquad \text{if } R \le L_0$$
$$= 0 \qquad \text{otherwise}, \tag{25}$$

where ψ is defined by $R=L_0 \sin \psi /2$. The resulting line shape, proportional to

$$\sum_{ij} e^{i\vec{Q}\cdot(\vec{R}_j-\vec{R}_i)} \tilde{p}_{ij}, \tag{26}$$

depends strongly on temperature and on island diameter.

4. Inelastic scattering

Inelastic scattering from adsorbates involves four types of possible excitations: |16,17|

1) modified substrate phonons
2) collective excitations (phonons) in the overlayer
3) adatom-solid vibrations
4) modified internal excitations of the adsorbed molecules (for polyatomic adsorbed species)

The most conventional probe, i.e. infrared absorption, is essentially limited, for technical reasons, to wavelengths shorter than 10 μ , i.e.

to frequencies above ~ 30 TH_Z $\sim 1500°K$ which implies that virtually only excitations of type 4) can be studied this way. |17|

The most important and fashionable probe nowadays is electron energy loss spectroscopy |16|. This extends down to much lower frequencies, say $\sim$ 5 TH_Z $\sim 250°K$ and can study, therefore, excitations of types 3) or 4) equally well. However, due to low angular resolution, collective excitations of types 1) and 2) cannot be studied with EELS.

These can only be studied with atom scattering (in the case of exfoliated graphite, also neutron scattering) but the experiments are still to be done, apart from the preliminary experiments by Mason and Williams |9|. Atom scattering has another advantage, that it can be extended easily to even lower frequencies and there is no problem in studying excitations below 1TH $\sim$ 50°K.

The present theory |11| should include automatically vibrations of types 1), 2) and 3) (but not 4, since structureless atoms are considered). Let me assume the correlation of displacements to depend only on the cells and not on the detailed structure within each cell; let me also consider only single-phonon scattering (or single-quantum scattering for discrete modes). Then the inelastic scattering probability can be written (for energy loss Δ)

$$dP_{inel} = \frac{k|k_{iz}|}{8\pi^3\hbar N} A d\Omega\, d\Delta \sum_{ij} e^{i\vec{Q}\cdot(\vec{R}_j-\vec{R}_i)} \ll H_{ij}\, \Phi_{ij} \gg \tag{27}$$

where H_{ij} is defined as above and

$$\Phi_{ij} = 2\int_{-\infty}^{+\infty} e^{-it\Delta/\hbar}\, W_{ij}(t;A)dt \ . \tag{28}$$

W_{ij} can be written in terms of momentum transfers $\hbar\vec{q}_{ni}$ (n labels the substrate ion or adsorbed atom that receives the momentum and j labels the crystal cell on which the beam atom falls) as

$$W_{ij}(t;A) = \tfrac{1}{2}\sum_{lm} \vec{q}_{li}\vec{q}_{mj} : <\vec{u}_l(0)\vec{u}_m(t)> \ . \tag{29}$$

In the single cell approximation, let $\vec{U}_j$ be the displacement of the center of the mass of the substrate ions underlying site j, and $\vec{u}_j$ be the displacement of the adatom at site j (if any). Then vibrations of type 1) (modified substrate phonons) may be separated from vibrations of types 2) and 3) (phonons in the adsorbate, and adatom-solid vibrations) by assuming

$$< \vec{U}_i(\vec{u}_j-\vec{U}_j)> = 0, \quad \forall\ i,j \ . \tag{30}$$

Let

$$\vec{A}_j = A^{-1}K^-\vec{U}_j \ , \qquad \vec{C}_j = A^{-1}K^+(\vec{u}_j-\vec{U}_j), \qquad \rho = 1 - K^+/K^- . \tag{31}$$

$$dP = \frac{k|k_{iz}|}{8\pi^3\hbar NA}\, d\Omega d\Delta\ \vec{q}\vec{q}:\sum_{ij} e^{i\vec{Q}\cdot(\vec{R}_j-\vec{R}_i)}\{\overleftrightarrow{T}_{ij}^{\,substrate} + \overleftrightarrow{T}_{ij}^{\,overlayer}\} \tag{32}$$

where

$$\overleftrightarrow{T}_{ij}^{\,substrate} = (|1-\Theta\rho|^2 + \Theta^2|\rho|^2\tilde{p}_{ij})\int_{-\infty}^{+\infty} e^{-it\Delta/\hbar} \langle\vec{A}_i^*(0)\vec{A}_j(t)\rangle\, dt\ , \tag{33}$$

$$\overleftrightarrow{T}_{ij}^{\,overlayer} = \Theta^2(1+\tilde{p}_{ij})\int_{-\infty}^{+\infty} e^{-it\Delta/\hbar} \langle\vec{C}_i^*(0)\vec{C}_j(t)\rangle\, dt\ . \tag{34}$$

In $\overleftrightarrow{T}_{ij}^{\,overlayer}$ the case $j\neq i$ should be distinguished from the case $j=i$. The case $j\neq i$ implies correlation between different adatoms, may describe phonons in the overlayer, and is small in some cases (for example H on graphite where the adatoms interact much more with the graphite than with each other; or any low density phase).

The remaining terms, $j=i$, describe individual vibrations of one adatom. These may be represented as Einstein oscillators (with typically two different angular frequencies, $\omega_{\parallel}$ and ω_z). The corresponding contribution to the scattering probability density is

$$dP_{Einstein} = \frac{\hbar k|k_{iz}|}{16\pi^3 A\mu}\, \Theta\{\Theta|K_o^+|^2 + (1-\Theta)|K_o^-|^2\}\, e^{-2W^+}\{\frac{Q^2}{\omega_{\parallel}}[(\langle n_{\parallel}\rangle+1)\delta(\Delta-\hbar\omega_{\parallel}) + \langle n_{\parallel}\rangle\delta(\Delta+\hbar\omega_{\parallel})] + \frac{q_z^2}{\omega_z}[(\langle n_z\rangle+1)\delta(\Delta-\hbar\omega_z) + \langle n_z\rangle\delta(\Delta+\hbar\omega_z)]\}\, d\Omega d\Delta, \tag{35}$$

where μ is the reduced mass, $\langle n_{\parallel}\rangle$ and $\langle n_z\rangle$ are given by Bose-Einstein statistics, and $K_o^{\pm}$ is defined by

$$K^{\pm} = e^{-W^{\pm}}\, K_o^{\pm}. \tag{36}$$

References

1. D.L.Smith and R.P. Merrill, J. Chem. Phys. 52 (1970) 5861
2. B.F. Mason and B.R. Williams, J. Chem. Phys. 56 (1972) 1895; Surf. Sci. 45 (1974) 141
3. T. Engel, J. Chem. Phys. 69 (1978) 373
4. K.H. Rieder and T. Engel, Phys. Rev. Letters 43 (1979) 373
5. D. Hayward, presented at I Seminar on Atom-Surface Scattering (Saclay,1978)
6. B. Poelsema, G. Mechtersheimer and G. Comsa, Proc. 4th Intern. Conf. on Solid Surfaces (Cannes, 1980) p. 834
7. J. Lapujoulade, Y. Le Cruer, M. Lefort, Y. Lejay and E. Maurel, Phys. Rev. B22 (1980) 5740
8. T.H. Ellis, S. Iannotta, G. Scoles and U. Valbusa, Phys. Rev. B 24 (1981) 2302.

9. B.F. Mason and B.R. Williams, Phys. Rev. Letters 46 (1981) 1138.

10. A.C.Levi, Nuovo Cim. 54B (1979) 357

11. A.C.Levi, R.Spadacini and G.E.Tommei, Surf.Sci. in press

12. A.C.Levi, R.Spadacini and G.E.Tommei, to be published

13. P.Beckmann and A.Spizzichino, The Scattering of Electromagnetic Waves from Rough Surfaces, Pergamon Press, Oxford, 1963

14. N.Garcia, V.Celli and M.Nieto-Vesperinas, Optics Comm. 30 (1979) 279

15. N.Garcia and N.Cabrera, presented at Rarefied Gas Dynamics Conference (Charlottesville, Va., 1980)

16. H.Froitzheim, in Electron Spectroscopy for Surface Analysis, ed. by H. Ibach, Topics in Current Physics, Vol. 4. Springer, Berlin, Heidelberg, New York, 1977, p. 205

17. See, however the excellent low-frequency studies by the Turin group: F. Boccuzzi, S. Coluccia, G. Ghiotti, C. Morterra and A. Zecchina, J. Phys. Chem. 82 (1978) 1298; F. Boccuzzi, E. Borello, A. Chiorino and A. Zecchina, Chem. Phys. Letters 61, (1979) 617; F. Boccuzzi, S. Coluccia, G. Ghiotti and A. Zecchina, Chem. Phys. Letters 78 (1981) 388.

Part III

Spectroscopy of Surface Optical Excitations

Surface Elementary Excitations

M. Šunjić

Department of Physics, Faculty of Science
University of Zagreb, P.O. Box 162
41001 Zagreb Croatia, Yugoslavia

1. Introduction

The presence of the surface (or interface) in a crystal modifies the elementary excitations of the infinite crystal, but it also leads to the creation of the new elementary excitations characteristic for the particular surface or interface, and to the onset of the single-particle or collective excitations of the species (atoms, molecules) adsorbed on the surface. In particular, some of the bulk modes are suppressed and replaced by the *surface modes*, the continuum of the bulk modes can become a discrete band of excitations in finite crystals, and these excitations may acquire a finite lifetime. The modification of the bulk modes and the onset of the new, surface elementary excitations have consequences for some microscopic and also macroscopic properties of the solid surfaces. For example, the electrostatic potentials - classical image potential and van der Waals potential in the physisorption - have their quantum-mechanical origin in the interaction of external charges with the long-wavelength surface modes, as will be discussed later.

The purpose of these lectures is to give an overall review of the surface elementary excitations in metals and in ionic crystals in the long-wavelength limit where the continuum approximation on the solid is valid. Other non-polar lattice excitations in solids and the lattice dynamical theory of the surface phonons, where the discrete nature of the solid is treated properly, are omitted because this topic is treated in detail in Professor Benedek's lectures.

In Sec.2 the ground state electronic structure of the metallic surfaces is discussed in the jellium model, in particular the local electronic densities and the electronic potentials. Sec.3 develops a continuum (or classical) theory of bulk and surface collective excitations in polarizable media, which can be applied both to the long-wavelenth optical phonons in the ionic crystals and to the plasmons in metals. In Sec. 4 we review the microscopic theory of surface plasmons in the Random Phase Approximation and discuss its numerical results. Sec. 5 briefly describes the optical spectroscopy of surface modes, using the gratings and the method of Attenuated Total Reflection. Sec. 6 deals in somewhat more detail with the powerful method of Electron Energy Loss Spectroscopy (EELS) of surface excitations, including both the semiclassical theory of multiple excitation processes and the quantum-mechanical (DWBA) theory of single-excitation processes, and also their application to the study of collective modes (optical phonons and plasmons) and adsorbed molecule localized vibrations. Finally, in Sec. 7 we discuss the quantum-mechanical derivation of the classical electrostatic potentials, and their extension to include the dynamical phenomena.

Apart from the standard texts dealing with the elementary excitations in solids, e.g. by KITTEL [1,2] or PINES [3], the reader is referred to a number of excelent review papers [4-7] for the details.

2. Electrons at metal surfaces

We should first outline some basic properties of the electrons near crystal surfaces in their ground state.

Though we know that the potential of an electron in a solid is a complicated many-body quantity, due to the screened ion potential and the electron - electron interaction, we often find it convenient to approximate it by a "jellium" model [1]. For a finite solid we thus start with the Sommerfeld model where the properties of the solid are defined by the empirical parameters: inner potential V_o, Fermi energy E_F, and work function $\phi = V_o - E_F$. Fermi energy is related to the only property of the jellium - its density:

$$n = (2mE_F/\hbar^2)^{3/2}/(3\pi^2) .$$

We shall assume that the electrons are confined in the z direction in a crystal slab of thickness L, which extends to infinity in the $\vec{R} = (x,y)$ directions. For simplicity we also assume the Infinite Barrier Model (IBM), i.e. that the potential V_o is infinite, so that the electronic wave functions are

$$\psi_{\vec{K}}(\vec{r}) = (2/AL)^{1/2} e^{i\vec{K}\cdot\vec{R}} \sin k_z z \tag{1}$$

where A is the surface normalization area, and

$$k_z = \frac{2\pi}{L} n_z, \quad n_z = 1,2, \ldots$$

is the discrete wave vector in the z direction.

Contrary to the infinite solid case, we have to define the *local* density of states (DOS) as

$$n(\vec{r},E) = \sum_i |\psi_i(\vec{r})|^2 \delta(E - E_i) , \tag{2}$$

defined as the number of states in the region around $\vec{r}$ at the fixed energy E. The quantum numbers are now $i = (\vec{K},k_z)$. Inserting (1) we find

$$n(z,E) = n^B(E) \{ 1 - \frac{\sin \tilde{z}}{\tilde{z}} \} ; \quad \tilde{z} = 2z\sqrt{2mE}/h \tag{3}$$

where

$$n^B(E) = \frac{AL}{2\pi^2} \left(\frac{2m}{\hbar^2}\right)^{3/2} \sqrt{E} \qquad (4)$$

is the *bulk density of states* which is reached for $z \gg \hbar(2mE)^{-1/2}$. The *total density of states* is given by

$$n(z) = \int_0^{E_F} n(E,z) \quad dE = \frac{k_F^3}{3\pi^2} \left\{1 + \frac{3\cos X}{X^2} - \frac{3\sin X}{X^3}\right\} \qquad (5)$$

where $X = 2k_F z$ defines the characteristic length. Both $n(z,E)$ and $n(z)$ exhibit characteristic Bardeen oscillations [8] as shown in Fig. 1.

The relative position of the potential barrier ("IBM edge") and the uniform positive background can be determined if we realize that the system should satisfy change neutrality:

$$\int_{-\infty}^{\infty} \left[n^-(z) - n^+(z)\right] \quad dz = 0 \qquad (6)$$

with $n^-(z)$ given by (5), and

$$n^+(z) = n \quad \Theta(z - d) \; . \qquad (7)$$

Inserting (5) and (7) into (6) gives

$$d = 3\,\pi/8k_F \quad .$$

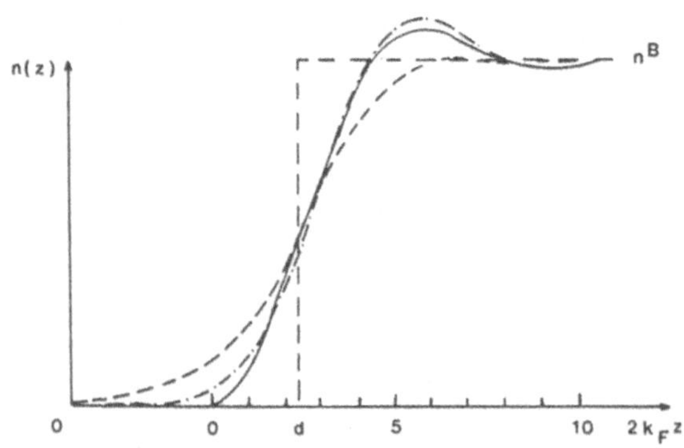

Fig. 1 Change density oscillations near a metal surface with infinite barrier model (solid line) and self-consistent calculations by LANG [9] (r_s=2, broken line, and r_s=5, dash-dotted line).

From Fig.1 we see that there is a net *dipole layer* or *surface double layer* formed at the surface, with excess negative charge spilled over the positive background boundary at $z=d$, a net positive charge inside, and a further decreasing charge oscillation. This dipole layer contributes to the electrostatic surface potential. Comparison with more realistic models, e.g. finite potential barrier, or especially with the self-consistent calculations [9], show that this oscillatory behaviour of the charge is a real effect which persists in all models, though with less pronounced amplitudes. The electrons obviously spill outside the IBM edge, but at lower densities (e.g. $r_s=5$); the results are pretty similar when scaled with the appropriate Fermi wave vector.

Until now we have treated the potential as defined by some given parameters (V_o), while in reality it results from the electron-ion and electron-electron interaction, and it should be evaluated self-consistently; e.g. one could start from the Schrödinger equation with some effective potential $V_{eff}[n(\vec{r}),\vec{r}]$ that has to satisfy the self-consistency because it is density dependent. In principle it contains

$$V_{eff} = V_{core} + V_{xc} + V_{es} \quad .$$

V_{core} describes the electron scattering on the ion cores, including the exchange and correlation between valence and core electrons. This term is usually identical to the one in the bulk, and can be treated similarly [10].

V_{xc} is the exchange and correlation potential between the valence electrons, and V_{es} is the electrostatic potential, which are both expected to be strongly modified near the surface. Exchange and correlation, which give a large contribution, about two thirds of V_o, are essentially many-body phenomena which are properly described by a non-local and energy-dependent electron self-energy. However, it is convenient to find some local approximations. The most popular one for the exchange was introduced by Gáspár and Slater [11] :

$$V_{ex}(\vec{r}) = - \frac{e^2}{\pi} (3\pi^2 n(\vec{r}))^{1/3} \tag{8}$$

where $n(r)$ is the *local* density. Nowadays it is known as X_α *approximation* , because (8) is usually multiplied by a factor $\alpha \sim 1$ which optimizes the results.

Another is the *Wigner interpolation scheme* [12] for exchange and correlation:

$$V_{xc}(\vec{r}) = - \{0.984 + \frac{8.77x + 0.944}{(1.00 + 12.57x)^2}\} x \; ; \; x \equiv (n(\vec{r}))^{1/3}. \tag{9}$$

The use of these approximations near the surface leads to difficulties. The exchange energy vanishes far outside ($z \to \infty$, where $n(z) \to 0$) as it should, but so does the correlation energy. In reality we know that the correlation hole in this limit gradually distorts and stays at the surface, leading to the classical image potential,

$$V_{xc}(z) \longrightarrow -\frac{\varepsilon-1}{\varepsilon+1}\frac{e^2}{4z} \quad . \tag{10}$$

The local approximation like (8) cannot reproduce this (correlation) effect, which on the other hand can be discussed in the framework of the classical electrostatic theory.

In Sec. 7 we shall discuss *electrostatic potentials* in more detail, especially the part that is connected with long-range correlations in the solid, so now we only derive the interaction of the electron with the static charge distribution, like the one in Fig. 1.

We start from the Poisson equation connecting the potential $V_{es(r)}$ with the total charge density $n(\vec{r}) = n_{ion} + n_{val}$:

$$\Delta V_{es}(\vec{r}) = -4\pi n(r) \quad . \tag{11}$$

Using the surface net translation symmetry defined by inverse net vectors $\vec{G}_{\parallel}$, we can expand V_{es} and n, and their Fourier coefficients satisfy the Poisson equation:

$$\left(-\frac{d^2}{dz^2} + G_{\parallel}^2\right) V_{G_{\parallel}}(z) = 4\pi\, n_{G_{\parallel}}(z) \quad . \tag{12}$$

The $G_{\parallel} = 0$ terms represent the surface-averaged quantities, and the solution of (12) is

$$V_o(z) = V_o(\infty) - 4\pi \int_z^{\infty} (z'-z)n_o(z')dz' \tag{13}$$

because for $z_o \to \infty$ the variation of the potential has to vanish. Contribution to the work function is defined as

$$\phi_{dip} = V_o(\infty) - V_o(-\infty) = 4\pi \int_{-\infty}^{\infty} z' n_o(z')dz' \tag{14}$$

where we have taken into account the charge conservation (6).

3. Continuum models of bulk and surface elementary excitations

We shall first consider long-wavelength surface excitations and assume that it is possible to describe the electron gas or the polarizable lattice as the dielectric continuum with the dispersionless *bulk dielectric function*:

$$\varepsilon(\omega) = \varepsilon(k \to 0, \omega) \quad .$$

It can be shown [13] that in an isotropic dielectric continuum, corresponding e.g. to the *ionic crystals* NaCl or ZnO,

$$\varepsilon(\omega) = \frac{\omega^2 - \omega_L^2}{\omega^2 - \omega_T^2} \qquad \begin{aligned} \omega_L^2 &= \omega_o^2 + \tfrac{2}{3}\,\omega_p^2 \\ \omega_T^2 &= \omega_o^2 - \tfrac{1}{3}\,\omega_p^2 \end{aligned} \quad . \tag{15}$$

Here ω_o is the mechanical restoring frequency, ω_p is the ion plasma frequency:

$$\omega_p = \left(\frac{4\pi e^{*2}}{M\,V}\right)^{1/2} \qquad \begin{aligned} &e^* \text{ - effective ion charge} \\ &M \text{ - reduced ion mass} \end{aligned} \quad , \tag{16}$$

ω_L and ω_T are the frequencies of longitudinal and transverse optical phonons, respectively. For simplicity we have assumed the point ion model, i.e. neglected the electronic polarizability of the ions.

Collective excitations of the electron-gas plasmons can also be described by the dielectric function; usually one considers the Random Phase Approximation result [2,3] :

$$\varepsilon_{RPA}(\vec{k},\omega) = 1 - V_k R(\vec{k},\omega) \tag{17}$$

where R is the Lindhard response function, and

$$V_k = 4\pi e^2/k^2 \tag{18}$$

is the Coulomb potential. In the long-wavelength limit

$$\varepsilon_{RPA}\ (k\to o,\omega) = 1 - \omega_p^2/\omega^2 \tag{19}$$

where

$$\omega_p = (4\pi e^2 n/m)^{1/2}$$

is the classical plasma frequency of the charges with the density n.

Formally [1] one can obtain (19) from (15) by identifying $\omega_T \to 0$ (no transverse force in the electron gas) and $\omega_L \to \omega_p$. In deriving the elementary excitations in this limit we start from the *Maxwell's equations* [4,5,7]:

$$\begin{aligned} &\operatorname{div}\vec{D} = 4\pi n && \operatorname{rot}\vec{E} = -\frac{1}{c}\dot{\vec{B}}\ , \\ &\operatorname{div}\vec{B} = 0 && \operatorname{rot}\vec{H} = \frac{1}{c}\dot{\vec{D}} + \frac{4\pi}{c}\vec{j}\ , \end{aligned} \tag{20}$$

and make the following approximations:

(i) Assume that are no free charges (n=0) or currents in the solid;

(ii) The medium is linear and non-magnetic: $\vec{B} = \vec{H}$, $\vec{D} = \varepsilon\vec{E}$.

(iii) In the long-wavelength limit the dielectric function is given by (15) or (19).

(iv) The fields are periodic with the time dependence $e^{-i\omega t}$.

Then the two equations in (20) combine into

$$\text{rot rot } \vec{E} - \frac{\omega^2}{c^2} \varepsilon(\omega) \vec{E} = 0 \quad . \tag{21}$$

We first look for the *bulk solutions* which have the spatial dependence

$$\vec{E} \sim \hat{e}\, e^{i\vec{k}.\vec{r}} \quad .$$

The equation (21) becomes

$$\vec{k} \times (\vec{k} \times \vec{E}) + \frac{\omega^2}{c^2} \varepsilon(\omega) \vec{E} = 0 \ .$$

The *longitudinal modes* with the polarization

$$\hat{e}||\vec{k} \ , \text{ i.e.} \quad \hat{e} \times \vec{k} = 0 \quad ,$$

then satisfy the relation

$$\varepsilon(\omega_L) = 0$$

which defines the longitudinal excitation frequency ω_L. The *transverse modes* have

$$\hat{e} \perp \vec{k}, \text{ i.e. } \hbar \times (\vec{k} \times \hat{e}) = k^2 \ ,$$

so they satisfy

$$k^2c^2/\omega_T^2 = \varepsilon(\omega_T) \tag{22}$$

which is the dispersion relation for the transverse mode frequencies ω_T.

For the dielectric functions (15) and (19) we can plot this dispersion of the bulk modes which are called *bulk polaritons*: coupled or hybridized excitations of photons and bulk optical phonons (Fig.2a) or plasmons (Fig.2b). The lower branch in Fig.2a

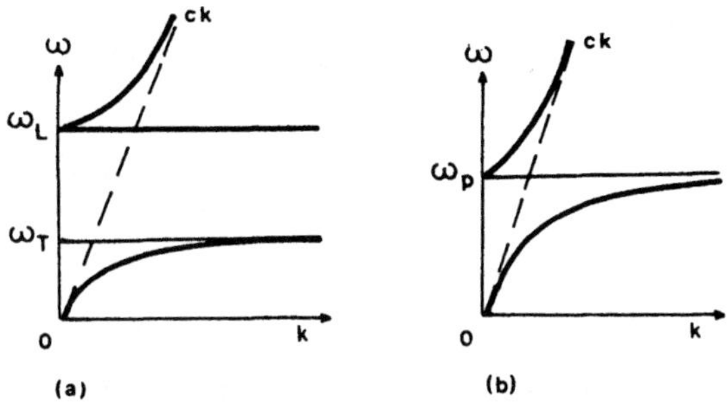

Fig. 2 Dispersion curves of bulk polaritons in (a) a polar dielectric and (b) a metal.

starts as a photonlike (i.e. transverse) mode and turns gradually into a TO phonon. The higher branch contains a longitudinal contribution which does not couple to photons, but stays an LO phonon, and a transverse contribution which at large k approaches the photon line ck.

In a metal (Fig.2b) the lower mode is reduced to the zero frequency, so we have only the higher transverse polariton branch. Very often one neglects the retardation effects, i.e. the coupling to the photons, which is obviously important in the region

$$k_r \lesssim \omega_L/c, \ \omega_T/c \qquad \text{or} \qquad k_r \lesssim \omega_p/c \ ,$$

but for $k \gg k_r$ becomes negligible. This can be formally achieved by taking $c \to \infty$ limit in all previous equations. One then recovers the usual LO and TO phonons and bulk plasmons, with the degenerate frequencies ω_L, ω_T and ω_p, respectively.

Now we look for the *surface modes*, assuming the planar geometry,i.e. translational invariance in the $\vec{R}$ direction:

$$\vec{E}(\vec{r},t) = \vec{E}(z)\, e^{i\vec{K}\cdot\vec{R} - i\omega t} \ . \tag{23}$$

Inserting (23) into (21) gives

$$\left\{ \frac{d^2}{dz^2} - \left[K^2 - \frac{\omega^2}{c^2}\,\varepsilon(\omega)\right]\right\} \vec{E}(z) = 0 \ . \tag{24}$$

The solutions are

$$\vec{E}(z) = \vec{E}_o\, e^{-\alpha|z|} \tag{25}$$

where

$$\alpha = \left[K^2 - \frac{\omega^2}{c^2}\,\varepsilon(\omega)\right]^{1/2} \tag{26}$$

is the attenuation parameter, and we have discarded the solutions which diverge as $z \to \pm \infty$. Outside (in the vacuum, $\varepsilon = 1$) the attenuation parameter is

$$\alpha_o = \left[K^2 - \frac{\omega^2}{c^2}\right]^{1/2} . \qquad (27)$$

3.1 Radiative and nonradiative modes

It is instructive to study the character of the solutions (25) as we scan the ω - K plane:

- for α *real* they decay exponentially: localized or *nonradiative* modes;
- for $\alpha = i\beta$ *imaginary* they oscillate: extended or *radiative* modes.

As an example we shall take a thin slab of ionic crystal, with ε slightly generalized to include the finite ion polarizability [13] .

It is easy to convince oneself, from (26), (27) and (25), that the solutions have the behaviour indicated in Fig. 3. Proper surface modes are therefore found only in the region L_2.

3.2 The dispersion relations

The dispersion relations have to be obtained by the matching of the fields on the boundaries between the different media: in our case we require the continuity of the $\vec{E}_{\parallel}$ tangential components of the electric field, and $D_{\perp}$ normal component of the displacement.

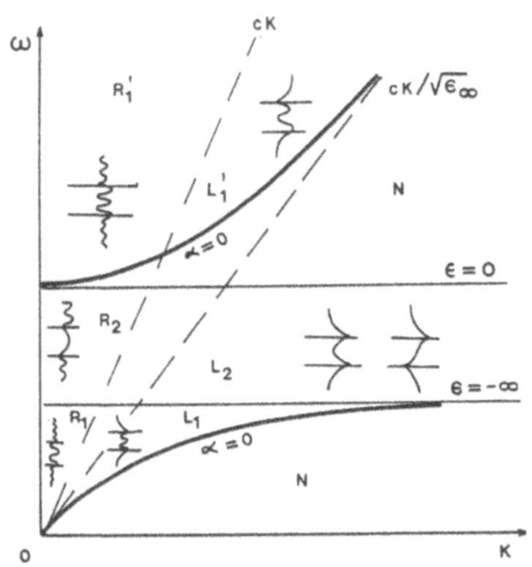

Fig. 3 Regions in the ω - K plane for a dielectric slab: R - radiative, L - localized and N - no solutions.

For the *semiinfinite solid* the solutions naturally split into two classes:

P - polarized or Transverse Magnetic (TM)mode: $\vec{E} = (E_x, 0, E_z)$
$\vec{B} = \vec{H} = (0, H_y, 0)$,

S - polarized or Transverse Electric (TE)mode: $\vec{E} = (0, E_y, 0)$
$\vec{B} = \vec{H} = (H_x, 0, H_z)$.

For the P mode the Maxwell's equations (20) give:

$$\frac{d}{dz} E_z + i\, kE_x = 0 \qquad \frac{d}{dz} E_x - i\, kE_z = \frac{i\omega}{c} H_y$$
$$\frac{d}{dz} H_y = -\frac{i\omega}{c} \varepsilon E_x \qquad i\, kH_y = -i\frac{\omega}{c}\varepsilon E_z\ . \tag{28}$$

Eliminating H_y and E_z gives:

$$\frac{d^2}{dz^2} E_x - \alpha^2 E_x = 0 \tag{29}$$

with α again given by (26).

The solutions again fall into two classes:

$\alpha^2 < 0 : E_x \sim e^{i\beta z}$, oscillatory bulk-like solution,

$\alpha^2 > 0 : E_x = C\, e^{\alpha z}$, transverse magnetic surface oscillation (TMSO).

The fields are therefore:

<u>inside, $z < 0$:</u>

$$E_x(z) = C\, e^{\alpha z}$$
$$E_z(z) = \frac{iK}{\alpha} C\, e^{\alpha z}$$
$$H_y(z) = -\frac{i\omega}{c}\frac{\varepsilon}{\alpha} C\, e^{\alpha z}$$

<u>outside, $z > 0$:</u>

$$E_x(z) = C'\, e^{-\alpha_o z}$$
$$E_z(z) = i\frac{K}{\alpha_o} C'\, e^{-\alpha_o z} \tag{30}$$
$$H_y(z) = -i\frac{\omega}{c\alpha_o} C'\, e^{-\alpha_o z}\ .$$

The matching conditions are:

$$\left.\begin{aligned} E_x(z=0^-) &= E_x(z=0^+) \\ H_y(z=0^-) &= H_y(z=0^+) \end{aligned}\right\} \text{(from } \vec{E}_{\parallel}\text{)}\ , \tag{31}$$

$$\varepsilon E_z(z=0^-) = E_z(z=0^+) \qquad \text{(from } D_\perp\text{)}\ .$$

These give $C' = C$ and $\varepsilon = -\alpha/\alpha_o$ which can also be written as

$$\left(\frac{\omega}{cK}\right)^2 = 1 + \frac{1}{\varepsilon(\omega)} \quad . \tag{32}$$

This is the dispersion relation for TMSO modes - or surface polaritons at a semiinfinite solid surface, shown schematically for a polar dielectric in Fig. 4a and a metal in Fig. 4b.

In a *metal* the bulk dielectric function is given by (19), and the dispersion relation (32) gives the solutions

$$\omega_{\pm}^2 = \frac{1}{2}\,\omega_p^2 + c^2K^2 \pm \left(\omega_p^4 + 4c^2K^2\right)^{1/2} \quad . \tag{33}$$

The ω_+ solution (Fig.4b) lies in the region of radiative modes and we shall discard it.

The proper surface mode ω_- lies always below the photon line. It starts as a photon-like mode at small K, and at large K approaches the *asymptotic surface plasmon frequency*

$$\omega_s = \omega_p/\sqrt{2} \quad . \tag{34}$$

The *nonretarded limit* can be obtained from (32) by letting $c \to \infty$:

$$\varepsilon(\omega_s) + 1 = 0 \quad . \tag{35}$$

This defines a single frequency ω_s, so that the nonretarded surface modes of a semiinfinite crystal, surface optical phonons and surface plasmons, show no dispersion.

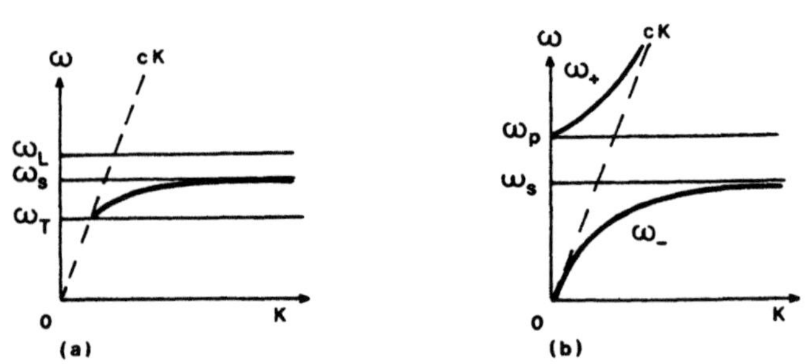

Fig. 4 Dispersion curves of surface polaritons in a semiinfinite polar dielectric (a) and a metal (b).

3.3 Matching conditions in the nonretarded limit

In the nonretarded limit we have to satisfy the Maxwell's electrostatic equations

$$\mathrm{div}\,\vec{D} = 0 \qquad \vec{D} = \varepsilon\vec{E}$$
$$\mathrm{rot}\,\vec{E} = 0 \tag{36}$$

and the boundary conditions. If we introduce the scalar potential $V(\vec{r})$, defined by

$$\vec{E} = -\,\mathrm{grad}\,V\,, \tag{37}$$

it has to satisfy the Laplace equation

$$\Delta V(\vec{r}) = 0\,. \tag{37'}$$

With the *planar geometry* we can obviously write the solution in the form:

$$V(\vec{r}) = V(z)\,e^{i\vec{K}\cdot\vec{R}} \tag{38}$$

so that the solutions of the Laplace equation are

$$V(z) = A\,e^{Kz} + Be^{-Kz}\,. \tag{39}$$

At the boundary, the continuity of $\vec{E}_{\parallel}$ and $D_{\perp}$ implies the continuity of $V(z)$ and $\varepsilon\frac{\partial V}{\partial z}$, respectively. For a boundary between the two media $\varepsilon_1(\omega)$ and $\varepsilon_2(\omega)$ we thus obtain:

$$A = B \qquad \varepsilon_1 A = \varepsilon_2 B$$

or the dispersion relation for the (nonretarded) *interface modes:*

$$\varepsilon_1(\omega) + \varepsilon_2(\omega) = 0\,. \tag{40}$$

3.4 Surface modes in a crystal slab

In a crystal slab with the boundary planes $z = \pm a$, because of the inversion symmetry at $z = 0$ the potential has to be of the form

$$\begin{aligned} V(z) &= B\,e^{-Kz}\,, && z > a \\ &= A(e^{Kz} \pm e^{-Kz})\,, && -a < z < a \\ &= \pm B\,e^{Kz}\,, && z < -a\,. \end{aligned}$$

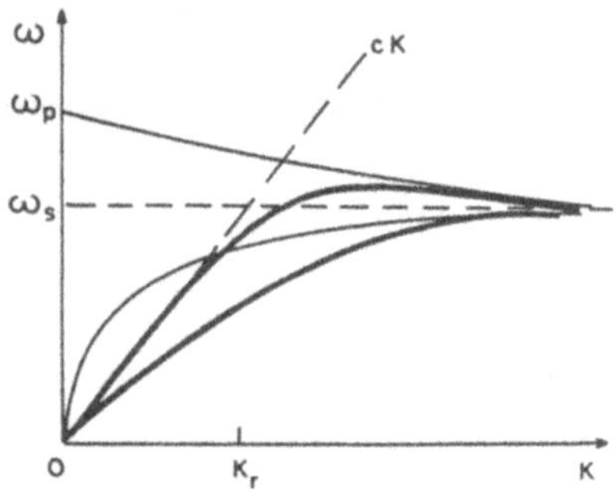

Fig. 5 Dispersion curves for surface polaritons in a metalic slab (thick lines) and in the nonretarded limit (thin lines).

The continuity conditions at z = a are

$$A = (e^{Ka} \pm e^{-Ka}) = B\, e^{-Ka} ,$$

$$\varepsilon K\, A\, (e^{Ka} \mp e^{-Ka}) = -B\, K\, e^{-Ka} ,$$

which give the dispersion relations

$$\varepsilon(\omega) = - [\tanh Ka]^{\pm 1} \tag{41}$$

for the two surface modes $\omega_{\pm}$.

For a *metallic slab* the solutions are

$$\omega_{\pm}(K) = \omega_s (1 \pm e^{-2Ka})^{1/2} \tag{42}$$

and are shown in Fig.5.

For an *ionic crystal slab* the solutions have the same shape as given in Fig.5, if we shift the points on the energy axis: $0 \rightarrow \omega_T$ and $\omega_p \rightarrow \omega_L$, as discussed earlier. Due to the presence of two surfaces we have found two solutions: ω_+ is due to an antisymmetric field and ω_- to a symmetric field distribution.

Surface modes have also been studied in crystals of other planar and nonplanar geometries, but we here refer the reader to the literature [4,5,7,14,15,16].

3.5 Quantization and the Hamiltonian formalism

It is often convenient to describe the elementary excitations and their interactions in the second quantized formalism in terms of the creation (a^+) and annihilation (a) operators, in the way completely analogous to the treatment of a collection of free harmonic oscillators [17] .

The relative atom or charge displacements $\vec{u}(\vec{r},t)$ and the induced polarization

$$\vec{P} = \frac{e}{v} \vec{u}$$

(where v is the unit cell volume) satisfy in the electrostatic limit the equation of motion

$$(\omega^2 - \omega_o^2 + \frac{1}{3} \omega_o^2)\vec{P} = - \frac{\omega_p^2}{4\pi} \vec{E} \ . \tag{43}$$

These equations of motion can be obtained from the Hamiltonian density [2] :

$$H = \frac{2\pi}{\omega_p^2} \dot{\vec{P}}^2 + (\frac{\omega_T}{\omega_p})^2 \vec{P}^2 - \frac{1}{2} \vec{P}\cdot\vec{E} \ , \tag{44}$$

where

$$\vec{\Pi} = \frac{4\pi}{\omega_p^2} \dot{\vec{P}}$$

is the conjugate momentum, and

$$[\vec{P}_i(\vec{r}), \ \vec{\Pi}_j(\vec{r}) \ ] = \delta_{ij} \ \delta(\vec{r} - \vec{r}) \ .$$

Expanding

$$\vec{P} = \sum_q (\hbar\omega_p^2/8\pi\omega_q)^{1/2} \vec{P}_q \ a_q + \text{h.c} \ . \ ,$$

$$\dot{\vec{P}} = i \sum_q (\hbar\omega_q\omega_p^2/8\pi)^{1/2} \vec{P}_q^* \ (a_q - a_q^*) \ , \tag{45}$$

where the creation (a^+) and annihilation (a) operators, representing the quanta of surface elementary excitations, satisfy the boson commutation relations

$$[a_q, a_{q'}^+] = \delta_{qq'} \ ,$$

we can rewrite the Hamiltonian (44) in the diagonal form:

$$H = \sum_q \hbar\omega_q \ (a_q^+ a_q + \frac{1}{2}) \ . \tag{46}$$

The interaction of the point charge outside the crystal with the polarization is

$$H_I(\vec{r}) = e \quad \phi \ (\vec{r}) \tag{47}$$

where the electrostatic potential is

$$\phi(\vec{r}) = \int \mathrm{div}\, \vec{P}(\vec{r}) |\vec{r} - \vec{r}'|^{-1} \, d\vec{r}' . \tag{48}$$

Using (45) for $\vec{P}$ we find the interaction Hamiltonian density:

$$H_I(\vec{r}) = \sum_q H_{Iq}(\vec{r})\, a_q + \mathrm{h.c.} \tag{49}$$

where the matrix elements can be calculated from (48) and (45) by expanding the potential (48) in the terms of the complete set of solutions of the Laplace equation, appropriate to the crystal geometry [18,15] .

4. Microscopic theory of surface electronic excitations

The basic equation which relates linearly the charge density fluctuation $\delta n(\vec{r},\omega)$ induced at $\vec{r}$ by an external charge density $\delta n_{ext}(\vec{r}',\omega)$ with frequency ω at $\vec{r}'$ is

$$\delta n(\vec{r},\omega) = \int d\vec{r}' \; \Lambda(\vec{r},\vec{r}';\omega) \left[\delta n(\vec{r}',\omega) + \delta n_{ext}(\vec{r}',\omega) \right] \tag{50}$$

where the kernel

$$\Lambda(\vec{r},\vec{r}';\omega) = \int d\vec{r}'' \, R_o(\vec{r},\vec{r}'';\omega)\, v(\vec{r}'' - \vec{r}')$$

contains the Coulomb potential v and the retarded irreducible polarization R_o, which in the Random Phase Approximation can be written as

$$R_o(\vec{r},\vec{r}';\omega) = 2e\Big\{ \sum_{\substack{E_i<E_F \\ \text{all } j}} - \sum_{\substack{E_j<E_F \\ \text{all } i}} \Big\} \frac{\psi_i^*(\vec{r})\psi_i(\vec{r}')\psi_j^*(\vec{r})\psi_j(\vec{r}')}{E_i - E_j - \omega + i\eta} . \tag{51}$$

Here ψ's are the electronic wave functions for the appropriate surface potential.

Assuming the translational invariance parallel to the surface we can Fourier transform (50). Self-sustaining charge density fluctuations then have to satisfy the integral equation

$$\lambda(\vec{K},\omega)\, \delta n_{\vec{K}}(z,\omega) = \int_{-\infty}^{\infty} dz' \, \Lambda_{\vec{K}}(z,z';\omega)\, \delta n_{\vec{K}}(z',\omega) \tag{52}$$

for the eigenvalue

$$\lambda(\vec{K},\omega) = 1 . \tag{53}$$

Equation (53) represents the dispersion relation for the normal modes; these are the surface plasmons if the associate charge density fluctuations $\delta n_{\vec{K}}(z)$ are localized at the surface.

WIKBORG and INGLESFIELD [19] have studied numerically these equations and found the results that we shall now discuss.

4.1 Surface plasmon dispersion

The potential was assumed to have a step at the surface (V_o= 15.83eV) corresponding to that of aluminium E_F = 11.64eV). Fig. 6 shows the calculated surface plasmon dispersion $\omega_s(K)$ and damping $\Gamma_s(K)$.

Apart from the detailed shape in the $K \to 0$ region, which seems to be sensitive to the shape of the potential barrier, two important conclusions can be drawn from this calculation:

i) The SP dispersion curve starts at ω_s for K = 0, which could be also obtained from purely classical arguments;

ii) The SP dispersion curve joins BP curve at the Landau cutoff wave vector K_c where they both merge with the pair-excitation continuum.

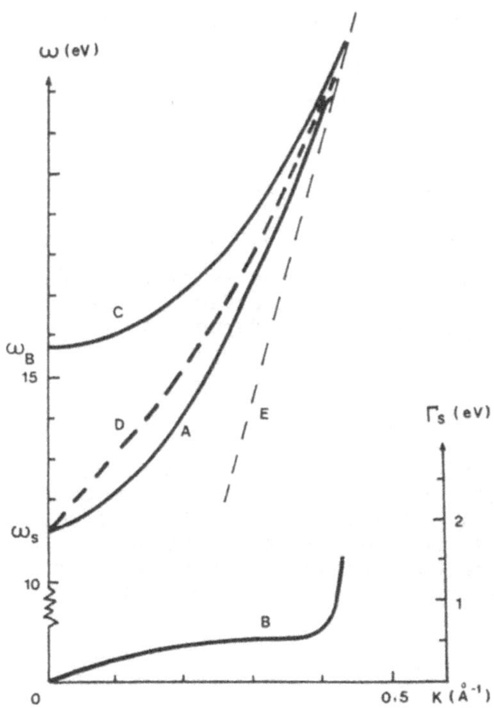

Fig. 6 A: Surface plasmon dispersion $\omega_s(K)$ in RPA; B: Surface plasmon damping $\Gamma_s(K)$; C: Bulk plasmon dispersion; D: Surface plasmon in SCIBM; E: Landau cutoff region. (From ref. [19])(Note the different scales for ω_s and Γ_s!)

The dispersion relation for small K can be written as an expansion:

$$\omega_s(K) = \omega_s \left[1 + A(K/k_F) + B(K/k_F)^2 + \ldots \right] . \tag{54}$$

The coefficient A can be related to the K=0 component of the density fluctuation [20,19],

$$A = k_F \int dz\, z \quad \delta n_{K=0}(z) \Big/ \int dz \quad \delta n_{K=0}(z) . \tag{55}$$

Surface plasmon damping in a step potential model, Fig.6, though small, extends to K=0, in contrast to the bulk case where the Landau damping starts at K_c. This is due to the presence of the surface which provides the necessary momentum for the excitation of electron-hole pairs with $K < k_c$ and $k_z \neq 0$.

Approaching K_c the damping increases dramatically, the surface plasmon ceases to be a well-defined excitation as it gradually merges with the bulk excitations. This can also be seen by looking at the electron density fluctuations associated with surface plasmons. In the classical dielectric model with a sharp step-density profile the fluctuations are confined to the z = 0 plane:

$$\delta n_{\vec{K}}(\vec{r}) = \delta n_o \, e^{i\vec{K}\cdot\vec{R}} \, \delta(z) . \tag{56}$$

In the quantum-mechanical case, with a continuous density profile (Fig.1), the density fluctuations also become broader functions $\delta n_{\vec{K}}(z)$, as calculated by WIKBORG and INGLESFIELD [19] and for larger K they develop an oscillatory part which gradually merges with the bulk density oscillations.

4.2 Semiclassical infinite barrier model (SCIBM)

Inspection of the polarization function R_o shows that it contains two types of contributions: one can be visualized as direct and another as due to the waves reflected from the surface. Assuming e.g. IBM wave functions (1) one gets the terms depending on

$$e^{\pm ik_z(z-z')} \qquad\qquad e^{\pm ik_z(z+z')} .$$

The last two terms contribute to the "interference terms", which e.g. modulate the electron density in (3) and (5). In the SCIBM we neglect them, so we get the sharp density step at the surface, and the integral equation (52) becomes easier to solve. RITCHIE and MARUSAK [21] first calculated surface plasmon dispersion in this model, where the implicit dispersion relation becomes

$$1 + \frac{K}{\pi} \int \frac{dk_z}{K^2+k_z^2} \; \frac{1}{\varepsilon(\vec{K},k_z;\omega)} = 0 \tag{57}$$

where ε is the bulk dielectric function for frequency ω and momentum $\vec{k} = (K, k_z)$.

The resulting SP dispersion is also plotted in Fig.6 for Al, where the dielectric function was taken to be the RFA (Lindhard) expression [2] .

5. Optical spectroscopy of surface excitations

Surface excitations do not couple directly to external photons because energy and momentum cannot be conserved in this transition. It can be easily checked that the photon and SP dispersion curves never cross (Fig. 4). Even if we try to lower the photon line, e.g. in a medium with the refractive index n, the dispersion curve of the new interface plasmon also gets modified: it gets suppressed below $\omega = cK/n$ line. Photons do couple to the radiative modes, with $\alpha = i\beta$ in the region R of Fig.3, but these modes are virtual, i.e. they cannot exist as normal modes because they radiate energy and thus decay.

5.1 Rough surfaces and gratings

In order to excite surface modes by photons we need an additional mechanism that would provide momentum. In a real surface *roughness* breaks the parallel momentum conservation. In an experiment we can do it in a controlled way by introducing a regular array of parallel gratings, spaced with the distance d.

A photon with an energy $\omega = kc$ incident at an angle α gets an additional parallel momentum,

$$k_m = \frac{\omega}{c} \sin\alpha + \frac{2\pi}{\alpha} m \qquad m = 0, \pm 1, \pm 2, \ldots$$

In the ω-K plane this gives a series of points where the photon-surface plasmon dispersion lines cross and therefore their coupling is possible. If the reflectivity of such a sample is measured as a function of α we find a number of dips. Their frequencies can be plotted to give a SP dispersion curve, as in the AL case [22,23] or In Sb case [24,25], agreeing very well with the theoretical curves.

5.2 Attenuated total reflection (ATR)

Attenuated Total Reflection (ATR) has been widely used to study surface excitations since it was introduced in 1968 by A.OTTO [26,27]for surface plasmons and in 1970 by R.RUPPIN [28] for surface optical phonons. The principle of the experiment is the following: The dielectric prism with the refraction index n, separated by an air gap from the metal surface, lowers the photon dispersion curve to $\omega = Kc/n$, but does not affect appreciably the surface modes. At total reflection angles θ,

$$1/n < \sin\theta < 1 ,$$

the evanescent photon waves from the prism, with parallel momenta

$$K_s = n(\frac{\omega}{c}) \sin \theta ,$$

couple to the surface plasmons in the region of K:

$$\frac{\omega}{c} < K < \frac{\omega}{c} n .$$

The resulting dips in the measured reflectivity for various angles θ give the SP dispersion curve [26,27,29] .

6. Electron energy loss spectroscopy of surface Excitations

6.1 Introduction

The observation of collective excitations - plasma oscillations - by electron loss spectroscopy dates from early fifties with the work by RUDBERG, RUTHEMANN, WATANABE and others [30].

The *surface plasmons*, predicted in 1957 by RITCHIE [31a] where first observed by POWELL and SWANN in 1959 [31b]. Subsequent history and application of this powerful technique is too rich to be reviewed here, and the interested reader should consult the lectures of Professor H.IBACH and the literature [32].

There are basically two experimental arrangements. In *transmission* experiments a beam of electrons with high kinetic energy E_o (say 50-80 keV) passes through a crystal slab losing energy E to both bulk and surface excitations, and one measures its intensity at the energy E_o - E.

In *reflection* experiments the electrons (of much lower energy) impinge at an angle θ, and one measures the reflected intensity as a function of loss energy E and final angle θ' . Here one mainly observes surface excitations, especially at very low energies and at grazing angle incidence, when the electron penetration is negligible.

Electrons couple to the electromagnetic field associated with the surface modes, so we can use them to study e.g., surface plasmons, surface optical phonons (Fuchs-Kliewer modes) or localized vibrations - adsorbed molecules, but they are not useful e.g. in the study of acoustic modes which do not produce polarization waves.

6.2 Experimental examples

Here we shall briefly discuss some typical experimental results. POWELL [33] measured loss spectra of electrons specularly reflected from liquid Al, for varying impact angles. At grazing incidence (Fig.3 from ref. [33]) we see multiple excitation of *surface plasmons (SP)* with the double and triple excitation processes most probable. As the beam direction approaches the normal to the surface, the SP excitation probability decreases. Also, the electrons start penetrating the sample, so the *bulk plasmon* peaks appear and for angles close to normal they dominate. IBACH [34] also measured a similar multiple exciation spectrum, but now the crystal is a polar dielectric ZnO which sustains *surface optical phonons*. The electron energy E_o = 7.5 eV is much lower than before, and the resolution has to be much better because the excitation energies are in the typical phonon range. It is interesting to observe a *gain* peak at the temperature T = 286K; the electron can absorb one or several of the thermally excited surface modes and appear at an energy higher than E_o.

As a third example we mention a high-resolution spectrum of electrons specularly reflected from nickel covered with adsorbed CO molecules [35,32]. The observed peaks in the spectrum correspond to the excitation of different vibrational modes, which are characteristic for the particular adsorption system. Some of them involve predominantly the adsorbed molecule, like the 239.5 meV peak at low coverage, which is due to the C-O stretching vibration, and some also involve the adsorbate atoms, like the low energy peaks at 59.8 meV and 44.5 and 81.5 meV, respectively. These are due to the two different modes of vibration of the molecule against the substrate, and therefore their identification can be used to study the adsorption process itself.

6.3 Interaction of electrons with collective surface modes

Here we shall briefly outline the Hamiltonian description of the electron interacting with collective surface excitations - surface plasmons and surface optical phonons. The details can be found in the original papers [18,36,37] and reviews 38,39 . Nonretarded theory usually suffices, because the relativistic effects become important for electron kinetic energies $E_o \rightarrow mc^2/2$ = 0.5MeV.

This can be checked by inspecting the energy conserving relation which gives the maximum in the transition probability for the excitation of $h\omega_K$ surface mode

$$h\omega_K = E_o - E_o^{final} \simeq \hbar\vec{K}\cdot\vec{v}_e \quad , \tag{58}$$

where we have neglected the small $K^2/2m$ recoil term on the r.h.s. Therefore the important surface mode momenta lie around

$$K \sim \omega_K/v_e << K_r = \omega_K/c$$

in the dispersion curve, Fig. 5.

The Hamiltonian of the system is

$$H = H_e + H_s + H_I \tag{58'}$$

where

$$H_e = T + U \tag{59}$$

describes the electrons with kinetic energy T in the crystal potential U (eventually including some other external field);

$$H_s = \sum_{i,K} \hbar\, \omega_{i,K} \left(a^+_{i,K}\, a_{i,K} + \tfrac{1}{2}\right) \tag{60}$$

is the Hamiltonian of the free surface modes with parallel wave vectors K and other quantum numbers i (e.g. parity), and H_I is the interaction of the point charge (electron) with the electrostatic potential ϕ, given by (47) or (49),

$$H_I(\vec{r}) = \sum_q \left[\Gamma_q(\vec{r}) a_q + \Gamma^*_q(\vec{r})\, a^+_q \right] . \tag{61}$$

For a semiinfinite solid ($q \equiv \vec{K}$),

$$\Gamma_{\vec{K}}(\vec{r}) = \frac{C}{\sqrt{AK}}\, e^{i\vec{K}\cdot\vec{R} - K|z|} . \tag{62}$$

A is the normalization area of the surface, and the constant C is for a *metal and point ion ionic crystal*

$$C^2 = \pi e^2 \hbar\, \omega_p^2 / 2\, \omega_K ,$$

and for a *general* (ε_o, ε_∞) *ionic crystal*

$$C^2 = \pi\, e^2 \hbar\, \omega_s \left(\frac{\varepsilon_o - 1}{\varepsilon_o + 1} - \frac{\varepsilon_\infty - 1}{\varepsilon_\infty + 1} \right) .$$

For the sake of completeness we also quote the analogous Hamiltonian for the interaction with the bulk modes inside a semiinfinite crystal:

$$H_I(\vec{r}) = \theta(z) \sum_{\vec{k}=(\vec{K},k_z)} \left[\frac{D}{\sqrt{V k^2}}\, e^{i\vec{K}\cdot\vec{R}} \sin k_z z\, b_{\vec{k}} + \text{h.c.} \right] \tag{63}$$

where b´s annihilate bulk plasmons, V is the unit volume, and

$$D^2 = 4\,\pi\, e^2\, \hbar\, \omega_p \qquad \textit{plasmons} ,$$

$$D^2 = 4\pi\, e^2 \left(\frac{1}{\varepsilon_\infty} - \frac{1}{\varepsilon_o}\right) \qquad \textit{LO phonons} .$$

6.4 Interaction of electrons with localized vibrations

Now we want to study the interaction of low-energy electrons, specularly reflected at metallic surfaces, with the dipolar electric fields of single adsorbed molecules. This system is described by the Hamiltonian

$$H = T + U + H_{vib} + V \; ; \tag{64}$$

T is the electron kinetic energy operator and U is the potential of the metallic surface that leads to the specular reflection. The eigenstates of T are plane waves

$$\phi_{\vec{Q}}(\vec{r}) = e^{i\,\vec{Q}\cdot\vec{r}}$$

where $\vec{Q} = (\vec{K}, \pm k\hat{z})$ are the electron momentum components parallel and normal to the surface, respectively. (We always define $k > 0$.) The eigenstates of $T + U$ are specularly reflected waves $\Psi_{\vec{Q}}(\vec{r})$, and they should be determined by solving a full low-energy electron diffraction (LEED) problem. However, we can write them in a rather general form outside the solid [40]:

$$\Psi_{\vec{Q}}(\vec{r}) = g_{\vec{Q}}(z)\; e^{i\vec{K}\cdot\vec{R}} \; ,$$
$$g_{\vec{Q}}(z) = \frac{i}{\sqrt{2\pi}} \left[\, e^{-ikz} + R(\vec{K},k)e^{-2i\delta(\vec{K},k)}e^{ikz}\right]\theta(z) \; , \tag{65}$$

where δ is the phase shift for the surface potential scattering, and the factor R, due to multiple scattering processes, absorptive part of U, etc., gives the reflectivity R^2. In writing (65) we have assumed translational invariance along the surface and the constant (eventually complex) potential outside the surface.

For the incoming electron energies in the energy gap region the reflectivity is very high, as is the case, for example, with copper and nickel (100) surfaces several eV above the vacuum level. Then we can approximate the potential in the simplest way by an infinite potential barrier. This gives $\delta = \pi/2$ and $R = 1$ and the function (65) becomes

$$\phi_{Q}(\vec{r}) = (2/\pi)^{1/2}\; e^{i\,\vec{K}\cdot\vec{R}}\; \sin kz \tag{66}$$

which is totally reflected plane wave (RPW). In (64) H_{vib} is the Hamiltonian of molecular vibrational modes, and we shall treat a single harmonic vibration of frequency ω_0:

$$H_{vib} = \hbar\,\omega_0\,(a^+a + 1/2) \; . \tag{67}$$

The contributions of different modes to the electron-loss spectrum are additive in the lowest (Born) approximation. The inelastic term V arises from electron coupling to dipolar fields of adsorbed molecules

$$H_{dip} = -e \left[\vec{\mu}_p + s\left(\frac{d\mu}{ds}\right) \vec{\mu}_o \right] \cdot \frac{\vec{r}}{r^3} \tag{68}$$

to which we have to add eventually a corresponding image term. We shall neglect the first term in (68), which comes from the permanent dipole and contributes only to the elastic scattering. The second term is due to the oscillating dipole in the $\vec{\mu}_o$ direction, and the vibrational normal coordinate is

$$s = (\hbar/2M\omega_o)^{1/2} \; (a+a^+) \; .$$

$\vec{r}$ is the position of the electron with respect to the molecule center of mass, and M is the reduced mass of the vibration. Therefor

$$V = -e \; (\hbar/2M\omega_o) \left(\frac{d\mu}{ds}\right) \frac{\vec{\mu}_o \cdot \vec{r}}{r^3} \; (a+a^+) \; . \tag{69}$$

6.5 Solution of the scattering problem

In both cases we were dealing with the Hamiltonian of the form

$$H = H_e + \sum_q \hbar\omega_q (a_q^+ a_q + \tfrac{1}{2}) + \sum_q \left[\Gamma_q(\vec{r}) a_q + \Gamma_q^*(\vec{r}) a_q^+ \right] . \tag{70}$$

There are two ways to solve it, i.e. to find the inelastically scattered electron spectrum:

i) *Quantum-mechanical perturbation approach*

The electrons are represented by distorted waves - solutions of the T + U Hamiltonian; the surface modes by the eigenstates $|n\rangle$ of the boson hamiltonian H_s. We have to calculate the transition probabilities for

$|n\rangle \to |n \pm 1\rangle$ single mode emission or absorption loss or gain

$\to |n+2\rangle$ double emission

and so on.

As mentioned earlier, the energy denominator in the transition probability is

$$\hbar\omega_K + E_f - E_i = \hbar\,\omega_K - \hbar\vec{K}\cdot\vec{v}_o + \hbar^2K^2/2m \; . \tag{71}$$

While we shall sometimes neglect the last (recoil) term on r.h.s. of (71), e.g. for high electron energies $E_o \gg \hbar\omega_K$, it is essential to keep it in the low energy limit, especially close to the excitation threshold $E_o = \hbar\omega_K$, as we shall see in the discussion of vibrational excitations.

ii) *Semiclassical (or trajectory) nonperturbative approach*

For $E_o \gg \hbar\omega_K$ the electrons can excite many surface oscillations, and these multiple processes can even become dominant, as can be seen in refs. [33] and [34]. One would therefore like to have a closed expression for the spectrum including all multiple excitation processes, which is feasible in the "trajectory approximation". One ascribes to the electron coordinate $\vec{r}_e$ the time development determinated by the Hamiltonian $H_e = T + U$, neglecting the effects of the subsequent electron-surface mode scattering. Examples are

$$\vec{r}_e(t) = \vec{v}_o t \qquad \text{in transmission,}$$
$$\vec{R}_e(t) = \vec{v}_{\parallel} t\ ,\ z_e(t) = -v_{\perp}|t| \qquad \text{in reflection at } z=0 \text{ plane.} \tag{72}$$

The meaning of the approximation can be seen by eliminating the electron term H_e from H; then the interaction Hamiltonian becomes

$$H_I(t) = \sum_q \{ \Gamma_q [\vec{r}_e(t)]\, a_q + \text{h.c.} \} . \tag{73}$$

For example, in the case $H_e = T$:

$$\vec{r}(t) = e^{iTt}\, \vec{r}\, e^{-iTt} = \vec{r} + i[T,\vec{r}]\, t = \vec{r} + \frac{\vec{p}}{m} t \tag{74}$$

where $\vec{r}$ and $\vec{p}$ are noncommuting operators. The trajectory approximation replaces in (74)

$$\frac{\vec{p}}{m} \rightarrow \frac{\langle\vec{p}\rangle}{m} = \vec{v}_e$$

neglecting small oscillations around the average velocity. Now the Hamiltonian takes the form

$$\tilde{H}(t) = H_s + H_I(t) \tag{75}$$

describing the driven field of surface oscillations. Γ's are time-dependent functions, and this problem can be exactly solved in the coherent state representation for the oscillation field [17].

6.6 Observation of surface optical phonons and surface plasmons

We can now use the results of the previous sections to write the electron loss spectrum A(E). By definition [36] it is equal to the boson field excitation (i.e. gain) spectrum due to the action of the electron from time $t_o = -\infty$. The exact boson function is (in the Schrödinger picture)

$$|\Psi_S(t,t_o)\rangle = D(t,t_o)\ |\Psi_S(t_o)\rangle \tag{76}$$

where

$$D(t,t_o) = \exp\left(-\frac{i}{\hbar} H_o t\right) \exp\left\{-\frac{i}{\hbar} \sum_i (I_i a_i + I_i^* a_i^+)\right\} \exp\left(\frac{i}{\hbar} H_o t\right)$$

is the displacement operator [17]. The loss spectrum is

$$A(E) = \sum_i |\langle\Psi_S(+\infty,-\infty)|\{i\}\rangle|^2\ \delta(E - E_{\{i\}} + E_o) \tag{77}$$

where $E_{\{i\}}$ is the total energy of the boson field {i} . By expressing the δ function in the integral form and after some operator manipulations [36] we find the result

$$A(E) = \frac{1}{2\pi} \int_{-\infty}^{\infty} e^{iEt}\ P(t)\ dt \tag{78}$$

where P(t) is the correlation function which can be calculated [36] :

$$P(t) = \hbar^{-2} \sum_i |I_i|^2 \ (e^{-i\omega_i t} - 1)\ . \tag{79}$$

We shall evaluate the coherent amplitudes I_i for the simplest case of a semiinfinite crystal, the electron incoming at an angle θ (velocities $v_\perp = v\cos\theta$, $v_\parallel = v\sin\theta$) and specularly reflected:

$$I_i = \int_{-\infty}^{\infty} dt \quad \Gamma_i[\vec{r}_e(t)] \quad e^{-i\ \omega_i t} \qquad (i \equiv \vec{K})\ . \tag{80}$$

From (62) and (72) we find $\Gamma_{\vec{K}}$, so that

$$I_{\vec{K}} = \frac{C}{\sqrt{AK}} \quad \frac{Kv_\perp}{(\omega_K - \vec{K}\cdot\vec{v}_\parallel)^2 + (Kv_\perp)^2}\ . \tag{81}$$

For a more detailed discussion of this and more general cases (thin crystal, non-specularly reflected electrons, bulk losses, transmission experiments) and the comparison with experiments the reader should consult the literature [36-39]. Some of the collective modes in a crystal can be brought into an excited state e.g. by the *finite temperature* T. Their subsequent deexcitation may be stimulated by some of the electrons in the beam, and these may emerge with energies higher than E_o, as shown in ref. [34]. In order to account for these processes we can generalize our theory to finite temperatures [38]. The resulting spectrum is again given by (78), except that now the correlation function is due to the statistical averaging:

$$P(t,T) = \exp\{\hbar^{-2} \sum_i |I_i|^2 (\operatorname{sh} x_i)^{-1} \left[e^{x_i}(e^{-i\omega_i t}-1) + e^{-x_i}(e^{i\omega_i t} - 1)\right]\} \qquad x_i \equiv \hbar\omega_i/kT . \tag{82}$$

When $P(t,T)$ is expanded and Fourier transformed, the second term in the square bracket contributes to the *gain peaks*, with the strength determined by the population factor e^{-x_i}, as discussed earlier.

6.7 Observation of adsorbed molecule vibrational modes

Now we shall use the quantum-mechanical perturbation approach to study the excitation of adsorbed molecule vibrational modes. Assumung that we may neglect higher order terms in the interaction V, the differential cross section in the distored wave Born approximation (DWBA) is

$$\frac{d^2\sigma}{d\Omega'\,dE'} = \frac{n(E')}{j_{in}} \frac{2\pi}{\hbar} |T|^2 \,\delta(E_i - E_f) \tag{83}$$

where E_i and $E_f = E' + \hbar\omega_o$ are the initial and final energies of the system, E' is the final electron energy, $n(E')$ is the density of final electron states, and j_{in} is the incoming electron current. The transition amplitude T is

$$T = \langle \chi_f^+ | U | \phi_i \rangle + \langle \chi_f^+ | V | \chi_i^- \rangle ; \tag{84}$$

$|\phi\rangle$ and $|\chi^{\pm}\rangle$ are products of vibronic (harmonic oscillator) and electronic (plane waves and RPW, respectively) wave functions satisfying appropriate boundary conditions. The first term in (84) contributes only to the elastic scattering, and we shall neglect it. The vibrational system is assumed to be initially in the ground state $|0\rangle$ and undergo transition into a singly excited state $|1\rangle = a^+|0\rangle$. Inserting (84) and (69) into (83) gives

$$\frac{d^2\sigma}{d\Omega' dE'} = \left(\frac{E'}{E}\right)^{1/2} \frac{m\,e^2}{M\hbar\,\omega_o\,E} \left(\frac{d\mu}{ds}\right)^2 |f(\vec{Q},\vec{Q}')|^2 \tag{85}$$

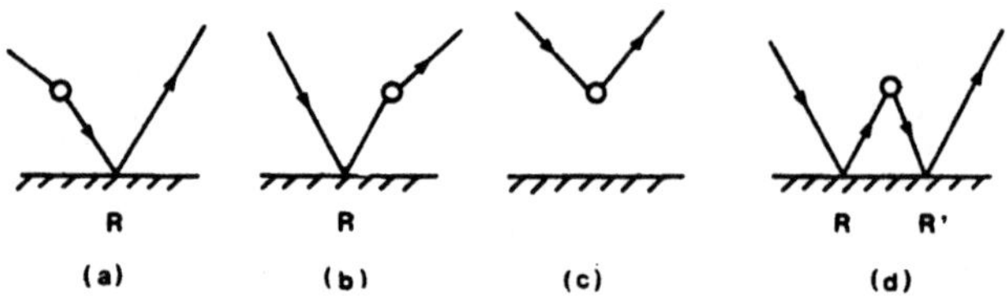

Fig. 7 Four contributions to the scattering amplitude.

where m is the electron mass and f is a dimensionless scattering amplitude:

$$f(\vec{Q},\vec{Q}') = \frac{Q}{2} \int \Psi^{*}_{\vec{Q}'}(\vec{r}) \frac{\vec{\mu}_o \cdot \vec{r}}{r^3} \Psi_{\vec{Q}}(\vec{r}) d^3r \quad . \tag{86}$$

Using the wave functions (65) in (86) gives four terms in scattering amplitude, which are represented schematically in (Fig.7). The largest contributions come from processes (a) and (b), because they include low-angle electron-dipole scattering, which cannot contribute to the observed current in processes (c) and (d). From (85) we see that the total inelastic cross section is proportional to

$$\bar{\sigma} = \int |f|^2 \, d\Omega' \tag{87}$$

which is a dimensionless function of E'/E and incident angle θ .

We can decompose an arbitrarily oriented dipole into components parallel and normal to the surface,and calculate the scattering amplitudes (86) for each component separately. In the high-reflectivity limit , the result can be written in the form

$$|f|^2 = |f_{\perp}|^2 \cos^2\theta_o + \frac{1}{2} |f_{\parallel}|^2 \sin^2\theta_o \tag{88}$$

where we have assumed a random distribution of dipole orientations parallel to the surface. Inserting the wave functions (65) into (86) and integrating, we can find the scattering amplitude $f_{\perp}$ and $f_{\parallel}$ [40] . For simplicity we shall quote the results in the high-reflectivity limit:

$$f_{\perp} = \frac{\Omega}{P_+} \{ K e^{-K|z_o|} - 2 (k+k')\sin[(k+k')z_o]\theta(z_o) \} \\ - (\text{same with } k' \longrightarrow -k') \ , \tag{89}$$

$$f_{\parallel} = i \frac{\Omega}{P_+} K \{ -e^{-K|z_o|} \operatorname{sgn} z_o + 2\cos[(k+k')z_o]\theta(z_o) \} \\ - (\text{same with } k' \rightarrow -k')$$

where

$$P_{\pm} = K^2 + (k \pm k')^2 \ .$$

Angular dependence of inelastic electron spectra can give information about the character of the interaction, e.g., the range of the potential.

As a first example, let us study the case of the dipole adsorbed perpendicularly at the surface ($z_0=0$). Figure 4 in ref. |40| shows the function $|f_{\perp}|^2$ for three characteristic angles: near normal ($\theta=5^{\circ}$), grazing incidence ($\theta=85^{\circ}$), and $\theta=45^{\circ}$, which is closer to the experimental situation [35]. Primary electron energies correspond, e.g. to 2.5 and 1.32 eV, respectively, if we assume that the vibration is a CO stretching mode with the frequency $\hbar\omega_0 = 0.25$ eV. The predicted angular distributions are reasonably narrow, except for near normal incidence, which means that the forward scattering dominates. However, they are much broader than would be expected for a purely elastically scattered beam on a clean surface, in accordance with the relatively localized character of the molecular dipole potential. This also indicates that the corrections to the Born approximation results may be important in cases when the low-energy electron reflectivity depends strongly on the angles of incidence. The main peak grows in intensity at higher electron energies, similar to the elastic cross section divergencies in scattering on gas molecules; it is somewhat shifted from the specular direction toward lower polar angles. This shift increases when approaching the normal incidence, and the second maximum develops. In real speace, these maxima are connected and form distorted ringlike structures around the minimum close to the specular direction. These features result from our choice of the electronic wave functions which have a node at the surface, i.e., at the position of the dipole, thus suppressing the specular peaks. By removing the dipole slightly away from the surface (or similarly, if we used more realistic electron wave functions), we would increase the coupling to the specularly reflected electrons.

For a more complete discussion of the angular and energy dependence of the spectra, and especially the selection rules in the dipole scattering we refer to refs. [40]

We can add a few words about other possible scattering mechanisms. Excitation of molecular vibrational modes by electron impact has been extensively studied in the gas phase [41-43]. However, most of the attention has been focused on resonant excitation processes, where the incoming electron gets trapped in the molecular potential well, temporarily forming a negative-ion state. This filling of the previously unoccupied molecular orbital strongly excites molecular vibrations, and multiple vibrational structures were observed at resonant electron energies [41]. Such shape resonances, connected with the electron capture in the molecular ground-state potential, are well known in most diatomic molecules, and occur at typical energies of, e.g., 1.7 eV for $^2\pi$ resonance in CO, or 2.3 eV for $^2\pi_g$ resonance in N_2.

However, in these lectures we were only looking at the direct coupling of the incoming-electron electric field with the (predominantly dipolar) field associated with adsorbed-molecular vibrations. There are many cases where this and not the resonant-scattering model is applicable [44]. In some experiments the electrons have energies below the actual free-molecule resonance energy and the electron energy increase does not lead to any dramatic changes in the observed loss spectra at the resonance energy. In particular, no multiple excitations, which usually occur when the scattering proceeds via a resonance, are observed. Second, shape resonances in molecules such as CO are connected with the electron capture in the lowest unoccupied molecular orbital. This is exactly the orbital that can be expected to participate in the molecule bonding to the substrate, and thus be most distorted with respect to the gas ohase, and even be partly populated by the "back donation" of charge from the substrate. In any case, in order to settle experimentally the question of the excitation mechanism, it is necessary to explore the observable predictions (in particular, angular and energy dependence) of the different models in as much detail as possible. The deviations from the simple dipolar scattering can be expected to show up at higher electron energies and in off-specular directions.

7. Dynamical screening - image potential

Now we shall continue the discussion of the electrostatic potentials near surfaces. The classical image potential can be visualised as arising from the coupling of an external charge Q at distance z from the surface with the charge density fluctuations in the solid. For a fixed charge(velocity $v = o$) in front of a semiinfinite solid, and in the long-wavelength limit this is described by the Hamiltonian [36]

$$H = E_o + \sum_{\vec{K}} \left[\hbar\omega_{\vec{K}} \, (a_K^{+} Q_K + \tfrac{1}{2}) + (\Gamma_K \, e^{i\vec{K}\cdot\vec{R} - Kz} \, a_K + \text{h.c.})\right] \qquad (90)$$

where we can take the constant electron energy E_o to equal zero.

Without interaction, the surface plasmons are (at temperature T=o) in their ground state $|0>$. When the interaction is switched on, we have to diagonalize the Hamiltonian (90), which can be achieved by replacing

$$a_K \to \tilde{a}_K = a_K + f_K, \quad \text{where} \qquad f_K = \frac{\Gamma_K}{\hbar\omega_K} \, e^{-i\vec{K}\cdot\vec{R} - Kz} \, . \qquad (91)$$

The Hamiltonian is diagonal in these new operators:

$$H = \sum_{\vec{K}} \hbar\omega_K \, (\tilde{a}_K^{+} \, \tilde{a}_K + \tfrac{1}{2}) + \Delta E(z) \qquad (92)$$

where

$$\Delta E(z) = - \sum_{\vec{K}} \hbar\omega_K \, |f_K|^2 \qquad (93)$$

is the total energy gain of the surface plasmon field, which is equal to the ground state energy shift of the external charge, and corresponds to the classical image potential. We can evaluate it for our dispersionless plasmon limit,

$$\Delta E(z) = - \pi Q^2 \int \frac{d^2K}{(2\pi)^2} \frac{e^{-2Kz}}{KA} = - \frac{Q^2}{4z} (1 - e^{-2K_c z}) \, . \qquad (94)$$

We have integrated up to some surface plasmon cutoff wave vector K_c, which obviously prevents the classical divergence at z=0.

We can see the meaning of this result better if we analyze the SP states in the diagonalized Hamiltonian (92). The substitution (91) corresponds to the canonical transformation

$$H \rightarrow \tilde{H} = D \, H \, D^{-1}$$

where

$$D(f) = e^{S(f)} \qquad S(f) = f^* a - f a^+ \qquad (95)$$

is the displacement operator. The ground state becomes

$$|0> \rightarrow |f> = D|0> = e^{-\frac{1}{2}|f|^2} \sum_n \frac{f^n}{n!} (a^+)^n \, |0> \qquad (96)$$

which is the coherent state with the amplitude f. It corresponds to a Poisson distribution of $|n>$ states :

$$|f> = e^{-\frac{1}{2}|f|^2} \sum_n \frac{f^n}{\sqrt{n!}} |n> \, . \qquad (97)$$

The advantage of this derivation of the image potential from the external charge coupling to the charge fluctuations in a solid is that it can be generalized to include the *dynamical* and *quantum-mechanical corrections* to the classical result. One of these was already introduced by restricting the SP fluctuations to $K < K_c$, which eliminated the classical short-range divergency at z = 0.

Now we shall study how the *motion* of the charge modifies the potential. In the semiclassical model the particle coming from

$z = +\infty$ at normal incidence with velocity v will excite the SP field to the coherent state

$$|\{I_K(z,\infty)\}\rangle$$

where the coherent amplitudes [17] are

$$I_K(z,\infty) = \int_{-\infty}^{t=-z/v} \Gamma_K \, e^{Kvt} \, e^{i\omega_K t} \, dt \ . \tag{98}$$

The energy shift $\Delta E(z,v)$ is, from (93) and (98),

$$\Delta E(z,v) = -\frac{Q^2}{2}\int dK \, e^{-2Kz} \,/\, [1 + (Kv/\omega_K)^2] \ . \tag{99}$$

We see that the potential is dynamically reduced, as compared to the adiabatic case ($v = 0$), and the relevant parameter is Kv/ω_K. Obviously this reduction arises because the SP fluctuations cannot follow the motion of the particle and screen it instantaneously.

From (98) and (99) we also see the dependence of the potential on the history of the particle: we find different results when the particle is brought to the distance z from different points and with different velocities. The connection with the induced potential is established only if the SP fields in the initial state are in their ground state, i.e. when the particle arrives from $z = +\infty$.

One could also study the effect of the particle recoil on the induced potential, the problem of self-consistent motion of the particle in the induced velocity-dependent potential and so on [45,46] . But even this brief account shows the advantages of discussing surface problems in terms of surface elementary excitations.

References:

1. C.Kittel: *Introduction to Solid State Physics* (Wiley,1971)
2. C.Kittel: *Quantum Theory of Solids* (Wiley,1963)
3. D.Pines: *Elementary Excitations in Solids* (Benjamin,1963)
4. E.N. Economou and K.L. Ngai: Adv. Chem. Phys. 27, 2654 (1974)
5. K.L. Kliewer and R. Fuchs: Adv. Chem. Phys. 27, 355 (1974), and references therein.
6. R.F. Wallis: Effects of Surfaces in Lattice Dynamics, in *Dynamical Properties of Solids*, ed. by G.K. Horton and A.A. Maradudin (North-Holland 1975), Ch.7
7. A. Ruppin and R. Englman: Rep. Prog. Phys. 33, 149 (1970)
8. J.Bardeen , Phys.Rev. 49,653(1936)
9. N.D.Lang, Solid State Commun. 7, 1047(1969). See also N.D.Lang, Solid State Physics 28,225(1973)
10. See e.g. V.L.Moruzzi, J.F.Janak and A.R.Williams; *Calculated Electronic Properties of Metals* (Pergamon, New York, 1978)

11. For a review see L.Hedin and S.Lundqvist, Solid State Physics 23, 1(1969)
12. E.P.Wigner, Trans.Faraday Soc. 34,678(1938)
13. M.Born and K.Huang: *Dynamical Theory of Crystal Lattices* (Oxford U.P.,1954)
14. E.N. Economou, Phys.Rev. 182,539(1969)
15. R.Brako, J.Hrnčević and M.Šunjić, Z.Physik B 21,193(1975)
16. A.Ronveaux, A.Moussiaux and A.A.Lucas, Can.J.Phys.55,1423(1977)
17. See e.g. E.Merzbacher: *Quantum Mechanics* (Wiley,1970),Ch.15
18. A.A.Lucas, E.Kartheuser and R.G.Badro, Phys.Rev.B2,2488(1970)
19. J.Inglesfield and E.Wikborg, Solid State Commun.14,661(1974). See also E.Wikborg and J.Inglesfield, Physica Scripta 15,37 (1977) for a review of their RPA results.
20. J.Harris and A.Griffin, Can.J.Phys.48,2592(1970)
21. R.H.Ritchie and A.L.Marusak, Surface Sci. 4,234(1966)
22. Y.T.Teng and E.A.Stern, Phys.Rev.Lett.19,511(1967)
23. R.H.Ritchie, E.T.Arakawa, J.J.Cowan and R.N.Hamm, Phys.Rev.Lett. 21,1530(1968)
24. N.Marschall, B.Fischer and H.J.Quiesser,Phys.Rev.Lett.27,95 (1971)
25. B.Fischer, N.Marschall and H.J.Quiesser, Surface Sci. 34,50 (1973)
26. A.Otto, Z.Physik 216,398(1968)
27. E.Kretschmann, Z.Physik 241,313(1971)
28. R.Ruppin, Solid State Commun. 8,1129(1970)
29. N.Marschall and B.Fischer, Phys.Rev.Lett.28,811(1972)
30. For a survey of early experimental work see *Advances in Electronics and Electron Physics*, ed. by L. Marton (Academic Press, New York, 1955), Vol.7
31.a) R.H.Ritchie, Phys.Rev. 106,874(1957)
b) C.J.Powell and J.B.Swann, Phys.Rev. 115,869(1959),ibid.116,81 (1954), ibid.118,640(1960)
32. H. Ibach (ed.): Electron Spectroscopy for Surface Analysis, in *Topics in Current Physics*, Vol.4 (Springer, Berlin, Heidelberg, New York 1977)
33. C.J.Powell,Phys.Rev.175,972(1968
34. H.Ibach,Phys.Rev.Lett. 24,1416(1970)
35. S.Andersson, Solid State Commun. 20,229(1976)
36. M.Šunjić and A.A.Lucas,Phys.Rev. B3,719(1971)
37. A.A.Lucas and M.Šunjić,Phys.Rev.Lett.26,229(1971)
38. A.A.Lucas and M.Šunjić,Progr.Surface Sci. 2,73(1972)
39. D.L.Mills, Progr.Surface Sci. 8,143(1977)
40.a) D.Šokčević,Z.Lenac,R.Brako and M.Šunjić,Z.Physik B28,273(1977)
b) Z.Lenac,M.Šunjić,D.Šokčević and R.Brako, Surface Sci. 80,602 (1979)
41. G.J.Schultz, Rev.Mod.Phys.45,423(1973)
42. K.Takayanagi,J.Phys.Soc.Jpn.21,507(1966)
43. H.S.W.Massey, Electron Collisions with Molecules and Photo-Ionization, in *Electronic and Ionic Impact Phenomena*, H.S.W.Massey, E.H.S.Burhop and H.B.Gilbody, Eds. (Oxford,England,1969),Vol.II,Chap.11
44. J.W.Davenport,W.Ho and J.R.Schrieffer, Phys.Rev.B17,3115(1978)
45. M.Šunjić,G.Toulouse and A.A.Lucas, Solid State Commun. 11,1629 (1972)
46. See G.D.Mahan, in *Collective Properties of Physical Systems*, ed. by B.Lundqvist and S.Lundqvist (Academic Press, (1973),p.164

Surface-Enhanced Raman Scattering

A. Otto

Physikalisches Institut III, Universität Düsseldorf
D-4000 Düsseldorf 1, Fed. Rep. of Germany

On "rough" surfaces of silver Raman signals of adsorbates are enhanced up to six orders of magnitude. The main aspects of this phenomenon are described and a short introduction to the ongoing controversial discussion is given.

1. Introduction

Surface enhanced Raman scattering from adsorbates on metals is an interesting, but still controversial and not well understood subject. It is not in the mainstream of the topics of a summer school on dynamical gas surface interaction. Therefore, only a short introduction into the phenomenon of SERS and some lines of the current discussion on SERS are given in section 2 (taken from ref. 1). For the interested reader, section 3 gives a list of "review" articles of SERS.

2. The phenomenon of surface enhanced Raman scattering, "roughness" and local field effects.

Surface enhanced Raman scattering (SERS) was first detected for pyridine C_5NH_5 (a benzene molecule, in which one CH group is replaced by N) adsorbed at silver electrodes |2-4|. However if one starts with a smooth silver electrode, e.g. a silver film evaporated at room temperature, a strong signal form the vibrational lines of pyridine appears only after an electrochemical oxidation reduction cycle, a so-called "activation cycle". Such a cycle consists of the transformation of the topmost metallic silver layers into Ag^+ ions (in aqueous Cl^- electrolytes a solid AgCl film is formed on the electrode) and the reduction of Ag^+ ions to redeposited metallic silver in the subsequent step. There are many experimental indications, that this second step is the important part of the "activation". Examples of this type of "SERS activation" are presented in figures 1 and 2. Figure 1 shows at the top the low intensity Raman spectrum of water, in the middle the additional weak signal from pyridine dissolved in water and at the bottom the signal from the electrode surface after activation. In this case, the lines from the monolayer of adsorbed pyridine are about 50 times stronger than the lines observed for the dissolved pyridine, though the laser power has been reduced by a factor of 9. The overall average enhancement, that is the ratio of the intensities from a molecule adsorbed at the activated electrode versus a dissolved molecule is estimated to be about 10^6. Fig.2 shows the analogous behaviour for a silver electrode in a cyanide aqueous electrolyte. It has been shown by radioactive tracer experiments with CN^- that both the unactivated and the activated surface is covered with about one monolayer of cyanide |5|. SERS from cyanide is only observed after activation. There

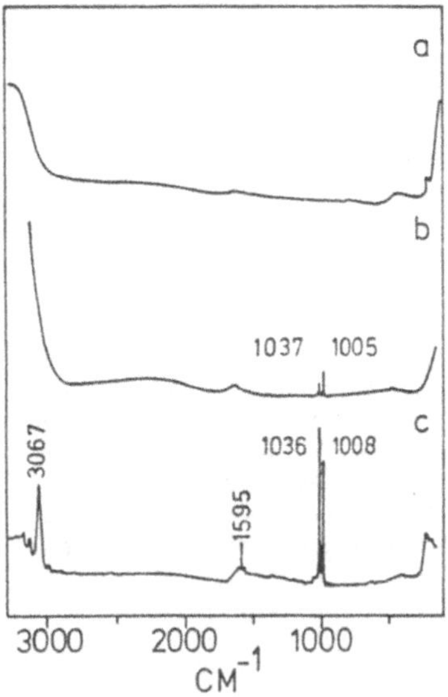

Fig.1 (a) Raman spectrum of a 0.1 M KCl electrolyte; (b) Raman spectrum of a 0.1 M KCl1, 0.05 M pyridine electrolyte, with 450 mW laser power. Lines at 1037 and 1005 cm^{-1} are due to pyridine. (c) Surface endine electrolyte after a dissolution and re-deposition of a layer of about 250 A thickness on a silver electrode. Laser power was 50 mW. With respect to (b), intensities are plotted 10 times demagnified. After Jeanmaire and Van Duyne |3|.

are other ways to produce a "SERS-active" silver surface. Mechanically polishing simply does it| 6,7| . A silver film, evaporated in ultrahigh vacuum onto a cooled substrate (120 K) and kept at this temperature is "SERS active" | 8,9| . A silver film, evaporated at room temperature, or warmed up from 120 K to room temperature is not, or no longer |10,9| "SERS active". Fig.3 shows the comparison of Raman spectra of pyridine adsorbed on a "SERS-active" silver film, evaporated at about 130 K and kept at 130 K and on a silver single crystal (110) surface, cleaned by Ar^+ bombardment, annealed, showing a well defined LEED pattern and cooled to 130 K. The Ag(110) surface has been exposed to 10^4L(1L = 1 Torr sec) of pyridine, so

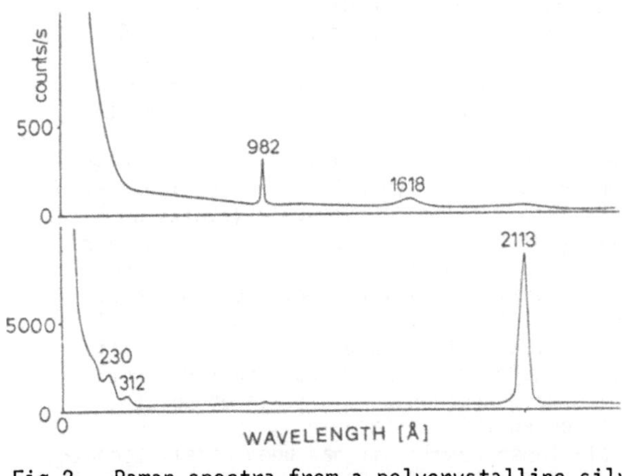

Fig.2 Raman spectra from a polycrystalline silver electrode at -1.0 V_{SCE} in 0.1 M Na_2SO_4, 0.01 M KCN aqueous electrolyte before any anodic sweep (above) and after switching for 5 s to +0.5 v (below). Laser wavelength:5145 A; abscissa is linear in wavelength, Stokes shifts of maxima are given in cm^{-1}. After ref.|5|.

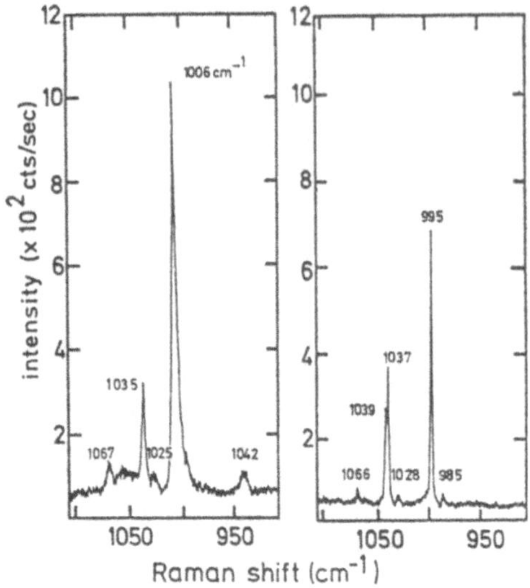

Fig.3 SERS spectrum of pyridine on a cold "SERS-active" silver film exposed to 1 L of pyridine (left) compared to normal Raman scattering from a thick film of pyridine on an Ag (110) surface exposed to 10^4L of pyridine at 120 K. After ref.|43|.

that a rather thick film of pyridine (5.10^3 monolayers) has been formed. The evaporated silver film has only been exposed to 1 L of pyridine. Even so, the Raman signal from the film is stronger than from the crystal, indicating 4 to 5 orders of enhancement for pyridine adsorbed on the film compared to pyridine on the (110) surface. It was possible to obtain a pyridine Raman signal after exposure of the cold silver film to only 3.10^{-2}L thus demonstrating the surprisingly high sensitivity of SERS under favorable conditions.

What makes a silver surface "SERS active"? In the case of electrode surfaces "activation" was possible without the presence of pyridine or cyanide. When pyridine or cyanide were added afterwards, the SERS spectrum was present as well. This indicates that it was a change in the topographical structure of the surface, or a change of the structure of the redeposites silver with respect to bulk silver which made an electrode "active" |5| . The experiments on cold and warm silver films under clean conditions in ultrahigh vacuum, without apparent coadsorbates, leave no other choice than to ascribe the "SERS activity" to a special surface topology, or to a change of the structure in the selvedge or in the bulk.

Scanning electron microscope pictures of heavily activated silver electrodes showed a rather coarse roughness on the 1000 Å scale (see for instance SEM pictures in ref.7). Silver island films of average thickness 50 Å were found to be SERS active |11| (for SEM pictures of SERS-active silver island films of average thickness of 90 Å see for instance |12| . Silver colloids were found to be SERS active |13| . Silver electrodes in tunnel junctions, deposited on rough CaF_2 films were found to be SERS active |14|. (For the 500 Å roughness scale of silver films evaporated on a CaF_2 film see electronmicrograph in ref.15). Rowe et al. |16| reported SERS from pyridine on silver surfaces, which were roughened to a scale of about 500-2000 Å by exposing a UHV cleaned Ag(100) surface to I_2 vapour and 4880 A laser light. Because of all these results it has been widely accepted that "surface roughness" is very important for SERS.
Under "surface roughness" one should also include possible subtle changes of the surface or selvedge topography, the crystallographic orientation of microcrystalline surfaces etc. The scale of the roughness which induces the "activity" is under debate. The scale may be of 500 to 2000 A

|16|, it may be "submicroscopic" (100 Å) as pointed out by Burstein and Chen |17| or "atomic" |5| . There is an intrinsic difficulty in separating "scales of roughness". "Submicroscopic" roughness may be present besides "microscopic" roughness (which means roughness which can be seen in a commercial reflection type scanning electron microscope). When there is roughness, there is always atomic scale roughness (steps, kinks, high index "open" planes, adatoms, clusters, surface voids) present (but not viceversa!) because of the atomic nature of solids. "SERS activity" was found for silver electrodes, where the scale of the roughness was below the detection limit: Pettinger et al. |18| observed SERS after recycling only 1 to 5 monolayers of epitaxial (111) silver films. It is unlikely that this results in a rms value of the roughness which is much bigger than the thickness of the dissolved silver film (one to five monolayers). Schultz, Janik Czachor and Van Duyne |7| prepared carefully mechanically polished electrodes. These were "SERS active" without the oxidation reduction cycle described above. Within the resolution of the SEM (about 250 Å) no roughness could be detected. There exists as yet no direct experimental value for the roughness of "SERS-active" silver films evaporated in UHV at low temperatures though one may expect a "bumpiness" on the scale of 50 Å, see ref.1. Udagawa et al. |19| reported a weak enhancement of about 440 on carefully prepared Ag(100) surfaces. This may be an indication of an enhancement mechanism on a smooth surface or an indication that there is still residual atomic scale roughness (as a silver surface is usually not better oriented as about 0.25 degrees with respect to the ideal crystallographic direction, a monocrystalline surface will always contain a surface concentration of atoms at steps with respect of atoms on terraces of 5.10^{-3} or greater, provided the steps are monoatomic).

Besides the relation between "roughness" and "SERS activity" there are other effects connected with SERS: in a SERS spectrum from silver the vibrational lines of an adsorbate are always superimposed on a nearly structureless inelastic background. This is apparent in fig.2. In the top spectrum there is a very low background of about 50 c/sec due to inelastic scattering from liquid water. After activation, the background has grown to about 300 c/sec. Compared to this new background, the $SO_4^=$ stretch mode at 982 cm^{-1} and the water bending mode at 1618 cm^{-1} are weak. This background is an intrinsic effect of "SERS-active" silver. It can be enormously enhanced by drastic mechanical polishing a polycrystalline silver slug |20| . It is present for silver films, evaporated in ultrahigh vacuum on substrates of 120 K, before any deliberate exposure to an adsorbate |9| . Silver films, which are not (or no more) "SERS active" do not display the background. The background is probably luminescence |21| due to a continuum of electronic excitation of the metal |22| . It shows a peculiar time dependence. Heritage et al. |21| observed in a picosecond Raman gain experiment on an activated silver electrode that the reflectivity of the probe beam (Stokes shifted with respect to the pump beam) was increased by the pump beam up to a delay time of 200 psec after switching off the pump beam. Chen et al. |23| found an "exceptionally strong" anti-Stokes background in second harmonic generation from SERS-active silver. For SERS-active silver films, the temporal behaviour of this background showed "a clear tail, several pulse widths (of 10 nsec each) long".

In SER spectra the Raman selection rules are relaxed, as detected for pyrazine on silver by Dornhaus et al. |24| and Erdheim et al. |25| and for benzene on silver by Moskovitz and DiLella |26| . The SER spectrum of pyrazine on a SERS-active silver electrode is shown in fig.4. The assignment and the comparison with the vibrational frequencies of the liquid phase of pyrazine is given in table 1. Free pyrazine $C_4N_2H_4$ has inversion symmetry and hence there is an exclusion rule of Raman and infrared activity

TABLE 1: Comparison of vibrational frequencies of pyrazine in aqueous solution (or in liquid pyrazine (1)) with SERS frequencies on a silver electrode at -0.4 V_{SCE}. Assignment and activity in the D_{2h} symmetry of the free molecule. After ref. |24|.

	aqueous,(1) liquid	SERS	D_{2h}	activity (D_{2h})
6a	615 vw	635 ?s	a_g	R
1	1017 vs	1018 vs	a_g	R
9a	1241 s	1242 s	a_g	R
8a	1594 s	1590 vs	a_g	R
2	3060 m	3050 m	a_g	R
16a	363 vw	362 w	a_u	-
	(393)			
17a	(950) (1)	966 vw	a_u	-
6b	677 vw	635 ?s	b_{1g}	R
		662 s		
		682 s		
3	1120 vw	1121 w	b_{1g}	R
8b	1529 m	1520 w	b_{1g}	R
7b	3041 - 3060 (1)	3031 vw	b_{1g}	R
10a	758 vw	744 m	b_{2g}	R
		(764) w		
4	703 s	700 m	b_{3g}	R
5	(922) vw	916 m-s	b_{3g}	R
16b	416 - 417 (1)	436 1	b_{1u}	IR
11	786 - 804 (1)	797 w	b_{1u}	IR
15	1063 - 1067 (1)	1069 w	b_{2u}/b_{3u}	IR
14	1342 - 1346 (1)	1340 vw	b_{2u}	IR
19b	1413 - 1418 (1)	1420 vs	b_{2u}	IR
20b	3066 - 3070 (1)	3060 ?	b_{2u}	IR
12	1021 - 1022 (1)	1038 vw	b_{3u}	IR
18a	1135 - 1148 (1)	1164 w	b_{3u}/b_{2u}	IR
19a	1484 - 1490 (1)	1485 s	b_{3u}	IR
13	3066, (3090) (1)	3183 ?	b_{3u}	IR

of the vibrations. The SER spectrum shows also the modes which are Raman inactive in the free molecule (pointed out by open dots in fig.4), some of them with surprisingly high relative intensity.

The SER spectra of benzene on a silver film evaporated at 11 K |26| is shown in fig.5. In contrast to pyrazine, the free benzene molecule has so-

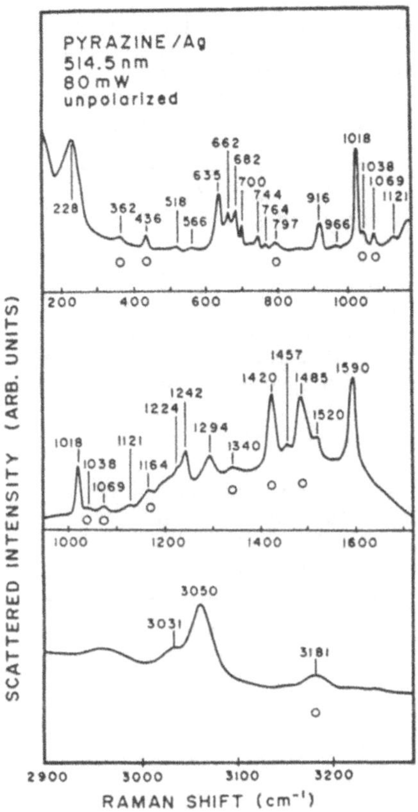

Fig.4 SERS spectrum of pyrazine adsorbed on an activated silver electrode at - 0.4 V_{SCE}. After ref.|24|. The open points mark lines which are IR active in free pyrazine.

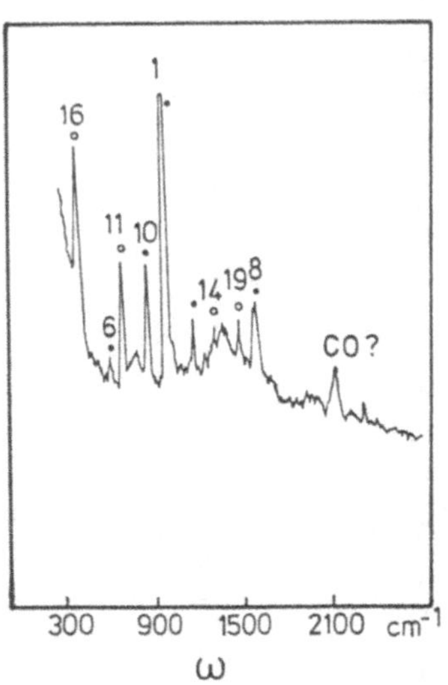

Fig.5 SERS spectrum of benzene adsorbed on silver, excited with laser light of 514.5 nm wavelength. Open (full) dots mark modes which in the free benzene molecule are IR active or silent (Raman active). After ref.|26|.
Numbers refer to the mode numbers in table II. The feature named CO? has not been assigned in ref.|26|.

called silent modes, neither Raman nor infrared active in first order, for instance the mode number 16 in table 2. This mode is seen with good intensity in fig.5. It is shifted by only 7 cm^{-1} from its position in the spectrum of the liquid. SER spectra are strongly depolarized, even for vibrational modes which do not depolarize, when the molecule is free. Overtones or combinational band are only very weak in SER |27| or not seen at all. The excitation spectra of SERS (intensity of various SERS lines versus exciting laser frequencies) do not show narrow resonances. An example is given in fig.6. For further references on excitation spectra and a recent discussion I refer the reader to |28| .

SERS activity has been found so far on silver, copper and gold (e.g. ref.29 and references therein) may be on aluminium |30,31| on Li |32| , on K 33 and on Pt, Hg, Ni (see ref. |1|). There is a big number of adsorbates, for which SERS has been observed on silver. However, there seems to be chemical specificity, in particular for water, as discussed in ref. | 1 | .

TABLE 2: Comparison of vibrational frequencies of liquid benzene and of benzene on a "SERS-active" silver film. Assignment and activity in the D_{6h} symmetry of the free molecule. After ref. |26| .
+: a printing error in ref. |26| has been corrected.

	liquid	SERS	D_{6h}	activity (D_{6h})
1	992	982	a_{1g}	R
2	3059	3062	a_{1g}	R
3	1346	-	a_{2g}	-
4	703	-	b_{2g}	-
5	989	?	b_{2g}	-
6	606	605	e_{2g}	R
7	3046	3047	e_{2g}	R
8	1596	1587	e_{2g}	R
9	1178	1177	e_{2g}	R
10	849	864	e_{1g}	R
11	670	697	a_{2u} +	IR
12	1008	-	b_{1u}	-
13	3062	?	b_{1u}	-
14	1309	1311	b_{2u}	-
15	1149	1149	b_{2u}	-
16	404	397	e_{2u}	-
17	966	?	e_{2u}	-
18	1036	1032	e_{1u}	IR
19	1479	1473	e_{1u}	IR
20	3073	?	e_{1u}	IR

The fairly general acceptance of the importance of "roughness" for SERS and the fact that the noble metals enhance well in those spectral ranges where they can support high quality surface plasmon resonances lead to an explanation of SERS which I will call from now on "classical enhancement" following Jha et al. |34| . The basic idea is the enhancement of the Raman scattered intensity by electro-magnetic resonances of the local field, both for the incident field and the frequency shifted emitted field. The solid surface is characterized by the boundary of a continuum with a dielectric constant $\varepsilon(\omega)$. In this way, one preempts the possibility that the electronic properties of a "SERS-active" sample be different at or near the surface compared to those of the bulk. Any specific adsorbate-metal interaction is neglected - the molecule "feels" the metal only through the enhancement of the local field. The response of the metal is not changed by the presence of the molecule. Without any further assumptions, the classical model cannot explain the background and the relaxation of Raman selection rules. In the sense of Jha et al., my definition of "classical enhancement" does not include possible mechanisms like the modulation of the surface potential barrier by the adsorbate |34| or the relaxation of selection rules in the strong field gradient at the metal surface |35| .

The discussion on the origin of SERS is not concerned with the question whether classical enhancement exists or not. The predicate "classical" implies that local field resonances must exist and there are many clear experimental verifications. The problem is whether the neglected points discussed above, do contribute to the enhancement and how much the local field resonances are damped by electronic properties not described by the bulk value of $\varepsilon(\omega)$. So the questions are |36| :

(i) Does the "classical enhancement" explain all the observed enhancement?

(ii) If not, to what extent are other than classical enhancement mechanisms involved?

(iii) Is SERS possible without "classical enhancement" for instance on transition metals, and how strong is the enhancement in this case?

There are clear and opposing statements in the literature of the year 1980:

"..... the dominant contribution to SERS is electromagnetic rather than chemical in origin" |16| .

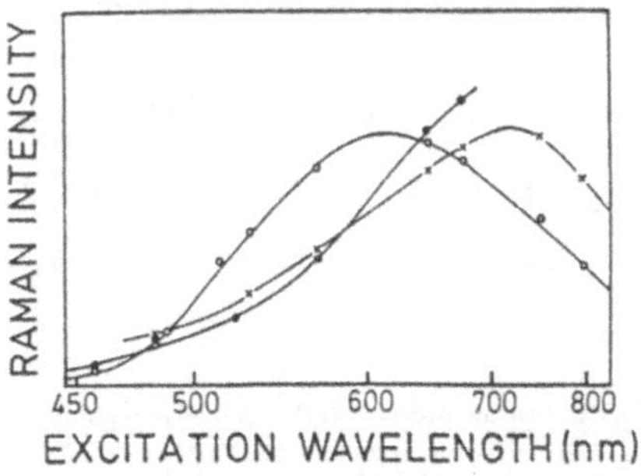

Fig.6 Raman excitation profiles for adsorbates at electrochemically activated silver electrodes for the 1008 cm^{-1} line of pyridine(x) for the 2114 cm^{-1} line of cyanide (•) and the 1000 cm^{-1} line of tryphenylphosphine (o). After ref.48.

"..... The overall enhancement is the result of a combination of contributions, some of which may be quite specific to the particular metal-adsorbed molecule system" |17| .
"..... SERS involves a chemical bond which may be intimately associated with microroughness by bonding site availability" |37| .
"..... A strong Raman enhancement for an adsorbate on a silver surface is only possible when the adsorbate is bound to an atomic scale surface roughness" |5| .

The hypothetical "nonclassical" contributions to the observed enhancement is usually addressed as "the chemical effect" caused by molecule-metal interaction. Many possible mechanisms including charge-transfer excitation have been proposed (for instance refs. 38-42). It has also been proposed this "chemical effect" is particularly strong at special surface sites ("active sites") of atomic scale roughness |5,36,43-45| .

3. Further Reading

The phenomenon of SERS is described in a number of more or less comprehensive articles, which are listed in the sequence of appearance (refs.46-50, 36,1) and by a collection of about 20 papers in "Surface enhanced Raman Scattering", eds. R.K. Chang, T.E. Furtak, Plenum, Oct. 1981.

References

1. A. Otto, in "*Light Scattering in Solids III*", M Cardona, G. Güntherrodt (eds), Topics in Applied Physics, Vol. 51 (Springer, Berlin, Heidelberg, New York 1982)
2. M. Fleischmann, P.J. Hendra, A.J. McQuillan, Chem. Phys. Letters 26, 123 (1974).
3. D.L. Jeanmaire, R.P. VanDuyne, J. Electroanal. Chem. 81, 1 (1977).
4. M.G. Albrecht, J.A. Creighton, J. Am. Chem. Soc. 99, 5215 (1977).
5. J. Billmann, G. Kovacs, A. Otto, Surf. Science 92, 153 (1980).
6. A. Otto, Surf. Science 75, L392 (1978).
7. S.G.Schultz, M. Janik-Czachor, R.P. VanDuyne, Surf. Science 104, 419, (1981).
8. T.H. Wood, M.V. Klein, J. Vac. Sci. Technol. 16, 459 (1979).
9. I. Pockrand, A. Otto, Sol. State Comm. 38, 1159 (1981).
10. T.H. Wood, M.V. Klein, Sol. State Comm. 35, 263 (1980).
11. C.Y. Chen, I. Davoli, G. Ritchie, E. Burstein, Surf. Science 101, 363 (1980).
12. H. Seki, J. Vac. Sci. Technol. 18, 633 (1981).
13. J.A. Creighton, C.G. Blatchford, M.G. Albrecht, J. Chem. Soc. Faraday II, 75, 790 (1979).
14. J.C. Tsang, J.R. Kirtley, J.A. Bradley, Phys. Rev. Letters 43, 772 (1979).
15. D. Beaglehole, O. Hunderi, Phys. Rev. B2, 309 (1970).
16. J.E. Rowe, C.V. Shank, D.A. Zwemer, C.A. Murray, Phys. Rev. Letters 44, 1770 (1980).
17. E. Burstein, C.Y. Chen, in Proc. VII Int. Conf. Raman Spectroscopy, W.F. Murphy (ed.) Ottawa, 1980 p. 346.
18. B. Pettinger, U. Wenning, D.M. Kolb, Ber. Bunsenges. Phys. Chem. 82, 1326 (1978).
19. M. Udagawa, Chih-Cong Chou, J.C. Hemminger, S. Ushioda, preprint.
20. A. Otto, J. Timper, J. Billmann, G. Kovacs, I. Pockrand, Surf. Science 92, L 55 (1980).
21. J.P. Heritage, J.G. Bergman, A. Pickzuk, J.M. Worlock, Chem. Phys. Letters 67, 229 (1979).

22. A. Otto in Proc. Int. Conf. on Vibrations in Adsorbed Layers, Jülich, June 78. Unchanged reprint: Surf. Science 92, 145 (1980).
23. C.K. Chen, A.R.B. De Castro, Y.R. Shen, Phys. Rev. Letters 46, 145 (1981).
24. R. Dornhaus, M.B. Long, R.E. Benner, R.K. Chang, Surf. Science 93, 240 (1980).
25. G.R. Erdheim, R.L. Birke, J.R. Lombardi, Chem. Phys. Letters 69, 495 (1980).
26. M. Moskovits, D.P. DiLella, J. Chem. Phys. 73, 6068 (1980).
27. B. Pettinger, Chem. Phys. Letters 78, 404 (1981).
28. I. Pockrand, J. Billmann, A. Otto, submitted to J. Chem. Phys.
29. I. Pockrand, submitted to Chem. Phys. Letters.
30. T.H. Wood, M.V. Klein, unpublished manuscript: Enhanced Raman Scattering from adsorbates on metal films in ultra high vacuum.
31. T. Lopez Rios, C. Pettenkofer, I. Pockrand, A. Otto, in praparation.
32. M. Moskovitz, private communication.
33. W. Schulze, private communication.
34. S.S. Jha, J.R. Kirtley, J.C. Tsang, Phys. Rev. B 22, 3973 (1980).
35. J.K. Sass, H. Neff, M. Moskovits, S. Holloway, J. Phys. Chem. 85, 621 (1981).
36. A. Otto, Appl. Surf. Science 6, 309 (1980).
37. T.E. Furtak, G. Trott, B.H. Loo, Surf. Science 101, 374 (1980).
38. E. Burstein, Y.J. Chen, C.Y. Chen, S. Lundqvist, E. Tosatti, Sol. St. Comm. 29, 567 (1979).
39. H. Ueba in *Surface Enhanced Raman Scattering*, ed. R.K. Chang, T.E. Furtak, Plenum, Oct. 1981.
40. G.W. Robinson, Chem. Phys. Letters 76, 191 (1980).
41. K. Arya, R. Zeyher, Phys. Rev. B 24, 1852 (1981).
42. B.N.J. Persson, Chem. Phys. Letters 82, 561 (1981).
43. A. Otto, I. Pockrand, J. Billmann, C. Pettenfoker in *Surface Enhanced Raman Scattering*, R.K. Chang,T.E. Furtak eds., Plenum, Oct. 1981.
44. A. Otto, submitted to Phys. Rev. B.
45. H. Seki, J. Vac. Sci. Technol. 18, 633 (1981).
46. R.P. VanDuyne in *Chemical and Biological Applications of Lasers 4*, ed. C.B. Moore (Academic Press, New York, 1979).
47. T.E. Furtak, J. Reyes, Surf. Science 93, 351 (1980).
48. C.Y. Chen, I. Davoli, E. Burstein in Proc. US-USSR Symp. Theory Light Scattering in Condensed Matter, eds. E. Burstein, C.Y. Chen, S. Lundqvist (Plenum, New York, 1979).
49. J.A. Creighton in *Vibrational Properties of Adsorbates*, ed. R. Willis, Springer Series in Chemical Physics, Vol. 15 (Springer, Berlin, Heidelberg, New York 1980)
50. E. Burstein, C.Y. Chen in Proc. VIIth Int. Conf. Raman Spectroscopy. W.F. Murphy, ed., Ottawa, 1980.

Calculation of the Phonon Spectrum of the Ni(111) Surface Covered with Oxygen

V. Bortolani, A. Franchini, F. Nizzoli and G. Santoro

Istituto di Fisica dell'Università di Modena, Gruppo Nazionale di Struttura della Materia del Consiglio Nazionale delle Richerche
I-41100 Modena, Italy

1. Introduction

In recent years there has been a noticeable development in the experimental techniques to detect surface phonons. Detailed information on the long wavelength acoustic part of the spectrum has been obtained with Brillouin scattering for clean and coated surfaces |1,2|. The other techniques used are the inelastic scattering of atoms |3,4| and the high resolution electron energy loss spectroscopy |5| (EELS). These two topics are discussed at length in this School both in their general aspects and in their fields of application. In particular EELS have been employed in the past for the study of vibrations of adsorbed layers of molecules in the investigation of the physics and chemistry of surfaces. More recently this technique has been used to observe surface phonons in metals.|6,7| In these notes we focus our attention on this particular subject.

The detection of surface vibrations with EELS generally occurs through the dipole coupling. For metal surfaces this dipole can be provided by a fraction of the monolayer of an adsorbed species such as oxygen. Therefore in this case there are problems of interpretation of the experimental loss spectra. In fact one has to discriminate between the modes of the substrate and the modes induced by the adsorbate.

In order to solve this problem we show here that a realistic surface phonon calculation is needed. We consider the case of Ni(111) surface covered with oxygen for which very accurate experimental data are available.|6| Our results indicate that the observed peaks relate to the phonons of the substrate for the Rayleigh wave but otherwise are due to the adsorbate layer.

2. Dipole Coupling

The long range part of the potential produced by a regular distribution of dipoles placed on the surface of a medium of dielectric constant ε is given by

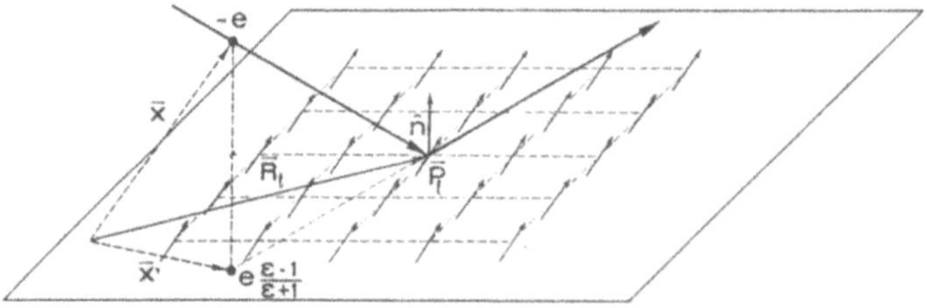

Fig.1 Geometry of the electron-dipole scattering. The image charge is shown.

$$\phi(\vec{x}) = -\sum \vec{P}_1 . \vec{E}_1(\vec{x}) \quad . \tag{2.1}$$

Here $\vec{P}_1$ is the dipole moment at the site $\vec{R}_1$ and $\vec{E}_1(\vec{x})$ is the electric field produced by the dipole acting on the electron in the $\vec{x}$ position as illustrated in Fig.1. If we consider the electric dipole produced by the excitation of a surface optical phonon of wave vector $\vec{Q}_{\parallel}$ we can write

$$\vec{P}_1 = \vec{P}_{\vec{Q}_{\parallel}} \exp(i\vec{Q}_{\parallel}.\vec{R}_1) \quad . \tag{2.2}$$

By using the image charge method 8 and performing the 1-summation in Eq.(2.1) the interaction potential results in

$$\phi(\vec{x}) = \frac{4\pi n_0 \varepsilon}{1+\varepsilon} e^{i\vec{Q}_{\parallel}.\vec{x}} e^{-|\vec{Q}_{\parallel}|z} (\vec{P}_{\vec{Q}_{\parallel}}.\hat{n} - \frac{i}{\varepsilon} \frac{\vec{Q}_{\parallel}.\vec{P}_{\vec{Q}_{\parallel}}}{|\vec{Q}_{\parallel}|}) \tag{2.3}$$

where n_o is the surface density of dipoles and n is the unit vector normal to the surface. For metals $|\varepsilon| >> 1$ so that only the first term of Eq.(2.3) is effective. By solving the time dependent Schroedinger equation outside the crystal, for this potential, one obtains the scattering cross section.

This expression, apart from angular factors, reads

$$\frac{d\sigma}{d\Omega} \propto <|\vec{P}_{\vec{Q}_{\parallel}}.\hat{n}|^2> \quad . \tag{2.4}$$

The brackets mean a thermal average over the phonon field. The dipole moment at the site $\vec{R}_1$ is defined by

$$P_{1\alpha} = \sum_{\beta} e_{\alpha\beta}(1_z) u_\beta(1) \tag{2.5}$$

where $e_{\alpha\beta}(1_z)$ is the dipole moment effective-charge tensor and $u_\beta(1)$ is the displacement operator given by

$$u_\beta(1) = \sum_n \frac{w_\beta(\vec{Q}_{\parallel}, 1_z, \omega_n)}{(2m_{1_z} \omega_n N_s)^{\frac{1}{2}}} e^{i\vec{Q}_{\parallel}.\vec{R}_1} (a_n + a_n^{\dagger}) \quad ; \tag{2.6}$$

$\vec{w}(\vec{Q}_{\parallel}, 1_z, \omega_n)$ are the polarization vectors of the surface phonons on the 1_z-th plane, N_s is the number of surface unit cells and m_{1_z} is the mass of the atoms in the 1_z-th plane.

The a_n and a_n^+ are the usual annihilation and creation phonon operators. Inserting Eq.(2.6) and Eq.(2.5) in the Eq.(2.2), one derives the explicit expression for $\vec{P}_{\vec{Q}_{\parallel}} . \hat{n}$. Performing the thermal average and retaining only single phonon processes we obtain the following expression for the cross section:

$$\frac{d\sigma}{d\Omega} \propto \frac{N(\omega)}{\omega} \left| \sum_{1_z} \sum_{\substack{\text{unit} \\ \text{cell}}} \frac{e_{zz}(1_z) w_z(\vec{Q}_{\parallel}, 1_z, \omega)}{\sqrt{m_{1_z}}} \right|^2 \tag{2.7}$$

where $N(\omega)$ is the Bose factor. For a metallic surface with an adsorbed layer one can assume that the dipole is entirely due to the charges of the two outermost layers and $e_{zz}(l_z=0)=-e_{zz}(l_z=1)$. Within this assumption and in the high temperature limit Eq.(2.7) reduces to

$$\frac{d\sigma}{d\Omega} \propto \frac{k_B T}{\omega^2} \left| \sum_{\text{unit cell}} \left[\frac{w_z(\vec{Q}_\parallel, l_z=0, \omega)}{\sqrt{m_{l_z=0}}} - \frac{w_z(\vec{Q}_\parallel, l_z=1, \omega)}{\sqrt{m_{l_z=1}}} \right] \right|^2 . \tag{2.8}$$

This is the expression that relates the EELS spectra with the polarization vectors of the surface phonons.

3. Bulk Phonons

For transition metals a first principles microscopic approach to the bulk problem is very hard. In fact one has to use a non-diagonal dielectric matrix in order to account for the presence of the d electrons that cause the interatomic potential to be anisotropic. The introduction of the surface makes the problem even more complicated and at the present there are no calculations based on this approach. For this reason we solve the dynamical problem of Ni in the framework of a force constant parametrization where the effects introduced by the surface can be easily accounted for.

We will consider central pairwise interactions and three body angular forces to simulate the anisotropy of the potential. In this scheme the force acting on an atom can be written as the sum of two terms. The first one, which represents the effect of the central part of the interatomic potential V, is

$$\vec{F}^c_n = -\sum_j \left[\alpha_{nj}(\vec{u}_n - \vec{u}_j) + (\alpha_{nj} - \beta_{nj}) \frac{\{\vec{R}^{oj}_n \cdot (\vec{u}_n - \vec{u}_j)\}\vec{R}^{oj}_n}{|\vec{R}^{oj}_n|^2} \right] ; \tag{3.1}$$

$\vec{R}^{oj}_n$ labels the equilibrium position between the atom "n" and the atom "j". The α_{nj} are the tangential force constants related to bending forces and the β_{nj} are the radial force constants responsible for the stretching between atoms. They are related to the central part of the potential $V(|\vec{R}^{oj}_n|)$ by:

$$\alpha_{nj} = \frac{1}{|\vec{R}^{oj}_n|} \frac{\partial V(|\vec{R}^{oj}_n|)}{\partial |\vec{R}^{oj}_n|} , \tag{3.2}$$

$$\beta_{nj} = \frac{\partial^2 V(|\vec{R}^{oj}_n|)}{\partial |\vec{R}^{oj}_n|^2} . \tag{3.3}$$

The $\vec{u}_m$ are the atomic displacements. The second contribution to the force arises from the angular part of the potential. It can be written as

$$
\begin{aligned}
\vec{F}_n^a = -\frac{1}{3}\sum_{j,k}\Bigg[& \frac{\eta_{njk}}{|\vec{R}_n^{0j}|}\psi(njk)\vec{\phi}(njk) - \frac{\eta_{jnk}}{|\vec{R}_j^{0n}|}\psi(jnk)\vec{\phi}(jnk)+ \\
& +\gamma_{njk}\{-\frac{1}{|\vec{R}_n^{0j}|^2}\big[(\vec{r}_n^{0j}.\vec{u}_n^j)\vec{r}_n^{0k}+(\vec{r}_n^{0k}.\vec{u}_n^j)\vec{r}_n^{0j} - \\
& -3(\vec{r}_n^{0j}.\vec{u}_n^j)(\vec{r}_n^{0j}.\vec{r}_n^{0k})\vec{r}_n^{\,j}+(\vec{r}_n^{0j}.\vec{r}_n^{0k})\vec{u}_n^j\big] + \\
& + \frac{1}{|\vec{R}_n^{0j}||\vec{R}_n^{0k}|}\big[\vec{u}_n^k-(\vec{r}_n^{0j}.\vec{u}_n^k)\vec{r}_n^{0j}-(\vec{r}_n^{0k}.\vec{u}_n^k)\vec{r}_n^{0k} + \\
& + (\vec{r}_n^{0j}.\vec{r}_n^{0k})(\vec{r}_n^{0k}.\vec{u}_n^k)\vec{r}_n^{0j}\big]\} - \gamma_{jnk}\{-\frac{1}{|\vec{R}_j^{0n}|^2}\big[(\vec{r}_j^{0n}.\vec{u}_j^n)\vec{r}_j^{0k}+ \\
& + (\vec{r}_j^{0k}.\vec{u}_j^n)\vec{r}_j^{0n}+(\vec{r}_j^{0n}.\vec{r}_j^{0k})\vec{u}_j^n-3(\vec{r}_j^{0n}.\vec{r}_j^{0k})(\vec{r}_j^{0n}.\vec{u}_j^n)\vec{r}_j^{0n}\big] + \\
& + \frac{1}{|\vec{R}_j^{0n}||\vec{R}_j^{0k}|}\big[\vec{u}_j^k-(\vec{r}_j^{0n}.\vec{u}_j^k)\vec{r}_j^{0n}-(\vec{r}_j^{0k}.\vec{u}_j^k)\vec{r}_j^{0k} + \\
& + (\vec{r}_j^{0n}.\vec{r}_j^{0k})(\vec{r}_j^{0k}.\vec{u}_j^k)\vec{r}_j^{0n}\big]\}\Bigg]
\end{aligned}
\tag{3.4}
$$

where

$$\vec{r}_n^{0m} = \frac{\vec{R}_n^{0m}}{|\vec{R}_n^{0m}|} , \tag{3.5}$$

$$\vec{u}_n^m = \vec{u}_n-\vec{u}_m , \tag{3.6}$$

$$\vec{\phi}(ijk) = \vec{r}_i^{0k}-(\vec{r}_i^{0j}.\vec{r}_i^{0k})\vec{r}_i^{0j} , \tag{3.7}$$

$$\psi(njk) = \frac{1}{|\vec{R}_n^{0j}|}\vec{\phi}(njk).\vec{u}_n^j+\frac{1}{|\vec{R}_n^{0k}|}\vec{\phi}(nkj).\vec{u}_n^k . \tag{3.8}$$

γ_{njk} and η_{njk} are the angular force constants defined by

$$\gamma_{njk} = \frac{\partial W(\vec{R}_n^0,\vec{R}_j^0,\vec{R}_k^0)}{\partial(\cos\theta_{njk}^0)} \tag{3.9}$$

and

$$\eta_{njk} = \frac{\partial^2 W(\vec{R}_n^0,\vec{R}_j^0,\vec{R}_k^0)}{\partial(\cos\theta_{njk}^0)^2} . \tag{3.10}$$

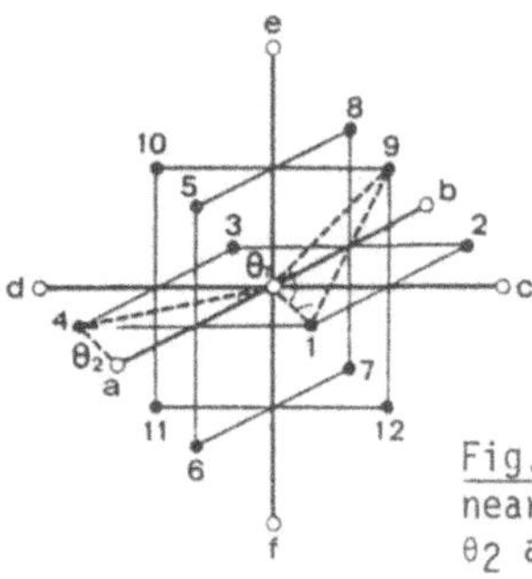

Fig. 2 First nearest neighbours (numbers) and second nearest neighbours (letters) in a fcc lattice. θ_1 and θ_2 are related to the angular forces.

Here θ°_{njk} is the angle between the vectors $\vec{R}^{\circ j}_n$ and $\vec{R}^{\circ k}_n$. In the previous expressions of the force [Eq.(3.1) and Eq.(3.4)], the sums run over all the atoms in the crystal. However the potential and the force constants decay rapidly with distance. For the case of Ni in which we are interested, a satisfactory parametrization can be achieved by retaining only second neighbour interactions. In Fig.2 are shown the positions of the first and second neighbours of fcc Ni together with the angles θ°_{njk} considered in the calculations. We note that for fcc structures the forces due to the γ_{njk} are zero for symmetry reasons. We have to determine six force constants. By imposing the equilibrium condition on the crystal we have $\alpha_1=-\alpha_2$ so that only five of them are independent.

The equations of motion written in terms of the polarization vectors $v_\beta(\vec{q},\omega)$ are

$$\sum_\beta \{ D_{\alpha\beta}(\vec{q}) - \omega^2(\vec{q})\delta_{\alpha\beta}\} v_\beta(\vec{q},\omega) = 0 \quad . \tag{3.11}$$

$D_{\alpha\beta}$ is the dynamical matrix related to the force constants previously introduced. The analytic expressions of the elements of the dynamical matrix are too long to be reported here but can be derived from a cumbersome calculation.

To determine the force constants we have written the expressions of the phonon frequencies along the symmetry lines of the Brillouin zone and we have performed a least squares fit of the experimental values |9|. With this procedure we have obtained for Ni the following values of the force constants:

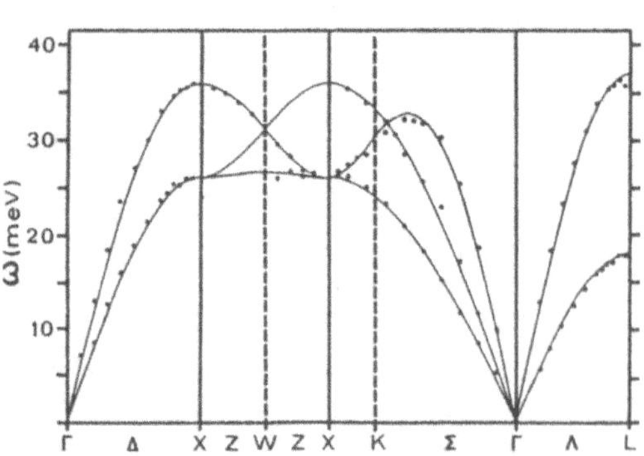

Fig. 3 Experimental (dots: from |9|) and theoretical (full line) phonon frequencies of nickel.

$$\alpha_1=-6.28\ 10^2\ \text{erg/cm}^2\ ;\quad \beta_1=3.99\ 10^4\ \text{erg/cm}^2\ ;\quad \eta_1=11.5\,10^{-14}\ \text{erg}$$

$$\alpha_2=6.28\ 10^2\ \text{erg/cm}^2\ ;\quad \beta_2=1.95\ 10^4\ \text{erg/cm}^2\ ;\quad \eta_2=-8.3\,10^{-12}\ \text{erg}\ . \tag{3.12}$$

The results of this fit are shown in Fig.3. The solid lines represent the calculations performed with these parameters. As one can see, this short range angular force model allows reproducing with good accuracy the bulk spectrum.

4. Surface Phonons

To evaluate surface phonons one should in principle deal with a semiinfinite medium. |10| This is actually true for long wavelength surface acoustic phonons |11| for which the penetration depth is very large. In this manner one avoids the interference effects occurring between the modes of the two surfaces present in a finite slab. The case of a semiinfinite crystal makes the theory very complicated. However we are interested here in long wavelength optical surface phonons only. In this case the surface vibrations are strongly localized at the surface so that the crystal can be treated as a slab |12| containing a limited number of atomic planes. Furthermore we remind that the observed (111) surface of Ni does not present any reconstruction or relaxation |13| . For this reason we can consider our slab as formed by a stacking of ideal (111) planes.

For this system the dynamical problem can be cast in the form

$$\sum_{l'_z}\sum_{\beta}\left[D_{\alpha\beta;l_z l'_z}(\vec{Q}_{\parallel})-\omega_j^2(\vec{Q}_{\parallel})\,\delta_{\alpha\beta}\,\delta_{l_z;l'_z}\right]w_\beta^j(\vec{Q}_{\parallel},l'_z,\omega_j)=0 \tag{4.1}$$

where l_z and l'_z label the N atomic planes, and α,β are the cartesian indices. The $\omega_j(Q_\parallel)$ are the frequencies of the j=1,..., 3N normal modes at fixed $\vec{Q}_\parallel$. The $w^j(Q_\parallel,\ l_z,\ \omega_j)$ are the polarization vectors of the slab modes.

The dynamical matrix is determined in terms of the angular model discussed in the previous section. In general the force constants in the surface region should be different from those of the bulk because of the different atomic charge distribution. This is particularly true for bcc metals and for semiconductors where one observes an electronic instability of the surface giving rise to strong effects of relaxation and reconstruction.|14| As already mentioned these effects are not present for Ni so that it seems reasonable to assume unmodified force constants also in the surface region. In the slab calculation we use this approximation and the coefficients of Eq.(3.12).

The 3N eigenvalues $\omega_j(\vec{Q}_\parallel)$, for $Q_\parallel$ along the symmetry lines of the two dimensional Brillouin zone (2DBZ) of the (111) surface of an fcc crystal, are sketched in Fig.4. There are frequencies associated with travelling modes which form a quasi continuum and frequencies associated with localized modes lying in energy gaps. If these gaps extend over the whole 2DBZ they are called absolute gaps, otherwise they are relative gaps. There are also regions forbidden to phonons of a given symmetry which form the symmetry gaps. The states decaying into the crystal are the surface states. Among them there is the Rayleigh wave whose frequency goes to zero in the $\bar{\Gamma}$ point. In the quasi continuum one finds states formed by travelling and localized waves, the so-called resonant modes. By increasing the number of planes, as one can see from Fig.4, the quasi continuum of states becomes a continuum, that is the bulk phonon spectrum projected on the surface. The penetration depth of the surface modes with a given $\vec{Q}_\parallel$ is related to the inverse of the

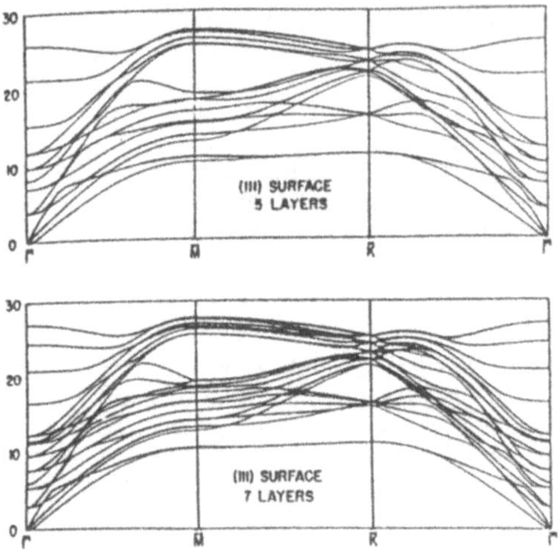

Fig. 4 Surface phonon frequencies for the (111) face of a model fcc crystal.

energy measured from the energy minimum (threshold) of the bulk branches having the same symmetry and the same $\vec{Q}_{\parallel}$. All the surface modes playing a role in the EELS experiments have frequencies lying far away from the respective thresholds so that they are strongly localized. This justifies the slab approach.

In our calculations we have used a slab of N = 21 ideal (111) planes which is quite sufficient to avoid interference effects between the two surfaces. The frequencies ω_i obtained by solving Eq.(4.1) are reported in Fig.5. There are various localized modes. The Rayleigh wave (S_1) is polarized in the sagittal plane defined by $\vec{Q}_{\parallel}$ and the normal to the surface. The mode S_2 in $\overline{M}$ shows the same kind of polarization. Close to the $\overline{K}$ point are present the modes: S_3 (shear horizontal), S_2 and S_4 (longitudinal). The S_5 mode is localized at the first internal plane in a small region aroung the $\overline{K}$ point and is polarized in the sagittal plane. The dashed curves are surface resonances having the same polarization as the surface states from which they originate. The resonance MS_3 occurs along the whole $\overline{T}$ direction and remains well localized on the surface.

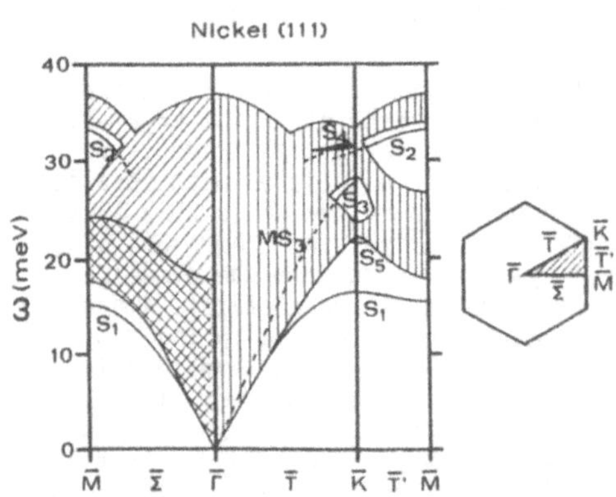

Fig. 5 Calculated phonon frequencies for a slab with 21 ideal (111) layers of Ni. The 2DBZ is also shown. The shaded regions represent the projected bulk phonons of different symmetry.

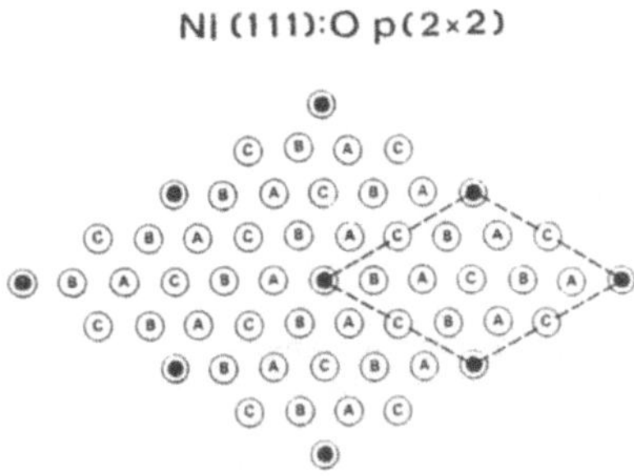

Fig. 6 Geometry of the Ni(111):O p(2x2) surface. Black dots represent the oxygen atoms. The atoms A refer to the surface of the substrate, the atoms B to the first layer and the atoms C to the second layer

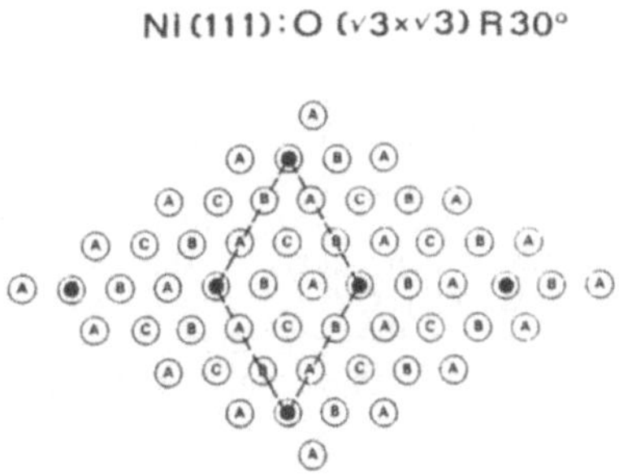

Fig. 7 The same as fig. 6 for the (√3x√3) R30° geometry.

5. Interpretation of the EELS Experiments

As previously discussed the main mechanism contributing to the EELS spectra is the surface dipole scattering. For metal surfaces one needs an adsorbate to produce such a dipole. With a fraction of monolayer of adsorbate the surface unit cell is enlarged and the corresponding 2DBZ reduced. In other words points at the border of the ideal 2DBZ are folded in the $\bar{\Gamma}$ point and can be seen with EELS. In fact this technique in specular reflection conditions gives information around the $\bar{\Gamma}$ point.

In the experiments performed by Ibach |6| for Ni(111), oxygen has been adsorbed in two different geometries |13| , namely the p(2x2) and the (√3x√3)R30° shown in Fig.6 and 7. In both cases each oxygen atom is bound to three Nickel atoms in a c_{3v} hole. The 2DBZ for the two geometries is shown in Fig.8. For the p(2x2) case the $\bar{M}$ point is folded in $\bar{\Gamma}$ while for (√3x√3)R30° the same happens for the $\bar{K}$ point. First of all we consider the experimental results |6| shown in Fig.9a relative to the p(2x2) geometry. Three peaks are present. The one at higher frequency which is far outside the phonon spectrum of Ni is clearly due to the motion of oxygen relative

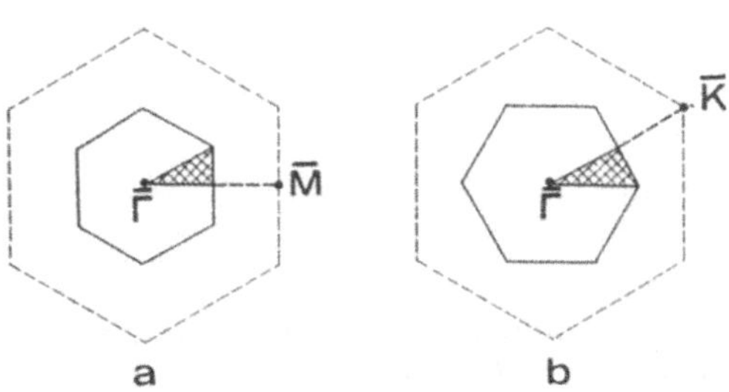

Fig. 8 Two-dimensional Brillouin zone for the (a) p(2x2) and (b) (√3x√3) R30° surfaces. The two-dimensional Brillouin zone of the ideal (111) surface is also shown (dashed lines).

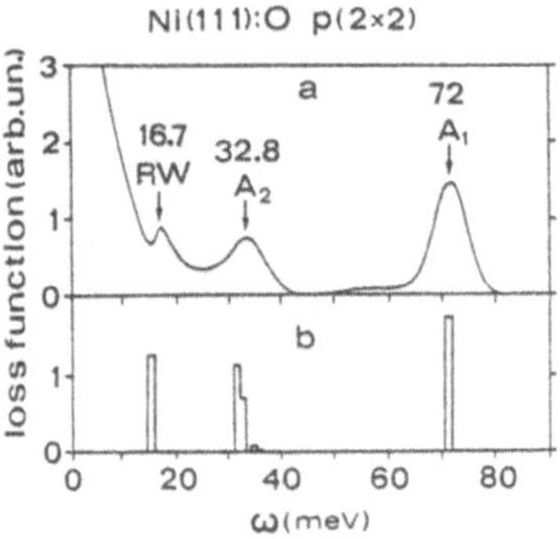

Fig. 9 Experimental energy loss spectrum for the Ni(111): O p(2x2) surface (a: from |6|) and calculated EELS spectrum from eq. (2.7) (b).

to the coordinated Ni atoms. The lowest peak can be interpreted in terms of the surface phonons of the clean surface. In fact close to this frequency there is the Rayleigh wave of the $\bar{M}$ point which is mainly polarized normal to the surface. The other peak at 30 meV inside the phonon spectrum of Ni cannot be explained in terms of the phonons of the clean surface. The surface modes of the $\bar{M}$ point near this frequency are longitudinal and do not couple with the impinging electron beam. It is hard to believe that the oxygen will be able to modify this polarization giving a transverse component to these modes.

To solve this problem we have to perform a full calculation of the phonon spectrum of the (111) surface of Ni with atoms of oxygen in the p(2x2) geometry. In our model |16| the oxygen interacts with the three nearest neighbour Ni atoms via a central Ni-O force and an angular Ni-O-Ni interaction. In order to determine the two force constants we consider the pyramidal molecule formed by the oxygen and the three coordinated Ni atoms. The dynamical problem can be solved by introducing the symmetry coordinates |17| shown in Fig.10. The two modes which transform according to the identity representation of the point group C_{3v} of the molecule have polarization given by linear combinations of the unit vectors depicted in Fig.10 for s_1 and s_2. These modes that we call A_1 and A_2 have a non-vanishing component of the Ni-O displacement normal to the plane of the Ni atoms. The s_{3a}, s_{3b}, s_{4a} and s_{4b}

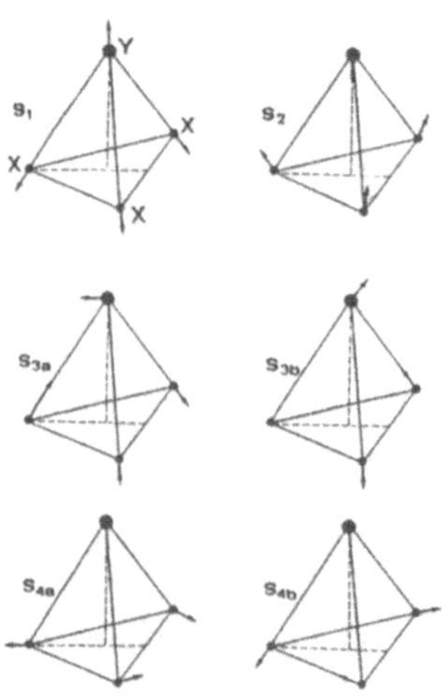

Fig. 10 Symmetry coordinates for the X_3Y pyramidal molecule.

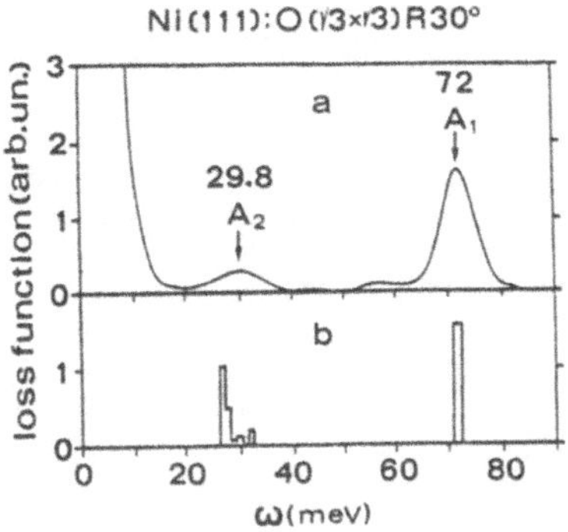

Fig. 11 Experimental energy loss spectrum for the Ni(111):O (√3x√3)R30° surface (a: from |6|), and (b) calculated EELS spectrum, from eq. (2.7).

transform according to a two-dimensional representation of C_{3v} and cannot mix with s_1 and s_2. The polarization of these modes is parallel to the Ni plane. By solving the dynamical problem for the A_1 and A_2 modes one finds that the frequency of A_1 is mainly due to the central Ni-O force constant, while that of A_2 depends strongly on the angular one. The frequency of A_1 turns out to be higher than that of A_2.

We fit the central force constant to the 72 meV experimental peak lying outside the frequency range of Ni. The angular force constant has been derived from the energy position of the loss peak seen at 30 meV in the experiments |6| performed with a fraction of disordered layer of oxygen on Ni(111). In this case the mean distance between oxygen atoms is large and their interaction is negligible. This situation is very close to that of the free pyramidal molecule.

With the parameters so derived we have evaluated the phonon spectrum of the covered surface. Our calculated curves for the normal component of the dipole moment at the surface, shown in Fig.9b, exhibit three structures. The lowest one corresponds to the Rayleigh wave of the ideal surface at the $\overline{M}$ point. The presence of the oxygen does not appreciably modify either the frequency and the polarization of this mode. The higher peak is the A_1 mode of the oxygen. The other peak is centered at 33 meV in good agreement with the position of the corresponding experimental loss. The evaluated polarizations are the same as those of the mode A_2 of the free molecule. We remark that in the calculated spectrum there is no evidence of the S_2 mode of the clean surface which is present in the $\overline{M}$ point at this frequency. Even with the presence of oxygen this mode maintains its longitudinal polarization. Furthermore the oxygen atom moves parallel to the surface so that there is no normal dipole moment.

A similar analysis has been carried out for the (√3x√3)R30° case. The experimental and calculated intensities are reported in Fig.11a and Fig.11b respectively. The same force constants derived before have been used. The two observed modes are related to the A_1 and A_2 modes of the molecule. In this

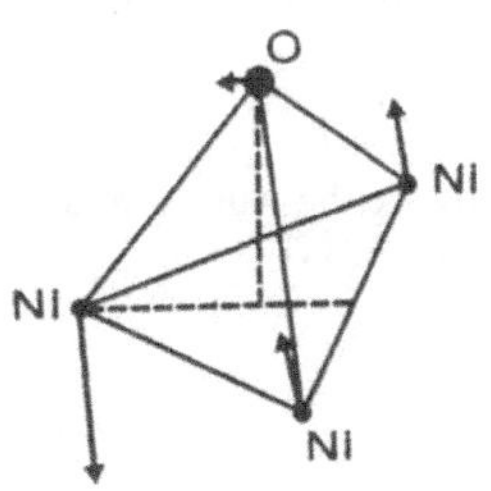

Fig. 12 Displacements of the oxygen and nickel atoms in the surface unit cell relative to the Rayleigh mode for the (√3x√3) R30° geometry.

case the Rayleigh wave does not appear in the spectra because the normal component of the surface dipole of the unit cell vanishes as shown in Fig.12. For the same reason also the S_2, S_3 and S_4 modes in $\bar{K}$ do not contribute.

We note that the evaluated frequencies of the A_2 peaks in the two considered geometries are slightly different each other in agreement with the experimental results. This shift can be understood by noticing that the indirect interaction between two oxygen atoms produced by the substrate is different for the two geometries.

In conclusion we have shown that a realistic calculation allows one to explain the loss spectra and to carefully identify the origin of the experimental structures in terms of either the modes of the substrate or the normal modes of the adsorbed species.

References

1. J.R. Sandercock, Sol. St. Comm. 26, 547 (1978);
   A. S. Pine, in Light Scattering in Solids, ed. by M. Cardona, Topics in Applied Physics, Vol. 8 (Springer, Berlin, Heidelberg, New York 1975), p.254; A. Marvin, V. Bortolani, F. Nizzoli and G. Santoro, J. Phys. C13, 1607 (1980).
2. W.L. Rowell and G.J. Stegeman, Phys. Rev. B18, 2698 (1978);
   J.R. Sandercock, F. Nizzoli, V. Bortolani, G. Santoro and A. Marvin, Phys. Rev. Lett. 43, 224 (1979).
3. G.Brusdeylins, B.R. Doak and J.P. Toennies, Phys. Rev. Lett. 46, 437 (1981); P. Cantini, R. Tatarek and G.P. Felcher, Phys. Rev. Lett. 37, 606 (1976).; B. Feuerbacher, to be published.
4. G. Benedek and N. Garcia, Surf. Sci. 80, 543 (1979).
5. H. Ibach, H. Höpster and B. Sexton, Appl. Surf. Sci. 1, 1 (1977).
6. H. Ibach and D. Bruchmann, Phys. Rev. Lett. 44, 36 (1980).
7. S. Anderson, Surf. Sci. 79, 385 (1979).
8. E. Evans and D.L. Mills, Phys. Rev. B5, 4126 (1972);
   A.A. Lucas and M. Sunjic, Phys. Rev. Lett. 26, 229 (1971);
   H.A.Pearce and N. Sheppard, Surf. Sci. 59, 205 (1976).
9. R.J. Birgenau, J. Cordes, S. Dolling and D.B. Woods, Phys. Rev. 136, 1359 (1964).
10. T.E. Feuchtwang, Phys. Rev. 155, 731 (1967).
11. V. Bortolani, F. Nizzoli, G. Santoro and E. Tosatti, Sol. St. Comm. 26, 507 (1978).
12. R.E. Allen, G.P. Alldredge and F.W. DeWette, Phys. Rev. B4, 1661 (1971).
13. J.E. Demuth and T.N. Rhodin, Surf. Sci. 45, 249 (1974).
14. M.K. Debe and D.A. King, Surf. Sci. 81, 193 (1979);
    P.J. Estrup, J. Vac. Sci. Technol. 16, 635 (1979);
    A. Fasolino, G. Santoro and E. Tosatti, in Modern Trends in the Theory of Condensed Matter. Proc. of the Karpacz Winter School of Theoretical Physics, ed. by A. Pekalski and J. Prystawa, Lecture Notes in Physics, Vol. 115 (Springer, Berlin, Heidelberg, New York 1980).
15. L.D. Landau and E. Lifchitz, Course of Theoretical Physics, Vol. 7 Elasticity Theory, 2nd ed. (Pergamon, New York 1970).
16. V. Bortolani, A. Franchini, F. Nizzoli and G. Santoro, to be published.
17. G. Herzberg, Molecular Spectra and Molecular Structure, vol.2 (Van Nostrand Inc., New York 1945) p. 155.

Part IV

Surface Phonon Spectroscopy by Atom Scattering

Phonon Interactions in Atom Scattering from Surfaces

J.P. Toennies

Max-Planck-Institut für Strömungsforschung, Böttingerstr. 4-8
D-3400 Göttingen, Fed. Rep. of Germany

Abstract

In the past very little was known about the dispersion curves of surface phonons and how they interact with gas atoms colliding with the surface. Recent molecular beam scattering experiments now provide the first detailed information on surface phonon dispersion curves and on the inelastic cross sections. In this brief review we first summarize some results from the theory of surface phonons and of inelastic cross sections. Next we describe the helium time of flight apparatus used to study the surface phonon dispersion curves of the alkali halides. Results are presented for LiF, NaF and KCl. Finally the influence of selective adsorption on the inelastic intensities is discussed.

1. Introduction

Figure 1 summarizes the various scattering processes, which can occur when an atom strikes a single crystal surface. Since typical light atoms and molecules (e.g. H_2, D_2, He, Ne) have DE BROGLIE wavelengths of the order of 1 Å, the scattering is dominated by the specular and diffraction peaks. These are adequately described in the rigid lattice approximation with only an intensity correction for inelastic effects, provided by the simple temperature

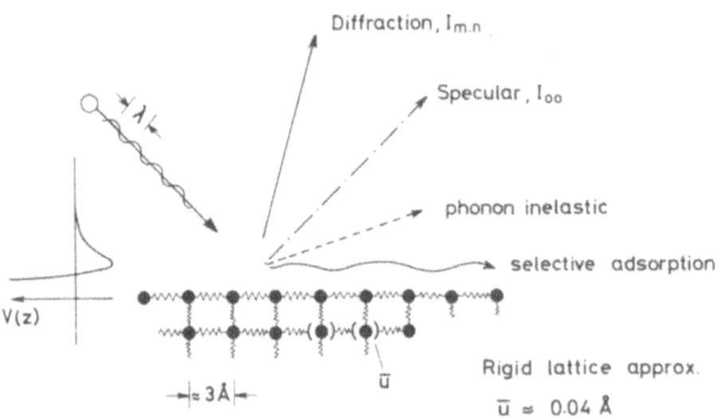

Fig. 1 Diagram showing the different collision processes, which can occur in the non-reactive scattering of a light atom with a DE BROGLIE wavelength comparable to the lattice dimensions. Since the lattice vibrational amplitudes are small, phonon inelastic scattering is expected to be improbable relative to elastic diffractive scattering.

dependent DEBYE-WALLER factor. The elastic resonance trapping of atoms in bound states on the surface, which is called selective adsorption, can also extensively modify these intensities at special angles and energies [1].

Inelastic cross sections, resulting from the interaction of the impinging atoms with the surface phonons and bulk phonons near the surface, have a relative probability of only about 10^{-3}. They provide a sensitive probe of the potential, which couples the motion of the atom with that of the lattice vibrations $V(\vec{R}, z, \vec{u})$, where $\vec{R}$, a vector along the surface, and z normal to the surface describe the locations of the surface atom, and $\vec{u}$ is the vibrational amplitude of a surface atom. Expanding we obtain

$$V(\vec{R}, z, \vec{u}) = V(\vec{R}, z)/_{u=0} + \frac{dV(R,z)}{du}/_{u=0}\, u + \dots .$$

The first term on the right is the corrugated elastic potential, which can be determined from the intensities of the elastic diffraction peaks. The second and higher order terms, which couple to the vibrations, can only be obtained from inelastic scattering measurements. An understanding of these coupling potential terms is fundamental for understanding the following important dissipative phenomena:

accommodation coefficients
sticking coefficients
transfer of energy from excited molecules near the surface of the crystal
transfer of energy from one site on the surface to other sites.

A complete knowledge of these elementary processes is of great importance for understanding catalysis. Another reason for studying inelastic atom scattering is that it provides a method for determining surface phonon dispersion relationships and consequently a test of theories of dynamical lattice properties of surfaces.

2. Review of Theoretical Studies of Surface Phonons

First recall that the bulk vibrations of an infinite crystal of an alkali halide (s=2) can be described by four dispersion curves describing the non-linear relationship between the frequency and wave vector of the 3s=6 normal modes of the phonons. As illustrated for the case of LiF in Figure 2a, the lowest two curves start at the origin and are due to the doubly degenerate transverse acoustical modes, which have the lowest frequencies, as well as the singly degenerate longitudinal acoustical mode of slightly higher frequency. The three optical modes have nearly flat dispersion curves with higher frequencies, which are finite at zero wave vector. The transverse optical mode is again doubly degenerate. The vibrational motions for the acoustical and optical branches are shown in Fig.2c. The dispersion relationships for the bulk shown in Fig.2a have been measured using neutron inelastic scattering [2] as indicated by the points in the curves. The solid curves are actually the results of a dynamical calculation. The agreement is very good.

In order to produce a surface, we have to cut the infinitely extended crystal. Because of the new boundary conditions, this will lead to a severe modification of the phonons. Essentially two new effects arise: (1) new phonon modes, which are localized at the surface, are added, and (2) the bulk modes will appear different at the surface since we will "see" only the projections of their wave vectors on the surface plane. We consider these two effects separately.

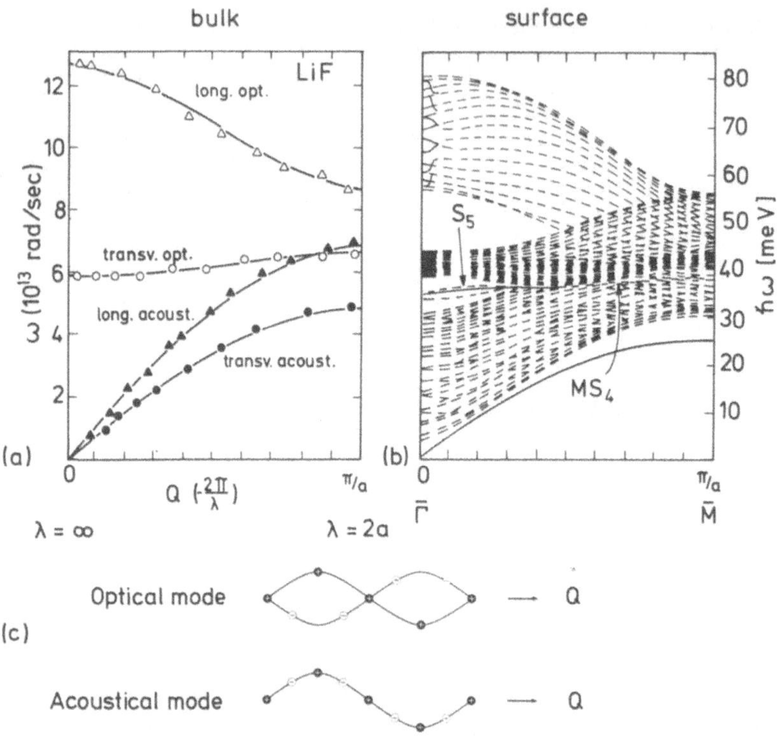

Fig. 2 Calculated dispersion curves for one direction (a) in the bulk and (b) on the surface (<100> direction) of a LiF crystal [2]. The points in (a) are from inelastic neutron scattering. In (b) we observe two types of vibrations: bulk modes projected onto the surface (dashed lines) and surface localized modes (solid lines). In (c) we show the motions in the optical and acoustical modes of two ions of equal mass.

The presence of vibrations localized at the surface was first pointed out by Lord RAYLEIGH on the basis of continuum theory [3]. The motions of the lattice atoms in RAYLEIGH waves are shown in Fig.3. Here the particles move in a saggital plane (SP), defined by the direction of motion and the surface normal. The deformation field extends into the crystal a distance proportional to λ the wavelength, and is composed of a longitudinal and vertical component, corresponding to compressional and shear distortions, whose amplitudes, as a function of reduced depth z/λ , are shown in Fig.3b. In general, RAYLEIGH waves have phase velocities (= ω/Q), which are less than the lowest bulk mode. The continuum theory is useful for classifying the different types of modes possible along different directions of a given crystal. Thus we find that for LiF there will not be a SP-RAYLEIGH wave along the <110> azimuthal direction, whereas NaF and KCl are expected to have SP-RAYLEIGH waves in all directions [4].

Only a theory which considers the individual atoms can account for dispersion. The result of such a calculation, based on the solution of the coupled motion of 15 slabs of a two-dimensional infinitely extended LiF crystal [5], can be compared with the bulk dispersion in Fig.2. The RAYLEIGH mode is clearly apparent below the lowest bulk mode. Striking is the appearance

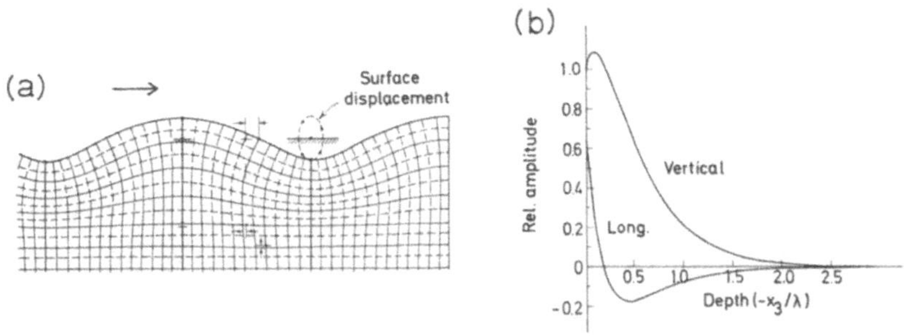

Fig. 3 Continuum theory motion of a lattice in a sagittal plane RAYLEIGH wave. In (a) the distortion of the crystal is shown, and in (b) the relative amplitudes of the vertical and longitudinal parts of the wave are shown.

of a broad band of modes above the lowest bulk modes. They appear as discrete curves since only a finite number of slabs were used, and in fact there is a near continuum of states for macroscopic crystals. If we carefully compare Fig.2a and 2b we see that these bulk bands can be considered to arise from the transverse acoustical mode in the bulk, which is projected onto the surface $\vec{Q}_{surface} = \vec{Q}_{bulk} \cdot \sin\Theta$, where Θ is the angle between $\vec{Q}_{bulk}$ and z. These are bulk modes at the surface, which, however, are not localized at the surface as are the RAYLEIGH waves. The same holds for the optical modes. Additional surface localized modes appear both in the gaps between the bulk modes projected on the surface and also in the bulk bands. These are shown in more detail for the perimeter of the irreducible section of the BRILLOUIN zone in Fig.4a [5]. The decay of the surface mode amplitudes with penetration into the surface is illustrated for the $\overline{X}$ point for NaCl in Fig. 4b. Fig.4c shows the surface BRILLOUIN zone for a rock salt crystal (001) face. One feature is of particular interest in Fig.4a. In going from the $\overline{M}$ point to the $\overline{X}$ point we note that at 30° there is an avoided crossing between the S_1 and S_7 curves. At this point the SP-RAYLEIGH wave switches from S_1 to S_7 and becomes inbedded in the bulk bands, while the S_1 mode, which is a shear horizontal (SH) mode, takes its place in the $\overline{\Gamma}$ - $\overline{X}$ direction. This is in agreement with the predictions of the continuum theory noted above. Other details on the many different types of surface modes, some of which are seen at higher frequencies in Fig.4, and their calculation by a different GREEN's function method, are discussed in the article by BENEDEK [6]. Finally we note that most calculations of surface vibrations have neglected relaxation, which refers to the small changes in lattice distances at the surface. For LiF these amount to at most 5% [7].

3. Kinematics of Inelastic Surface Scattering

These theoretically predicted dispersion curves present a challenge to the experimentalist. Obviously a scattering experiment is called for which is able to probe the region of large wave vectors. Neutrons are not suitable since they are insensitive to the surface. Information on the phonon density of states can nevertheless be obtained from neutron diffraction by comparing the inelastic scattering from microcrystals with that from powders with a much smaller surface area [8]. Photons, because of their large velocity, are only sensitive to surface waves of very large wavelengths [9]. Electrons can yield some information on optical surface phonons [10]. They have the disadvantage that they also penetrate deeply into the lattice and cannot effectively probe

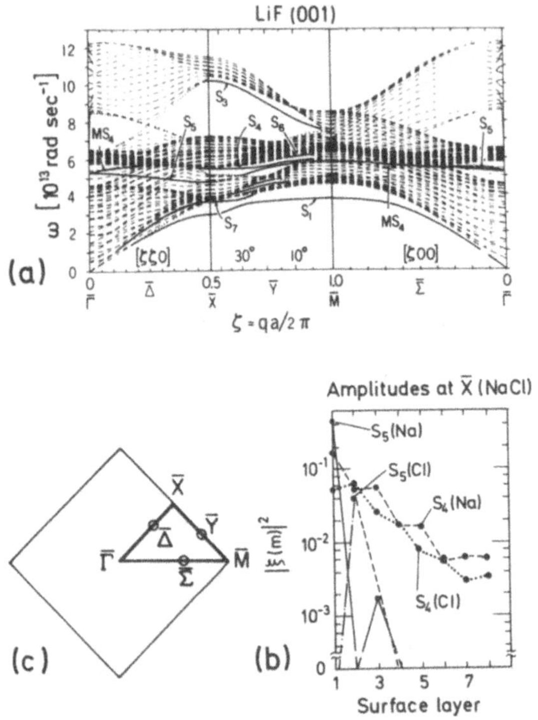

Fig. 4 Calculated surface phonon modes for a LiF surface [5]. In (a) the dispersion curves are shown for different directions on the surface. The part of the dispersion curve from $\bar{X}$ to $\bar{M}$ shows the frequencies at the edge of the BRILLOUIN zone boundary as indicated in (c). Part (b) shows the amplitudes at $\bar{X}$ of some surface optical modes imbedded in the continuum of bulk modes as calculated for NaCl.

the dispersion curves. Helium atoms on the other hand are ideal for many reasons as we shall see below. For now it is sufficient to note that helium atoms have similar energies and wavelengths as neutrons, but probe only the uppermost layer! This fact simplifies the kinematics of He surface scattering since the particle momentum along the surface is conserved. Fig.5 shows the usual geometry and definition of angles used in surface scattering, the conservation of momentum and energy equations and a typical EWALD construction. Note that these equations assume single phonon interactions. If bulk phonons are excited, then only their momentum projected onto the surface is measured. The EWALD diagram provides a "graphical" solution of the conservation equations. In Fig.5b we have only shown a solution for elastic scattering ($\omega = 0$, $\vec{Q} = 0$) which yields the angular directions for the specular and diffraction peaks. Consequently we expect inelastic events to be observable in the angular regions between the diffraction peaks.

Fig.6 shows a side view of the EWALD diagram for inelastic scattering along the <100> direction. In the example the incident wave vector k_i is 6 $Å^{-1}$, and the angle between incident and scattered beams is $\Theta_{SD} = 90°$ and $\Theta_i = 64.2°$. The detector thus sees all particles scattered in a direction $\Theta_f = 25.8°$. The point (1) indicates one of many possible inelastic events, in which a phonon has been *annihilated* and the atom has gained energy. The phonon momentum transfer $\vec{Q}$ to the atom is in the *forward* direction. (6) indicates another possibility, in which a phonon has been *created* and the atom has lost energy. The momentum transfer $\vec{Q}$ is in the *backward* direction. Note that $\vec{Q}$ is always measured to the nearest reciprocal lattice vector. Thus in the example of Fig.6 we would speak of phonons associated with the $(\bar{1},\bar{1})$ reciprocal lattice vector. Thus for given scattering angles Θ_i and

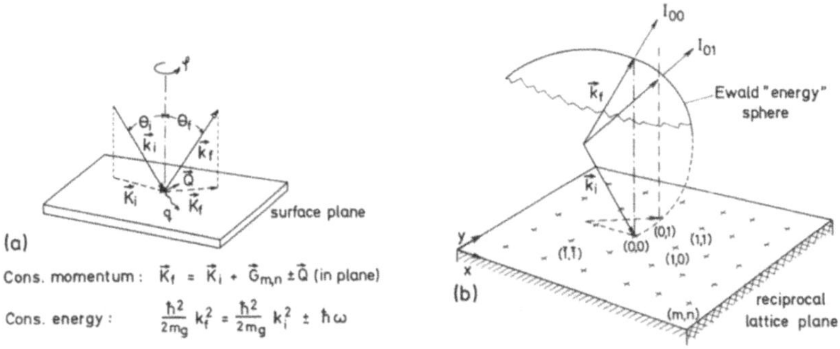

Fig. 5 In (a) the usual definition of angles and particle momenta and the conservation equations used in single phonon inelastic scattering are shown. In (b) the surface EWALD diagram is shown for the case of elastic scattering. Note that elastic events are limited to specific discrete directions indicated, for example, by I_{00} (specular) and I_{01} (a possible diffraction peak).

Θ_f there is a set of values $\{\omega,\vec{Q}\}_{app}$, seen by the detector. Only those events will be registered corresponding to $\{\omega,\vec{Q}\}_{app} \cap \{\omega,\vec{Q}\}_{surface\ phonons}$. To separate these different contributions, we must measure one additional quantity such as the energy of the scattered particles. This is most conveniently done by measuring the time of flight distribution.

A more useful representation of the EWALD diagram of Fig.6 is to project the $\{\omega,\vec{Q}\}_{app}$ onto an extended zone diagram containing the $\{\omega,\vec{Q}\}_{surface\ phonons}$. This is shown for many different Θ_i in Fig.7 for the geometry, shown in Fig.6 and assuming only RAYLEIGH phonons [11]. We call $\{\omega,\vec{Q}\}_{app}$ the scan curve. Thus in a time of flight spectrum we expect to observe maxima corresponding to the intersections of the scan curves with the dispersion curves. From the difference in energy, compared to the elastic peak, we can determine the $\hbar\omega$ of the phonon involved. From the scan curve we can then determine the wave vector of the surface phonon. Alternatively it is easy to derive from the conservation equations the following equation relating the total momentum transfer $\Delta\vec{K} = \vec{G} \pm \vec{Q}$ with the experimental quantities $h\omega$ and Θ_f for in-plane scattering [12]:

$$\frac{\hbar\omega}{E_i} = -1 + \left(1 + \frac{\Delta K}{K}\right)^2 \frac{\sin^2\Theta_i}{\sin^2\Theta_f} . \qquad (1)$$

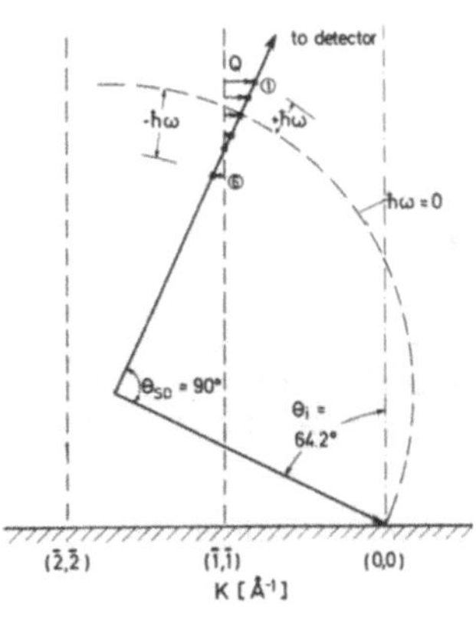

Fig. 6 Side view of an in-plane EWALD diagram along the <100> direction for an experiment, where the angle between incident and scattered beams $\Theta_{SD} = 90°$ and $\Theta_i = 64.2°$. The points ①...④, ⑥ indicate different inelastic events observable at this angle. Their resolution requires a measurement of the final velocity distribution.

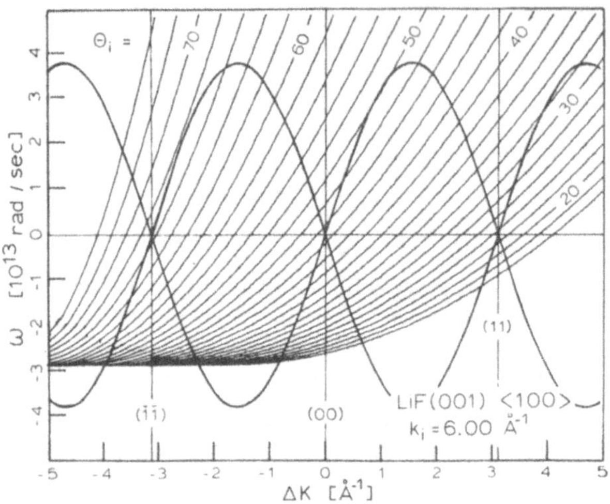

Fig. 7 Extended zone diagram showing scan curves for various Θ_i and for $\Theta_{SD} = 90°$ and in-plane scattering along the <100> direction with $k_i = 6.00$ Å^{-1} [11]. The heavy solid lines show the RAYLEIGH dispersion curves in the sine approximation. $\Delta K = -3.12$, 0 and +3.12 correspond to $(\bar{1},\bar{1})$ diffraction, specular scattering and (1,1) diffraction, respectively.

4. Dynamical Theory of Inelastic Scattering

Before describing the experiments, it is useful to review briefly what factors are thought to affect the intensity of the various possible inelastic events. Several theoretical estimates have been presented concerning the relative contributions of single phonon to two-phonon events. As is well known from neutron scattering, two-phonon events do not lead to sharp peaks in the energy loss spectra, but only contribute to a broadly smeared-out background. WEARE [13] has suggested the following criterium for a predominance of one-phonon processes

$$\frac{M_{beam}}{M_{surface}} \frac{E_{iz}\, T_s}{k_B\, \Theta_D^2} < 0.01 \quad , \tag{2}$$

where M_{beam} and $M_{surface}$ are the mass of the beam projectile and of the surface atom, respectively. E_{iz} is the normal energy component, T_s the surface temperature, k_B the BOLTZMANN constant, and Θ_D the surface DEBYE temperature. According to Equation (2), light atoms and low beam energies and surface temperatures are to be preferred in order to reduce the occurance of multiple phonon events. MEYER [14] has recently used a fairly realistic model to predict the contribution of two-phonon scattering to energy change spectra.

If only single phonons are involved, then the differential reflection coefficient is given by

$$\frac{d^2R}{d\omega\, d\Omega} = \frac{(\Delta k_z)^2}{2\Pi^3} \frac{k_f}{|k_{iz}|} |n^{\pm}|^2 \cdot \qquad (3)$$

$$g(\vec{K} - \vec{G}) \cdot \rho(\vec{Q}, \omega) \quad ,$$

where Δk_z is the change in the normal momentum, and k_{iz} is the initial normal momentum. The factor $n^{\pm}$ is a BOSE factor and is discussed below; $g(\vec{K} - \vec{G})$ is the coupling coefficient and is related to the T matrix. We will not discuss this factor further, but only note that it is proportional to the elastic scattering diffraction intensity $|A_G|^2$ of the "nearest" diffraction peak. $\rho(\vec{Q}, \omega)$ is the surface projected density of states, which is defined in the article by BENEDEK [6]. This formula, which was derived by LEVI [14] using a distorted wave BORN approximation, has been used by BENEDEK and GARCIA [16] to calculate the measured time of flight spectra described in the next section.

Fig.8 shows the surface projected density of states $\rho(\vec{Q}, \omega)$ of LiF <100> for SP polarization for 5 different values of $\vec{Q}$, as calculated by BENEDEK [17]. The surface modes show up as sharp spikes, whereas the bulk projected modes lead to broad bands.

The BOSE factor $n^{\pm}$ takes account of both the probability P_n of finding a phonon oscillator in state n, where

$$P_n = \frac{\exp(-(n+1/2)\frac{\hbar\omega}{k_BT})}{\sum_{m=0}^{\infty} \exp(-(m+1/2)\frac{\hbar\omega}{k_BT})} \quad , \qquad (4)$$

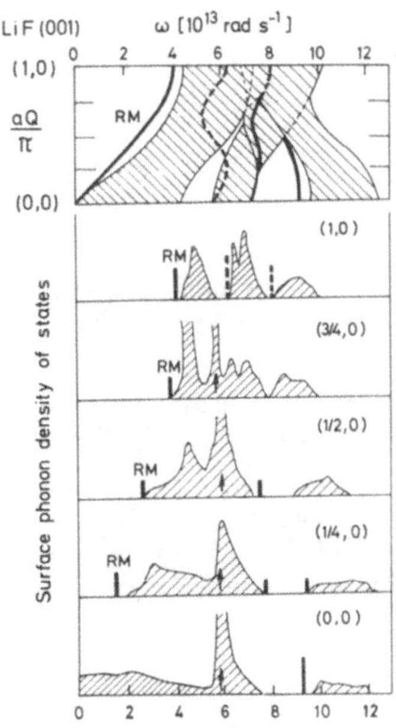

Fig. 8 Calculated surface projected phonon density of states of sagittal plane vibrations along the <100> direction of LiF [17]. The top part shows the dispersion diagram on its side. At the bottom the densities are shown for the end points and three intermediate values of aQ/π. The arrows indicate regions where a mixed surface-bulk mode (dashed line at top) leads to an enhancement in the density of states.

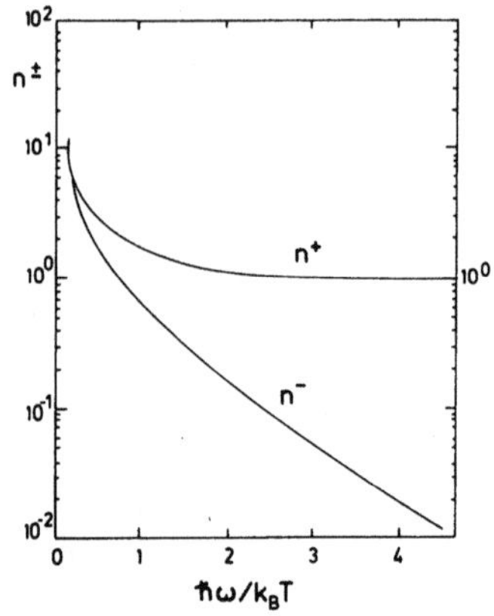

Fig. 9 Calculated BOSE factors for phonon creation n^+ and phonon annihilation n^-.

and the fact that for harmonic oscillators the upward (creation) transition probabilities for given initial state n are related to the $n = 0 \to 1$ transition probability σ_0 by

$$\sigma^+_{n\to n+1} = (n+1)\,\sigma_0 \quad , \tag{5a}$$

while the downward (annihilation) transition probabilities go as

$$\sigma^-_{n\to n-1} = n \cdot \sigma_0 \quad . \tag{5b}$$

Thus the averaged probability for phonon creations (index +) and annihilations (index -) is given by

$$n^\pm = \sum_{n=0}^{\infty} P_n \; \sigma_n^\pm \tag{6}$$

$$= \frac{1}{2}\,\sigma_0 [\coth \frac{\hbar\omega}{2k_BT} \pm 1] \; .$$

Fig.9 shows n^+ and n^- as a function of $\hbar\omega/k_BT$. At low temperatures, creation processes will predominate, while at temperatures for which $\hbar\omega/k_BT < 1$ both processes will become equally probable.

We conclude the first part of this lecture by noting that experimental results are needed to provide answers to the following questions: (1) How accurate are calculated surface phonon dispersion curves? What is the effect of surface relaxation and which theory is best? (2) What is the relative importance of surface and bulk phonons in energy loss and gain in gas surface collisions? Can optical modes be excited? (3) What is the relative importance of single and multiple phonon processes?

5. Inelastic Scattering Experiments

The experimental study of surface phonons has a long history, which is summarized in Table 1. The early experiments have provided valuable information on the relative role of single and multiple events and have suggested that RAYLEIGH modes are probably more important than bulk modes. However, all of the experiments have been hampered by insufficient velocity resolution and insufficient sensitivity to detect the signals from phonons at small wavelengths near the zone boundary.

These problems have been solved in the apparatus used by BRUSDEYLINS, DOAK and TOENNIES [11,25,30], which is shown schematically and to scale in Fig.10. The apparatus has a fixed angle of 90° between incident and scattered beams. This apparatus has two improvements over previous machines: (1) High stagnation pressures (200 atm) and small nozzles (5 microns) have been used to obtain nozzle beams with very high speed ratios S of about 200 . The relative velocity half-width is given by $\Delta v/V = 1.65/S$ and therefore is only 0.8% [31]. By using a small nozzle the gas load in the first stage is similar to other apparatus. The beam intensity is limited only by the size of the diffusion pump in this stage. In our apparatus a pump capable of handling a gas throughput of 15 Torr liters/s was used. (2) Extensive differential pumping is provided to reduce the helium pressure, which is nearly 10^{-3} Torr in the source chamber, to about 10^{-13} - 10^{-14} Torr helium partial pressure in the detector. Altogether 8 pumping stages are located between source and detector chambers. In this way it was possible to detect much more feeble signals, amounting to a beam pressure of only 10^{-14} Torr, than in conventional machines, where the detector is in the target chamber, which usually has a base pressure of 10^{-10} Torr. Of course, a greater detector sensitivity is desirable to compensate for the much longer flight path between target and detector, which is required to resolve the small energy changes.

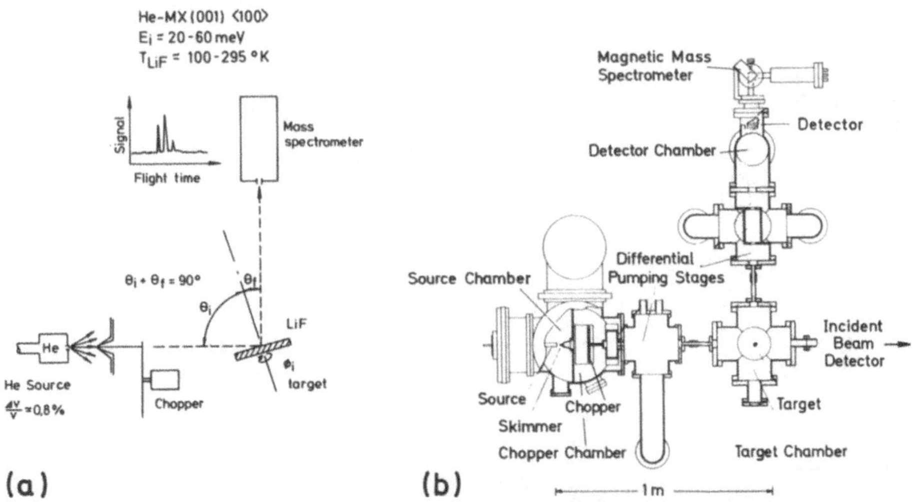

Fig. 10 Part (a) shows a schematic diagram of the apparatus of BRUSDEYLINS, DOAK and TOENNIES [25,30]. Part (b) shows a cross-section through the vacuum chambers used in the actual apparatus, which contains altogether 8 pumping stages between source and detector chambers.

Table 1 Low energy molecular beam inelastic scattering experiments

Authors and recent reference	atom	surface	technique
Subbarao, Yerkes and Miller (1969, 1972,1980) [18]	He -	Ag(111)	TOF
Mason and Williams (1971,1972,1974) [19]	He -	LiF(100) NaF(100)	tilting crystal
Fisher and Bledsoe (1972) [20]	He -	LiF(100)	TOF
Lapujoulade and Lejay (1975) [21]	He, Ne - Ar	Cu(100)	TOF
Cantini, Tatarek and Felcher (1976,1977) [22]	He, Ne -	LiF(100)	selective adsorpt.minima in tails of specular peak
Horne and Miller (1978) [12]	He -	LiF(100)	TOF
Feuerbacher, Adriaens and Thuis (1980) [23]	He -	LiF(100)	TOF pulsed beams
Allison and Feuerbacher (1980) [24]	H_2,D_2 -	LiF(100)	TOF pulsed beams
Brusdeylins, Doak and Toennies (1980) [25]	He -	LiF(100)	TOF $\Theta_{SD} = 90°$
Semerad and Hörl (1981) [26]	Ne -	LiF(100)	TOF $\Theta_{SD} = 90°$
Valbusa et al. (1981) [27]	Ne -	LiF(100)	TOF
Cantini and Tatarek (1981) [28]	He -	graphite	selective adsorption
Mason and Williams (1981) [29]	He - and Xe on	Cu(100) Cu	energy analyzer

beam energy	resolution Δv/v	ΔΘ	comments
23 meV	5%	-	evidence for RAYLEIGH modes
≈ 60 meV	7%	0.6°	see structures in angular distributions assigned to RAYLEIGH waves
58 meV	5%	±1.5°	evidence for inelastic scattering with $\overline{Q} = 0$
65 meV	5-10%	2°	evidence for one-phonon scattering with He, many phonons with Ne and Ar
60 meV	-	-	evidence for RAYLEIGH mode and some bulk contributions
63 meV	15%	-	evidence for RAYLEIGH mode with small Q
60 meV	10%	-	evidence for RAYLEIGH modes
60 meV	10%	-	see energy changes resulting from combined rotational and phonon excitation
20 meV	0.8%	0.1°	resolved RAYLEIGH modes out to zero boundary
60 meV	5%	0.3°	with Ne large multi-phonon contributions are affected by selective adsorption
60 meV	5%	0.6°	large multi-phonon contributions
43 meV	-	-	rough agreement with bulk TA mode
22.6 meV	3%	-	see RAYLEIGH mode and multi-phonon excitation of adsorbed Xe

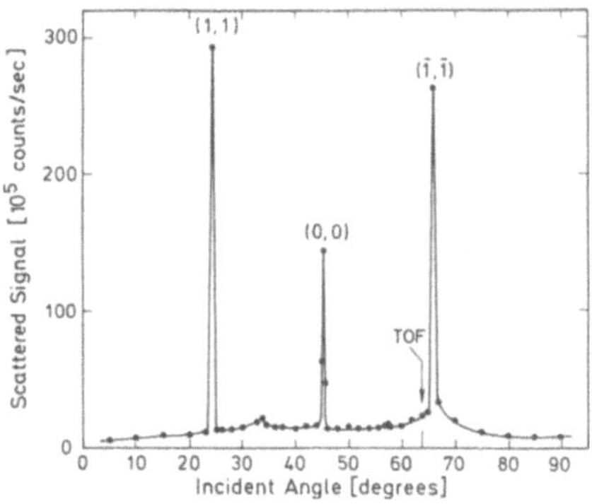

Fig. 11 Measured scattered intensity as a function of incident angle for scattering of He from LiF <100> in the apparatus shown in Fig.10. The beam energy was about 19 meV and $k_i \simeq 6.00$ Å^{-1}.

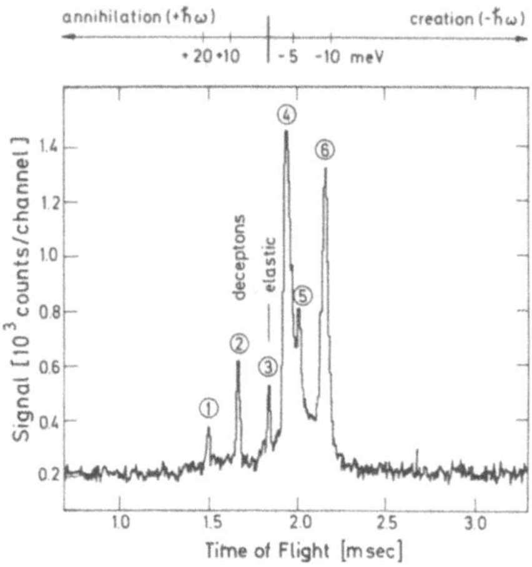

Fig. 12 Measured time of flight spectrum for the same conditions as in Fig.11, but at $\Theta_i = 64.2°$. From the scale at the top the corresponding energy changes can be estimated. The assignment of the six maxima are discussed in the text and illustrated in Fig.13.

Despite the restriction to a fixed angle between incident and scattered beams, the apparatus is capable of seeing all the diffraction peaks by rotating the crystal. This is illustrated in Fig.11, where we show the scattered intensity as a function of the angle of rotation of the crystal for LiF (001) <100> . The predominance of diffraction over the specular peak is not unexpected at this Θ_{SD} angle [32]. Note that there is a hint of structure between the specular and diffraction peaks. The origin of this structure is discussed below.

Recalling Fig.7 it is apparent that the fixed angle apparatus also is able to probe the RAYLEIGH phonons over the entire range of frequencies and over a wide range of momentum transfer. Fig.12 shows a time of flight spectrum taken at 64° corresponding to the EWALD diagram of Fig.6. From Fig.11 we see that this angle is very close to the $(\bar{1},\bar{1})$ diffraction peak. In Fig.12 we see that altogether 6 maxima have been resolved in the time of flight spectrum. By referring to the appropriate scan curve shown in Fig.7, also shown in Fig.13, we can assign these peaks in the following way: peak ① is due to annihilation of a RAYLEIGH mode near the zone boundary, peak ② is due to a "decepton", which is our designation for peaks arising from diffraction of those beam atoms whose velocities lie within the low intensity (10^{-3} of peak) tails of the incident beam velocity distribution [33]. These appear as sharp peaks since (as with a crystal monochromator) the detector acceptance angle projects out only a narrow range of velocities faster than the primary beam. This weak component is difficult to detect otherwise since in the direct beam the background from the major component would overwhelm this weak signal. Similar spurious peaks are observed in neutron scattering, where they are referred to as spurions. Peak (3) is due to a small amount of incoherent elastic scattering from crystal imperfections. It serves as a useful fiduciary mark. Peak ④ is due to the creation of a RAYLEIGH phonon

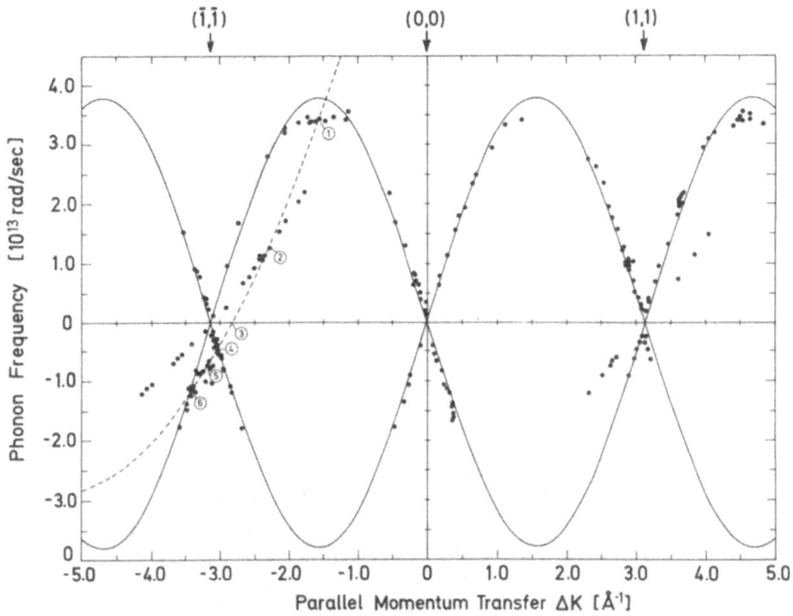

Fig. 13 The scan curve for Θ_i = 64.2° and the assignment of the different points of the time of flight spectrum are shown in an extended zone plot for the <100> direction of LiF. Also shown are the results of all measured time of flight spectra, which can be compared with the sine approximation for the RAYLEIGH dispersion curves.

with forward momentum; peak ⑤ is due to the creation of a bulk phonon with nearly zero tangential momentum; and finally, peak ⑥ is due to creation of a RAYLEIGH phonon with backward momentum.

Over 50 time of flight spectra were measured [11,25] for this crystal direction and the measured peak locations are all shown in an extended zone plot in Fig.13. Here we see that the deceptons appear near the $(\bar{1},\bar{1})$ and (1,1) peaks but not near the specular peaks, confirming our interpretation in terms of diffraction of a broad velocity component.

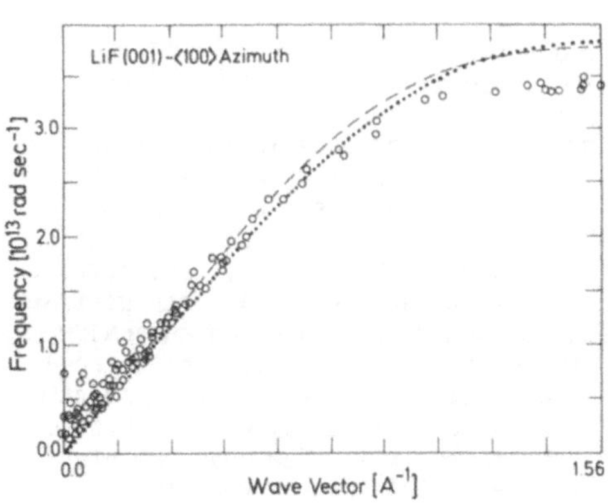

Fig. 14 Reduced zone plot showing all measurements without decepton points for LiF <100>. The dashed line shows the theoretical curves of CHEN, ALLDREDGE and DE WETTE [5] and the dotted line shows the theoretical curve of BENEDEK [17]. The agreement is very good, except near the zone boundary.

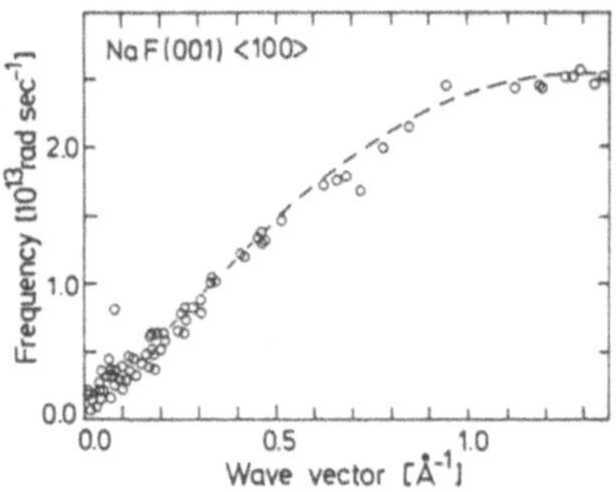

Fig. 15 Reduced zone plot showing all measurements without decepton points for NaF <100>. The dashed line shows the theoretical curves of CHEN, ALLDREDGE and DE WETTE [5].

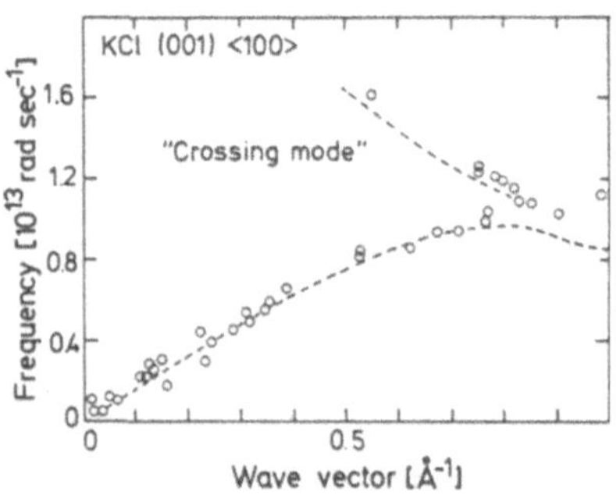

Fig. 16 Reduced zone plot showing all measurements for KCl <100>. The results show additional surface modes in the bulk continuum, which were predicted by BENEDEK and GALIMBERTI [35] to be due to a crossing mode (see text).

Fig.14 summarizes the results for this crystal orientation in a reduced zone plot and compares them with the theoretical predictions of DE WETTE and coworkers [5] and BENEDEK [17]. Note that we have deleted the "decepton" points. It is gratifying to find good agreement between theory and experiment over most of the range of wave vectors. Starting at about $Q = 1$ Å^{-1}, however, the measured points fall below the theoretical predictions. In this region the phonon frequencies are especially sensitive to small changes in the force constants because of the increase in relative amplitudes of adjacent atoms and the shallow penetration of the phonons as the wavelength decreases. This discrepancy has been attributed to relaxation [16,25], but could also be due to small errors in the force constants used in the calculations.

Similar experiments have been performed for LiF <110> and NaF <100> and KCl <100> [11]. Fig.15 and 16 show the measured dispersion relationships for NaF and KCl together with the available theoretical results. For NaF we find very good agreement with no evidence for a decrease in the frequency at the BRILLOUIN zone boundary. The KCl results are particularly interesting since they show the first evidence for a surface mode imbedded in the bulk modes at frequencies higher than the RAYLEIGH wave. This surface mode, which is peculiar to KCl, was predicted by BENEDEK and GALIMBERTI [35], who attributed it to the near equality in the masses of K and Cl. If both particles were identical, then the lattice spacing would be halved and the BRILLOUIN zone boundary would be shifted to twice the value in KCl. The corresponding RAYLEIGH mode thus appears reflected back upon itself in the momentum space of the KCl lattice. BENEDEK [34] has noted that it may be possible to see optical modes in the time of flight spectra for KCl. New measurements with an improved apparatus are required to see if this can be realized.

In summary, these experiments have demonstrated that surface phonon dispersion curves can be measured with high accuracy out to the BRILLOUIN zone boundary. The He atoms appear to excite mainly the acoustical RAYLEIGH surface modes. The measured dispersion curves are in good agreement with calculations, except near the zone boundary, where for LiF <100> relaxation appears to lower the frequencies. Bulk and optical phonons have a much smaller effect on the scattering. By using low energy beams ($E \lesssim 20$ meV) it has been possible to suppress multiple phonon processes.

6. The Role of Resonances in Phonon Interactions

As early as 1936, LENNARD-JONES, DEVONSHIRE [36] and STRACHAN [37] suggested that selectively adsorbed atoms will eventually "evaporate" from the surface after interacting with a surface phonon. Another process, which may influence the intensities of inelastically scattered particles, is called "kinematic focussing". This phenomenon, which was first predicted by BENEDEK [38], can be understood by examining the scan curves of Fig.7. There we observe that for certain angles the scan curves run tangent to the RAYLEIGH mode for annihilation events. Thus at these angles we expect a rainbow-like phenomenon to enhance the inelastic scattering. To explore the importance of these different mechanisms, we have looked more carefully at the angular distributions between the specular and diffraction peaks [30]. Fig.17 shows such an angular scan. The intensity scale has been expanded by about three orders of magnitude compared to Fig.11. The angular distribution reveals a large number of maxima and minima. At the bottom of the spectrum the shaded wedges show the expected angles and intensity profiles for kinematic focussing. In addition, we have indicated by vertical bars the angles at which we expect elastic selective adsorption to occur via the in-plane (1,1) and the out-of-plane (1,0) reciprocal lattice vectors. For the former we observe maxima corresponding to each of the known bound states n=0, 1, 2 and 3. For the latter we find minima also for each of the bound states.

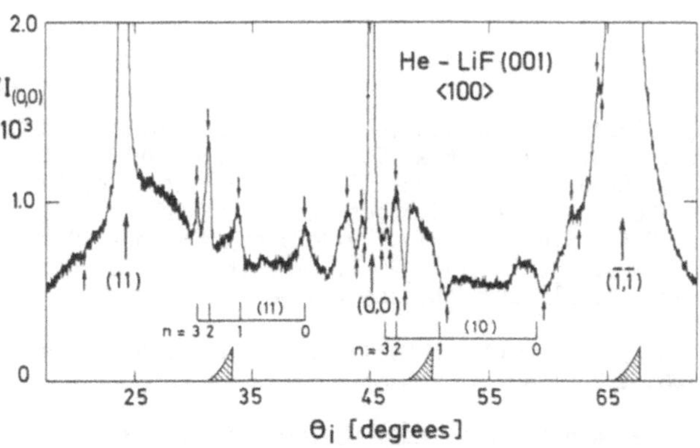

Fig. 17 Angular distribution as a function of incident angle measured with an unchopped beam on an expanded ordinate scale [30]. Some of the maxima and minima observed between the diffraction and specular peaks have been assigned to an intermediate selectively adsorbed state. The wedges at the bottom show the expected contributions from kinematic focussing.

The simple physical explanation is that in the case of the in-plane resonances the trapped helium atoms bounce along the surface until they interact with a phonon. The resulting change in energy and momentum brings them out of resonance and they desorb. Since there is a strong coupling to RAYLEIGH modes with momenta Q in the scattering plane, the atoms remain in the scattering plane and can strike the detector. Since selective adsorption is an elastic intermediate step, the overall inelastic process must follow the same scan curves as for the direct inelastic collisions. The minima are explained by the fact that in this case the atoms move in a direction out of the scattering plane. The chance that they will once again arrive at the detector is much smaller so that this resonant channel removes flux from the direct process. It is interesting to note that the observed intensities are not at variance with the WEARE [39] and CELLI [40] rules, based on quantum interference for the occurrence of maxima and minima in specular peaks due to selective adsorption, However, since we have found exceptions to these rules in furthergoing experiments [41] and since these rules have not yet been modified to take account of phonon interactions, we prefer the above simpler classical explanation.

The experiment in Fig.17 provides no evidence for a significant contribution from kinematic focussing [30]. We suspect that these structures may be much sharper than originally expected and can only be resolved with an improved angular resolution.

Finally we note that in more recent experiments with a new apparatus, which enables us to scan both initial and final angles, we have found evidence for additional resonant processes [41].

Acknowledgements

I am most grateful to R.B. DOAK for many enlightening discussions. I want to thank him und G. BRUSDEYLINS for permission to present here some data which have not been published previously.

References

1. For recent reviews of elastic and inelastic surface scattering see J.P. Toennies, Appl. Phys. 3, 91 (1974); M.W. Cole and D.R. Frankl, Surf. Sci. 70, 535 (1978); T. Engel, K.H. Rieder "Structural Studies of Surfaces with Atomic and Molecular Beam diffraction", Springer Tracts in Modern Physics, Vol. 19 (Springer, Berlin, Heidelberg, New York 1981); H. Hoinkes, Rev. Mod. Phys. 52, 933 (1980); P. Cantini and G. Boato, Advances in Electronics and Electron Physics (to be published).

2. G. Dolling, H.G. Smith, R.M. Nicklow, P.R. Vijayaragharan and M.K. Wilkinson, Phys. Rev. 168, 970 (1968).

3. G.W. Farnell, in "Physical Acoustics 6" eds. W.P. Mason and R.N. Thurston (Academic Press, New York) 1970 p. 109-166.

4. D.C. Gazis, R. Herman and R.F. Wallis; Phys. Rev. 119 (2) 533 (1960).

5. T.S. Chen, F.W. de Wette and G.P. Aldredge; Phys. Rev. B15, 1167 (1977).

6. G. Benedek; This volume.

7. G.C. Benson and T.A. Claxton; J. Chem. Phys. 48, 1356 (1968).

8. K.H. Rieder and W. Drexel; Phys. Rev. Lett. 34, 148 (1975).

9. J.R. Sandercock; Solid State Comm. 26, 547 (1978).

10. H. Ibach, in "Festkörperprobleme XI" (Vieweg, Braunschweig) 1971, p. 135-174; H. Ibach and D. Bruchmann; Phys. Rev. Lett. 41, 958 (1978).

11. R.B. Doak; Ph.D. Dissertation, MIT 1981.

12. See for example J.M. Horne and D.R. Miller; Phys. Rev. Lett. 41, 511 (1978) and G. Brusdeylins, R.B. Doak and J.P. Toennies; Phys. Rev. Lett. 44, 1417 (1980).

13. J.H. Weare; J. Chem. Phys. 61, 2900 (1974).

14. H.D. Meyer; Surf. Sci. 104, 117 (1981).

15. A.-C. Levi; Il Nuovo Cimento 54B, 357 (1979).

16. G. Benedek and N. Garcia; Surf. Sci. 103, 1143 (1981).

17. G. Benedek; Surf. Sci. 61, 603 (1976).

18. S.C. Yerkes and D.R. Miller; J. Vac. Sci. Technology 17, 126 (1980).

19. B.F. Mason and B.R. Williams; Proc. 2nd Intern. Conf. on Solid Surfaces, Japan. J. Appl. Phys. Suppl. 2, Pt. 2, 1974, p. 557.

20. S.S. Fisher and J.R. Bledsoe; J. Vac. Sci. Technology 9, 814 (1972).

21. J. Lapujoulade and Y. Lejay; J. Chem. Phys. 63, 1389 (1975).

22. P. Cantini, R. Tatarek and G.P. Felcher; Surf. Sci. 63, 104 (1977).

23. B. Feuerbacher, M.A. Adriaens and H. Thuis; Surf. Sci. 94, L 171 (1980).

24. W. Allison and B. Feuerbacher; Phys. Rev. Lett. 45, 2040 (1980).

25. G. Brusdeylins, R.B. Doak and J. Peter Toennies; Phys. Rev. Lett. 46, 437 (1981).

26. E. Semerad and E.M. Hörl; Österreichisches Forschungs-Zentrum Seibersdorf, Ber. OEFZS No. 4058, 1981.

27. U. Valbusa; This volume.

28. P. Cantini and R. Tatarek; Phys. Rev. B23, 3030 (1981).

29. B.F. Mason and B.R. Williams; Phys. Rev. Lett. 46, 1138 (1981).

30. The apparatus is described in G. Brusdeylins, R.B. Doak and J.P. Toennies, J. Chem. Phys. 75, 1784 (1981).

31. J.P. Toennies and K. Winkelmann; J. Chem. Phys. 66, 3965 (1977); G. Brusdeylins, H.-D. Meyer, J.P. Toennies and K. Winkelmann, Prog. Astronaut. Aeronaut. 51, 1047 (1977).

32. G. Boato, P. Cantini and L. Mattera; Surf. Sci. 50, 141 (1976).

33. W. Allison, R.F. Willis and M. Cardillo; Phys. Rev. Rapid Communications, to be published (1981).

34. G. Benedek; private communication at Erice, 1981.

35. G. Benedek and F. Galimberti; Surf. Sci. 71, 87 (1978).

36. J.E. Lennard-Jones and A.F. Devonshire; Proc. Roy. Soc. A158, 253 (1937).

37. C. Strachan; Proc. Roy. Soc. London A158, 591 (1937).

38. G. Benedek; Phys. Rev. Lett. 35, 234 (1975).

39. K.L. Wolfe and J.H. Weare; Phys. Rev. Lett. 41, 1663 (1978).

40. V. Celli; This volume.

41. G. Lilienkamp and J.P. Toennies, to be published.

Surface Phonons in Ionic Crystals

G. Benedek*

Department of Physics, University of Virginia
Charlottesville, VA 22901, USA

The first part of these lecture notes is a basic introduction to the theory of surface vibrations from the macroscopic point of view. The dispersion relations of acoustic Rayleigh waves and of surface optical phonon polaritons are respectively obtained from the elastic wave equation and from the Maxwell equations, combined with the appropriate boundary conditions. The microscopic approach to surface phonons is explained in the second part. An overview of the various methods used in surface lattice dynamics introduces the Green's function theory of surface vibrations in ionic crystals. The calculated dispersion curves are compared to the recent data obtained by inelastic scattering of atoms, with an illustration of the theory of one-phonon scattering processes.

1. Introduction

From a macroscopic point of view a solid surface is conceived as the boundary of a semiinfinite homogeneous continuum. The collective oscillations of microscopic variables, such as the positions of nuclear masses, and of ionic charges are regarded respectively as oscillations of the mass density (acoustic phonons) and of the ionic charge density (optical phonons). In this framework, the dynamical behaviour of the medium is fully accounted for by the macroscopic response functions of the crystal such as the elastic constants and the frequency-dependent dielectric susceptibility. Consequently, a surface acoustic phonon is seen as a vibrating strain field solving the elastic wave equation under surface boundary conditions. In a similar way, ionic charge collective surface excitations are described by the associated oscillating electromagnetic field fulfilling Maxwell equations and the continuity conditions at the surface boundary.

As far as a solid is regarded as a continuum, the eigensolutions localized at the surface in the non-dispersive regime obey a simple scale rule, according to which the penetration length into the solid has to be proportional to the wavelength along the surface. This general property (only approximately valid when dispersion effects and polariton coupling are taken into account) characterizes macroscopic surface excitations.

Such a macroscopic point of view holds only when the wavelength is much larger than lattice distances, and even in this case it is not at all exhaustive. In fact several microscopic surface excitations may appear at

---

* Permanent address: Gruppo Nazionale di Struttura della Materia del C.N.R., Istituto di Fisica dell'Universita, Via Celoria 16, I-20133 Milano, Italy

large $\vec{K}$ or exist even for $K \to 0$ as an effect of the lattice discreteness. In this case however the localization at the surface is usually strong (few lattice distances) and almost independent of the surface wavevector.

The first part of these notes outlines the macroscopic theory of surface vibrations. The surface elastic waves are explicitly derived for an isotropic medium (Sec. II) following the simple method reported in a review article by FARNELL [1]. Surface optical phonons are discussed in Sec. III. They are obtained as solutions of Maxwell equations including retardation effects (polaritons). In this section I have adopted the macroscopic approach illustrated in the recent book by MARADUDIN, WALLIS and DOBRZYNSKI [2,3].

The second part of the notes (Sec. IV) presents the microscopic theory of surface phonons in the framework of Green's function method. Conceptually the Green's function theory of surface lattice vibrations represents the natural extension of the macroscopic elastic wave theory, both treating the surface as a perturbation. Actually, by incorporating the differential elastic wave equation and the surface boundary conditions into a single integral equation, we have just the transcription of the ordinary dynamical problem in the Green's function language. The Green's function theory of surface excitations in the continuum limit is reported by GARCIA-MOLINER in a recent review article [4].

2. Surface Elastic Waves

2.1 Theory

The equation of motion for a continuous and homogeneous elastic medium reads

$$\rho \ddot{u}_\alpha = \Sigma_{\beta\gamma\delta} c_{\beta\alpha\gamma\delta} \nabla_\beta \nabla_\delta u_\gamma , \qquad (\alpha,\beta,\gamma,\delta = 1,2,3), \tag{1}$$

where ρ is the mass density, $\vec{u}(\vec{x})$ is the displacement vector field, $\vec{\nabla}$ the gradient operator and $c_{\alpha\beta\gamma\delta}$ the elastic tensor. For a semiinfinite medium occupying the half-space $x_3 < 0$, the displacement field at the free surface must fulfill the boundary condition

$$F_\alpha \equiv \Sigma_{\gamma\delta} c_{3\alpha\gamma\delta} \nabla_\delta u_\gamma = 0, \quad \text{for } x_3 = 0,\ \alpha = 1,2,3, \tag{2}$$

in order to have a vanishing external force F_α acting on the surface. Eq. (1) admits travelling wave solutions like

$$\vec{u}^{(n)} = \vec{e}^{(n)} \exp[\, ik\, (\vec{\alpha}\cdot\vec{x} - vt)], \tag{3}$$

where $\vec{e}^{(n)}$ is the unit polarization vector of the n-th eigensolution (n = 1,2,3). The components of $\vec{\alpha}$ are the direction cosines of the wavevector $\vec{k} = k\vec{\alpha}$, and v is the phase velocity. Inserting (3) into (1), we have the secular equation for the polarization vector

$$\Sigma_\gamma C_{\alpha\gamma} e_\gamma^{(n)} = \rho\, v^2\, e_\alpha^{(n)} , \tag{4}$$

where

$$C_{\alpha\gamma} \equiv \Sigma_{\beta\delta} c_{\beta\alpha\gamma\delta} \alpha_\beta \alpha_\delta . \tag{5}$$

The three eigenvalues of v^2 are solutions of the 3×3 determinantal equation

$$\det | C_{\alpha\gamma} - \delta_{\alpha\gamma}\rho v^2 | = 0 \; . \tag{6}$$

So far we have found the usual wave solutions for the infinite medium, where all the components of $\vec{\alpha}$ must be real, and the eigenvalues $v = v_L$, v_{T1}, v_{T2} are the velocities of the longitudinal and transverse acoustical waves. In isotropic media or along certain symmetry directions of anisotropic media the two transverse velocities are degenerate: $v_{T1} = v_{T2} = v_T$.

The surface intervenes through the boundary condition. In general all stationary bulk solutions $\vec{u}^{(n)}$ having an antinode at the surface (namely $\nabla_\delta u_\gamma^{(n)} = 0$ at $x_3 = 0$) fulfill trivially (2). There may be however other special solutions taking the linear form

$$\vec{u} = \Sigma_n \, a_n \vec{u}^{(n)} \tag{7}$$

which fulfill the boundary condition in a non-trivial way, provided that α_3 is allowed to be a complex constant assuming different values $\alpha_3^{(n)}$ for each wave component. When

$$\mathrm{Im} \, \alpha_3^{(n)} < 0 \tag{8}$$

we have physical solutions which are wavelike along the surface with parallel wavevector

$$\vec{K} = k \; (\alpha_1, \alpha_2) \quad , \qquad \alpha_1^2 + \alpha_2^2 = 1 \tag{9}$$

and decay exponentially in the interior of the solid according to the factor $\exp(-x_3 k \, \mathrm{Im} \, \alpha_3)$. These special solutions, localized near the surface, are called surface waves.

When (7) is inserted into (2), we get a system of three equations for the coefficients a_n

$$\Sigma_n \, [\Sigma_{\gamma\delta} c_{3\alpha\gamma\delta} \, e_\gamma^{(n)} \alpha_\delta^{(n)}] a_n = 0 \; , \quad (\alpha = 1,2,3) \; , \tag{10}$$

where $\vec{\alpha}^{(n)} \equiv (\alpha_1, \alpha_2, \alpha_3^{(n)})$. A non-vanishing solution a_n is found if

$$\det | \Sigma_{\gamma\delta} \, c_{3\alpha\gamma\delta} \, e_\gamma^{(n)} \alpha_\delta^{(n)} | = 0 \; . \tag{11}$$

Since $e_\gamma^{(n)}$ and $\alpha_\delta^{(n)}$ can be derived from (4-6) and written as functions of v^2, (11) turns out to be an implicit equation in the unknown squared velocity v^2 for the surface waves. Except for special crystal symmetries combined with special surface and K-vector orientations, the set of simultaneous equations (4,6,10,11) cannot be solved analytically. Surface wave velocities and polarizations are normally reported in a numerical form, even for cubic crystals. The reader can find several examples in FARNELL's review article [1].

2.2 Isotropic Crystals

For isotropic crystals, i.e. for cubic crystals whose elastic constants (hereafter given in Voigt notation) fulfill the isotropy condition

$$\eta \equiv 2\,c_{44}/(c_{11} - c_{12}) = 1\ , \tag{12}$$

the surface wave problem can be easily solved in an analytical way. It is quite instructive to show how the solution is derived. We choose $\vec{K}$ along x_1, so that $\vec{\alpha} = (1,0,\alpha_3)$. Eq. (6) becomes (after the substitution $c_{12} = c_{11} - 2c_{44}$)

$$\det \begin{vmatrix} c_{11} + c_{44}\alpha_3^2 - \rho v^2 & 0 & (c_{11} - c_{44})\alpha_3 \\ 0 & c_{44}(1+\alpha_3^2) - \rho v^2 & 0 \\ (c_{11} - c_{44})\alpha_3 & 0 & c_{44} + c_{11}\alpha_3^2 - \rho v^2 \end{vmatrix}$$

$$= [c_{44}(1+\alpha_3^2) - \rho v^2]^2\,[c_{11}(1+\alpha_3^2) - \rho v^2] = 0\ . \tag{13}$$

Taking into account that $c_{11}/\rho = v_L^2$ and $c_{44}/\rho = v_T^2$, we can readily find the eigenvalues of α_3 as functions of v^2

$$\alpha_3^{(1)} = \alpha_3^{(2)} = -i\ (1 - v^2/v_T^2)^{1/2}$$

$$\alpha_3^{(3)} = -i\ (1 - v^2/v_L^2)^{1/2}\ , \tag{14}$$

and the eigenvectors

$$\vec{e}^{(1)} - \frac{v_T}{v}\,(-\alpha_3^{(1)},0,1);\quad \vec{e}^{(2)} = (0,1,0);\quad \vec{e}^{(3)} = \frac{v_L}{v}\,(1,0,\alpha_3^{(3)})\ . \tag{15}$$

Hence we write (11) as

$$\det \begin{vmatrix} \rho\,\frac{v_T^3}{v}\,(2 - \frac{v^2}{v_T^2}) & 0 & 2\rho\,\frac{v_T^3}{v}\,\alpha_3^{(1)} \\ 0 & \rho\, v_T^2\alpha_3^{(2)} & 0 \\ 2\rho\,\frac{v_L v_T^2}{v}\,\alpha_3^{(3)} & 0 & -\rho\,\frac{v_L v_T^2}{v}\,(2 - \frac{v^2}{v_T^2}) \end{vmatrix} = 0 \tag{16}$$

which is satisfied by either

$$\alpha_3^{(2)} = 0,\ \text{i.e.}\quad v = v_T \tag{17}$$

or

$$(1 - \tfrac{1}{2}v^2/v_T^2)^4 = (1 - v^2/v_T^2)(1 - v^2/v_L^2)\ . \tag{18}$$

The first condition (17), tells us that the bulk transverse wave polarized parallel to the surface obeys the boundary condition in an essential way (since $\nabla_1 u_2 \neq 0$), and therefore has to be considered as a degenerate surface wave. The second condition (18), is an equation in the unknown v^2 which admits (for $v_T < v_L$) a single solution below v_T^2. In this case we have a true surface wave polarized in the plane $(x_1 x_3)$ (sagittal plane) known as Rayleigh wave (RW). Two other solutions of (18) above v_L^2 do not yield any surface wave as both $\alpha_3^{(1)}$ and $\alpha_3^{(3)}$ become real. When the spurious solution $v^2 = 0$ is eliminated, we obtain the famous Rayleigh wave cubic equation ($x_T \equiv v^2/v_T^2$):

$$\frac{1}{16}x_T^3 - \frac{1}{2}x_T^2 + (\frac{3}{2} - v_T^2/v_L^2)x_T + v_T^2/v_L^2 - 1 = 0 \, . \tag{19}$$

We can finally solve (10) and derive a_n. Then from (3), (7) and (15) we obtain the displacement field of Rayleigh waves

$$\begin{aligned} u_1 &= -N[(1-v^2/v_T^2)^{1/4} e^{ik\vec{\alpha}^{(1)}\cdot\vec{x}} - (1-v^2/v_L^2)^{-1/4} e^{ik\vec{\alpha}^{(3)}\cdot\vec{x}}] e^{-ikvt} \\ u_2 &= 0 \\ u_3 &= iN[(1-v^2/v_T^2)^{-1/4} e^{ik\vec{\alpha}^{(1)}\cdot\vec{x}} - (1-v^2/v_L^2)^{1/4} e^{ik\vec{\alpha}^{(3)}\cdot\vec{x}}] e^{-ikvt} \end{aligned} \tag{20}$$

where N is a normalization constant. The factor i in u_3 causes an elliptical polarization in the sagittal plane. At the surface ($x_3 = 0$) we have

$$(u_3/u_1)_{x_3=0} = i[(1 - v^2/v_L^2)/(1 - v^2/v_T^2)]^{1/4} \, . \tag{21}$$

Since $v_L > v_T$, the major axis is normal to the surface. For $v \to 0$ the polarization becomes circular, whereas for $v \to v_T$ the surface mode tends to be linearly polarized with a transverse polarization normal to the surface. The displacement field decays in the interior of the crystal with two components having different penetration lengths γ. The slowly decaying component penetrates a distance $\gamma = 1/k|\alpha_3^{(1)}|$; the fast decaying component has $\gamma = 1/k|\alpha_3^{(3)}|$. The inverse proportionality of γ to k descends also from the scaling argument. Moreover, the penetration tends to infinity as v tends to the transverse velocity v_T.

2.3 Anisotropic Cubic Crystals

The above properties of a Rayleigh waves, valid for the isotropic case, offer a guideline for a qualitative understanding of their behaviour in isotropic crystals with cubic symmetry. The bulk anisotropy implies anisotropy of Rayleigh waves with respect to the surface orientation and to the propagation direction on a given surface. If we consider the (001) surface of a cubic crystal, v varies as the propagation direction changes from (1,0) to (1,1), as shown in the upper plots of Fig. 1.

Here we use the notation (ζ,ξ) for surface directions as an abbreviation of $[\zeta,\xi,0]$. We note that the (001)-surface basis vectors have orientations $[1,0] \equiv (1,\bar{1})/\sqrt{2}$ and $[0,1] \equiv (1,1)/\sqrt{2}$ while $(1,0) \equiv [1,1]/\sqrt{2}$, which means that the surface basis vectors are rotated by 45 degrees with respect to the crystallographic directions.

Along (1,0) the RW velocity is always smaller than v_T whatever is η. This appears in the lower plot of Fig. 1, where v/v_T is given as function of the

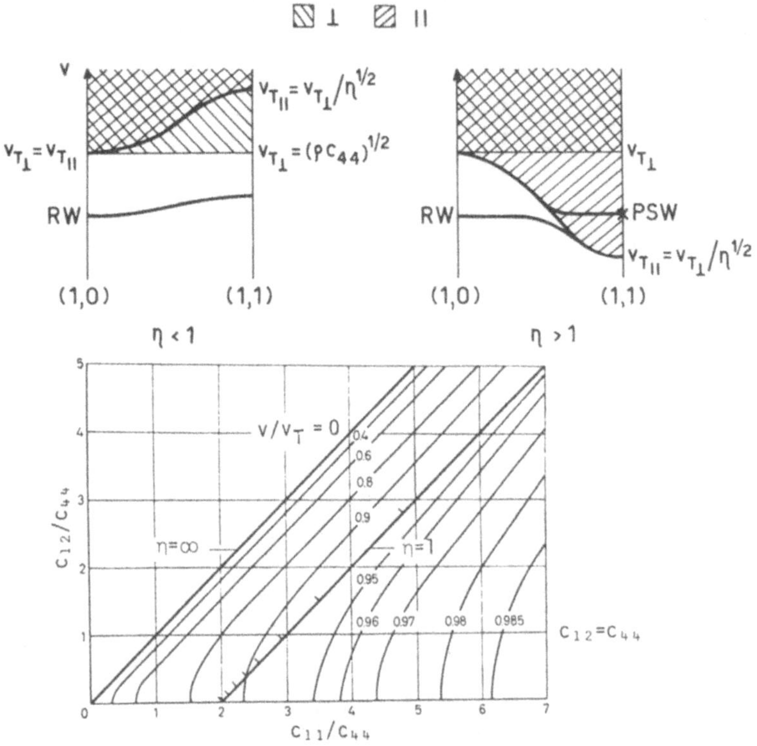

Fig.1 Rayleigh wave velocity on the (001) surface of cubic crystals as function of the propagation direction and of the crystal anisotropy (upper plots), and of the elastic constants for propagation along (1,0) (lower plot; from [1])

elastic constants. In addition to the isotropy line ($\eta = 1$) I show also the line corresponding to the Cauchy relation ($c_{12} = c_{44}$), which holds in solids where three-body interatomic forces are absent, and is approximately fulfilled in several ionic crystals. For $\eta < 1$, v/v_T is close to unity, whereas for $\eta > 1$ the RW velocity can be quite smaller than v_T.

There is another important difference between the cases $\eta < 1$ and $\eta > 1$. In the former the parallel transverse velocity v_T increases as direction is rotated from (1,0) to (1,1), and the RW exists everywhere. In the latter case v_T takes at (1,1) a value lower than at (1,0), and the RW dispersion curve (v as function of the propagation direction) may cross the dispersion of v_T and enter the continuum of bulk phase velocities. In this case the RW becomes a resonance and only at the symmetry direction (1,1) it recovers a pure sagittal surface wave character (pseudo-surface wave, PSW). Since the v_T dispersion corresponds to a degenerate surface wave, the crossing exhibits hybridization effects. Thus in the crossing region also the transverse parallel mode has a true surface character. However out of the symmetry directions the polarizations of the two hybridizing surface modes are no longer exactly parallel or sagittal.

3. Surface Polaritons

3.1 Introduction

From the macroscopic point of view, valid for a continuum, the collective oscillations of ionic charges in insulators (optical phonons) are accounted for by the oscillations of the associated electromagnetic field. Thus we have to solve Maxwell equations coupled to surface boundary conditions, the bulk behaviour being described by given frequency-dependent dielectric constant and magnetic permeability tensors, $\varepsilon(\omega)$ and $\mu(\omega)$, respectively. Electromagnetic waves propagating inside the medium obey the dispersion law,

$$k^2c^2 = \omega^2\varepsilon(\omega)\mu(\omega) \ . \tag{22}$$

In non-dispersive regimes, where ε and μ are constant, the natural excitations of the medium are just photons with a constant phase velocity $c/(\varepsilon\mu)^{1/2}$. In the highly dispersive regions where either $\varepsilon(\omega)$ or $\mu(\omega)$ has poles, k can vary over a wide range keeping ω almost constant; ω tends asymptotically to a pole ω_0 as $k \to \infty$. Such a flat branch of the dispersion law (22), having vanishing group and phase velocities, is associated with matter mechanical oscillations of the medium (phonons) and loses completely any photon character. Since one can go continuously along the electromagnetic dispersion curve from the pure photon limit to the limit of pure matter oscillations, we shall use the general name of polariton for any electromagnetic wave travelling in a dispersive medium. Optical phonons and photons are seen as limiting cases of polaritons.

3.2 Surface Phonon Polaritons

Consider a planar interface located at z = 0 between two non-magnetic isotropic media. They are characterized by their frequency dependent dielectric constant and conductivity $\varepsilon = \varepsilon_1(\omega)$ and $\sigma = \sigma_1(\omega)$ for $z > 0$, or $\varepsilon = \varepsilon_2(\omega)$ for $z < 0$, respectively. We start from the equation of motion for the electric field $\vec{E} = \vec{E}(x,y,z,t)$ as deduced from Maxwell equations,

$$-c^2 \vec{\nabla} \times (\vec{\nabla} \times \vec{E}) = \varepsilon\ddot{\vec{E}} + 4\pi \sigma \dot{\vec{E}} \tag{23}$$

with $\mu = 1$ and the assumption of charge neutrality

$$\vec{\nabla} \cdot \varepsilon\vec{E} = 0. \tag{24}$$

We look for special solutions, wavelike in two dimensions and localized at the surface:

$$\begin{aligned} \vec{E} &= \vec{E}_1^o \exp\left[-\kappa_1 z + i(\vec{K}\,\vec{R} + \omega t)\right] \qquad \text{for } z > 0 \\ &= \vec{E}_2^o \exp\left[+\kappa_2 z + i(\vec{K}\cdot\vec{R} + \omega t)\right] \qquad \text{for } z < 0 \end{aligned} \tag{25}$$

with $\vec{R} \equiv (x,y)$, $\vec{K} = (k_x,k_y)$ and Re κ_1, Re $\kappa_2 > 0$. From (4) and from the isotropy condition ($\varepsilon_{xx} = \varepsilon_{yy} = \varepsilon_{zz} \equiv \varepsilon$)

$$i\ \vec{E}\cdot\vec{K} = E_z \qquad (\kappa = \kappa_1, -\kappa_2) \ . \tag{26}$$

Solutions with $\vec{E}$ orthogonal to $\vec{K}$ imply $\kappa = 0$. Thus surface localized solutions must be sagittal with vector amplitudes

$$\vec{E}_1^o = E_1^o \, (\vec{K}/K \, , \, -iK/\kappa_1) \, , \tag{27}$$

$$\vec{E}_2^o = E_2^o \, (\vec{K}/K \, , \, +iK/\kappa_2) \, .$$

Inserting the trial solution (25) with (27) into (23) we recover the dispersion law (22) in the form

$$(K^2 - \kappa_j^2) \, c^2 = \omega^2 \, \varepsilon_j(\omega) \, , \quad (j = 1,2). \tag{28}$$

As usual the conductivity term has been incorporated into the dielectric term to give a complex-valued dielectric constant: $\varepsilon_j + 4\pi i\sigma_j/\omega \rightarrow \varepsilon_j$. However, in our examples the imaginary part of ε, yielding a damping of the polariton mode, can normally be neglected. The boundary conditions, consisting in the continuity through the interface of parallel $\vec{E}$ and normal $\vec{D}$ components, yield

$$E_1^o = E_2^o \tag{29}$$

$$\kappa_1/\kappa_2 = -\varepsilon_1(\omega)/\varepsilon_2(\omega). \tag{29'}$$

Combining (29') with (28) we get the surface polariton dispersion

$$K^2 c^2 = \omega^2 \, \frac{\varepsilon_1(\omega)\varepsilon_2(\omega)}{\varepsilon_1(\omega) + \varepsilon_2(\omega)} \, . \tag{30}$$

We note that (30) is simply obtained by setting in (22) $\eta = 1$ and replacing the inverse bulk dielectric constant $\varepsilon^{-1}(\omega)$ with the inverse surface dielectric constant

$$\varepsilon_s^{-1}(\omega) = \varepsilon_1^{-1}(\omega) + \varepsilon_2^{-1}(\omega). \tag{31}$$

Looking at (31) we note that the eventual poles of $\varepsilon_1(\omega)$ and $\varepsilon_2(\omega)$, corresponding to bulk transverse excitations in the two non-interacting media, are no longer poles of $\varepsilon_s(\omega)$. Rather, the poles $\omega = \omega_s$ of $\varepsilon_s(\omega)$, giving interface optical phonons, are solutions of the equation

$$\varepsilon_2(\omega) = -\varepsilon_1(\omega). \tag{32}$$

According to (29'), $\kappa_1 = \kappa_2$ in the limit $\omega \rightarrow \omega_s$. For a free surface, where one of the two media is the vacuum, say $\varepsilon_1 = 1$, surface phonons occur when

$$\varepsilon_2(\omega) = -1. \tag{33}$$

So far we have considered ε as a function of ω only, neglecting any k dependence possibly introduced by spatial dispersion. Actually, where dealing with surface polaritons associated with the transverse infrared-active optical (TO) phonon of a ionic crystal, the contribution of the TO phonon branch to the dielectric constant can be written as

$$\varepsilon(\vec{k},\omega) = \varepsilon_\infty \left[\, 1 + \frac{\omega_p^2}{\omega_T^2(\vec{k}) - \omega^2 - i\omega\tau^{-1}} \, \right] . \tag{34}$$

Here the frequency $\omega_T(\vec{k})$ of the TO phonon shows its explicit dependence on the wavevector $\vec{k} \equiv (\vec{K},k_z)$, and the relaxation time τ is also, in principle, a function of $\vec{k}$. The plasma frequency ω_p is given by

$$\omega_p^2 \equiv \omega_L^2 - \omega_T^2 = (1 - \varepsilon_\infty/\varepsilon_s)\omega_L^2, \tag{35}$$

where $\omega_T \equiv \omega_T(0)$; $\varepsilon_s = \varepsilon(0,0)$ is the static dielectric constant and ω_L is the $k = 0$ longitudinal optical (LO) phonon frequency, related to ω_T through the LYDDANE-SACHS-TELLER relation $\omega_L^2/\omega_T^2 = \varepsilon_s/\varepsilon_\infty$. The form of $\varepsilon(\vec{k},\omega)$ expressed by (34) is also valid for excitons, provided that ω_T is interpreted as the $k = 0$ excitation frequency, and $\omega_T(\vec{k})$ gives the exciton dispersion. In solving the surface polariton problem again we look for solutions where the third component of the wavevector $\vec{k}$ is imaginary, namely

$$\vec{k}_1 = (\vec{K},i\kappa_1) \text{ and } \vec{k}_2 = (\vec{K},-i\kappa_2). \tag{36}$$

But now the conditions for κ_1 and κ_2, obtained by replacing the trial solution into (23,24), are no longer explicit, since $\varepsilon_j(\vec{k}_j,\omega)$, with $j = 1,2$, depends on both $\vec{K}$ and κ_j. As shown by MARADUDIN and MILLS [2], in the simpler case of a free surface ($\varepsilon_1 = 1$) there are three independent values of κ_2 whose corresponding trial solutions fulfill (23,24). Like in the surface elastic wave problem, the boundary conditions are now satisfied only by a linear combination of the three trial solutions, and the resulting dispersion relation can be worked out only numerically.

However, some relevant physical aspects of surface phonon polaritons can be shown by assuming no dispersion and no damping of the optical branch, i.e., $\omega_T(\vec{k}) = \omega_T$ and $\tau^{-1} = 0$. Then, solving (30) with $\varepsilon_1=1$, the dispersion relation for surface polaritons turns out to be (Fig. 2)

$$\omega^2 = \tfrac{1}{2}[\omega_L^2 + (1 + \varepsilon_\infty^{-1})K^2c^2][1 - \sqrt{1-4\frac{(\omega_L^2+\varepsilon_\infty^{-1}\omega_T^2)K^2c^2}{(\omega_L^2+(1+\varepsilon_\infty^{-1})K^2c^2)^2}}\,] \quad . \tag{37}$$

The surface optical phonon frequency is obtained for $K\to\infty$:

$$\omega_s^2 = \frac{\omega_T^2 + \varepsilon_\infty\omega_L^2}{1 + \varepsilon_\infty} = \omega_T^2\,\frac{1 + \varepsilon_s}{1 + \varepsilon_\infty} . \tag{38}$$

This is the famous FUCHS and KLIEVER (FK) mode [5,6] found for a non-dispersive semiinfinite dielectric medium. We see that ω_s is related to ω_T through a sort of LYDDANE-SACHS-TELLER relation for the surface, with $\omega_T < \omega_s < \omega_L$, and that no surface polariton exists below ω_T.

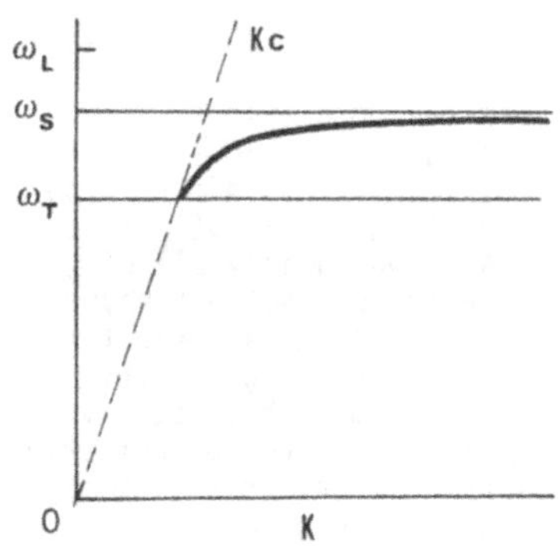

Fig. 2 Dispersion curve of surface optical phonon polaritons in a semi-infinite dielectric medium.

Setting $Kc = \omega_T$ in (35) one obtains $\omega^2 = \omega_T^2$. For $\omega < \omega_T$ the phase velocity of the photon-like wave is larger than c and tends to $(1+\varepsilon_s^{-1})c$ for $K \to 0$. Actually, κ_2 diverges at $\omega = \omega_T$ and becomes imaginary for $\omega < \omega_T$. Since κ_1 as well as ε_2 stay real, (29') is no longer fulfilled. As an exercise, the reader could discuss the existence and find the dispersion relation of polaritons localized at the interface between two ionic solids whose optical phonon frequencies are $\omega_{T,1}$, $\omega_{L,1}$ and $\omega_{T,2}$, $\omega_{L,2}$.

An important remark: The above results, frequently used in discussing the optical properties of crystal surfaces, are not at all rigorous, since the effects coming from LO and TO branch dispersion have been neglected. As a consequence, the frequency ω_S of the electromagnetic mode as given by (38), and corresponding to the root of $\varepsilon(0,\omega) = -1$, is expected to be not in good agreement with the prediction of lattice dynamical calculations in the long wave limit. In real crystals the region between ω_T and ω_L is partially or even entirely filled by the continuous bands of TO and LO bulk frequencies $\omega_T(\vec{k})$ and $\omega_L(\vec{k})$, with $\vec{k} = (0,k_z)$ and k_z running all over the allowed values in the 3D Brillouin zone. Thus the FK mode, falling in most cases into the LO bulk band, actually corresponds to a broad resonance, whose frequency may be quite apart from the prediction of (38). However in the special case when a sufficiently large gap exists between TO and LO bands, the FK mode might be a localized frequency in the gap.

An example, discussed by MARADUDIN et al. in [2], is found in the lattice dynamical calculation of surface modes in NaCl(001) using a rigid-ion model. According to these authors, the FK mode occurs just below the bottom of the LO band, at a frequency which is quite lower than that predicted by (38). In several other crystals, where ω_S falls into the LO band [7], the FK mode loses much of its surface character, particularly when, at long wave, it is deeply penetrating into the bulk. In this case, however, the electric field is also penetrating deeply into the vacuum outside the crystal, since $\kappa_1 \sim \kappa_2$, allowing for a selective coupling with incident photons (e.g., in the attenuated total reflection method developed by OTTO [8,9] and in Raman scattering [10]) or electrons (in high-energy [11] or low-energy [12] scattering experiments). On the other hand, neutral molecular beams, which interact exclusively with the topmost surface layers and are therefore the best probe for surface phonons, have no appreciable coupling with FK modes. Actually lattice dynamical calculations of surface phonon densities *projected* onto the first layer of the surface do not show any dynamical resonance associated with in-band FK modes [13].

4. Surface Lattice Dynamics

4.1 Introduction

When the wavelength of surface waves is so small to be comparable to the interatomic distances of the discrete crystal lattice, the previous continuum theory is no longer valid. This occurs for angular frequencies of the order of 10^{11} s^{-1} or higher. This limit is quite above the frequencies of surface waves used in devices of technological interest, but nearly all of the phonons involved in surface thermodynamic and response functions have higher frequencies, typically of the order of 10^{13} s^{-1}. In this respect the lattice dynamical approach is needed and a detailed knowledge of the interatomic force constants is required. It should be added that usually a crystal surface is not simply a step discontinuity in an otherwise perfectly homogeneous medium, but is accompanied by a local relaxation and sometimes by a reconstruction, which means local changes of elastic constants and density. In these cases, as well as in the presence of adsorbed layers of foreign atoms, the lattice dynamical approach is needed also in the long-wave limit.

Surface lattice dynamics has been studied from different points of view. A method, conceptually analogous to the elastic wave theory and frequently used in the past for simple crystal models with short range forces, is that based on the construction of a trial solution for the semiinfinite lattice [14]. The extension to crystals with long-range interactions of any kind has been given by FEUTCHWANG [15], and successfully used for metal surfaces by BORTOLANI et al. in Modena [16]. The hystorical importance of this method is bound to the discovery of microscopic optical modes in diatomic crystals, frequently referred as to Lucas modes (LM) [17].

Another method, introduced and extensively used by the Austin group (ALLDREDGE, ALLEN, DE WETTE, CHEN et al.) [18,19], consists in the direct calculation of eigenvalues and polarization vectors of a slab formed by a sufficiently small number of layers. This method yields all the variety of acoustic and optical surface modes all over the surface Brillouin zone (SBZ), provided that their penetration length is much smaller than the slab thickness. The above authors have shown that ten to twenty layers are normally sufficient to give all microscopic surface modes with excellent precision, which makes the direct method of practical use and quite convenient for most purposes.

Two limitations should be mentioned, however. One is represented by the poor statistics in the calculation of the momentum-selected surface phonon densities, due to the small number of layers. Such densities, entering several surface response functions, are more conveniently derived by a Green's function technique. Another difficulty in the slab calculation is the identification of resonances. To some extent this can be done by examining the loci of hybridization of crossing bulk mode dispersion curves. In principle the Green's function method provides the most natural way to determine resonant surface modes.

The Green's function theory of surface vibrations, treating the surface as a perturbation which modifies the spectrum of bulk vibrations, was introduced by LIFSCHITZ and ROSENZWEIG in the early forties [20]. Its formal advantage is the reduction of a big problem to a small one: it enables one to work in the perturbation subspace rather than in the large slab space. In practice the Green's function method has a drawback in the severe computational difficulties coming from the singular nature of the surface-projected Green's functions, which partly nullify the advantage of dealing with small matrices [13]. In this section I present first the direct method in order to introduce standard nomenclature of surface lattice dynamics and to provide a basis for the main subject, the Green's function theory of surface lattice vibrations in crystals with general long-range interatomic forces.

4.2 Dynamics of a Thin Slab

We express the equilibrium positions of crystal atoms as

$$\vec{x}(\vec{\ell}\kappa) = \vec{x}(\vec{\ell}) + \vec{x}(\kappa), \tag{39}$$

where

$$\vec{x}(\vec{\ell}) = \ell_1\vec{a}_1 + \ell_2\vec{a}_2 + \ell_3\vec{a}_3 \tag{40}$$

denotes the positions of the unit cells of the periodic lattice; $\vec{a}_j$ are the basis vectors and ℓ_j are integer numbers. Each unit cell contains s atoms whose positions are $\vec{x}(\kappa)$, with $\kappa = 1,2,\ldots,s$. In the presence of a surface it is possible to select $\vec{a}_1$ and $\vec{a}_2$ parallel to the surface and to write

$$\vec{x}(\vec{\ell}\kappa) = \vec{x}(\vec{L}) + \vec{x}(\ell_3\kappa), \tag{41}$$

where

$$\vec{x}(\vec{L}) = \ell_1\vec{a}_1 + \ell_2\vec{a}_2, \qquad \vec{L} = (\ell_1,\ell_2) \tag{42}$$

Here ℓ_3 denotes the atomic layers parallel to the surface. In a slab-shaped lattice with N_L layers, $\ell_3 = 1,2,\ldots,N_L$. The two-dimensional reciprocal lattice associated with the periodic structure of the surface is

$$\vec{G} = g_1\vec{b}_1 + g_2\vec{b}_2 \ , \tag{43}$$

with g_1 and g_2 integer numbers, and

$$\vec{b}_1 = 2\pi \frac{\vec{a}_2 x(\vec{a}_1 x\vec{a}_2)}{|\vec{a}_1 x\vec{a}_2|^2} \quad , \qquad b_2 = 2\pi \frac{\vec{a}_1 x(\vec{a}_2 x\vec{a}_1)}{|\vec{a}_2 x\vec{a}_1|^2} \ . \tag{44}$$

In the harmonic approximation [21] the interaction among the crystal atoms is represented by the set of force constants

$$\Phi_{\alpha\beta}(\vec{L},\vec{L}';\ell_3\kappa,\ell_3'\kappa') = \Phi_{\alpha\beta}(\vec{L}-\vec{L}';\ell_3\kappa,\ell_3\kappa') = \\ = \frac{\partial^2 U}{\partial u_\alpha(\vec{L},\ell_3\kappa)\partial u_\beta(\vec{L}',\ell_3'\kappa')} \ , \tag{45}$$

defined as the second-order derivative of the crystal potential energy with respect to the atomic displacement vectors $\vec{u}(\vec{L},\ell_3\kappa)$. Since the periodicity of the lattice with cyclic boundary conditions is preserved along the surface, the force constants depend only on the difference $\vec{L} - \vec{L}'$. According to (45)

$$\Phi_{\alpha\beta}(\vec{L}-\vec{L}';\ell_3\kappa,\ell_3'\kappa') = \Phi_{\beta\alpha}(\vec{L}'-\vec{L};\ell_3'\kappa',\ell_3\kappa). \tag{46}$$

The force constants are usually derived from a model microscopic description of the two- and many-body interatomic potentials. For a two-body potential $\phi_{\kappa\kappa'}(r)$ the force constants are also symmetric with respect to the cartesian indices $\alpha\beta$ and the atomic labels, separately, since from (45)

$$\Phi_{\alpha\beta}(\vec{L}-\vec{L}';\ell_3\kappa,\ell_3'\kappa') = \frac{r_\alpha r_\beta}{r^2}\left(\frac{\partial^2\phi_{\kappa\kappa'}}{\partial r^2} - \frac{\partial\phi_{\kappa\kappa'}}{r\partial r}\right) + \delta_{\alpha\beta}\frac{\partial\phi_{\kappa\kappa'}}{r\partial r}, \tag{47}$$

with

$$\vec{r} = \vec{x}(\vec{L},\ell_3\kappa) - \vec{x}(\vec{L}',\ell_3\kappa') \ , \qquad r = |\vec{r}| \ . \tag{48}$$

Since the crystal potential is invariant under a displacement field corresponding to an arbitrary rigid-body translation or rotation, the force constants must satisfy the conditions [21]

$$\Sigma_{L'\ell_3'\kappa'} \ \Phi_{\alpha\beta}(\vec{L}-\vec{L}';\ell_3\kappa,\ell_3'\kappa) = 0 \qquad \forall\ \alpha,\beta \tag{49}$$

(translational invariance condition: TI)

$$\Sigma_{L'\ell_3'\kappa'}[\Phi_{\alpha\beta}(\vec{L}-\vec{L}';\ell_3\kappa,\ell_3'\kappa')x_\gamma(\vec{L}'\ell_3'\kappa')$$

$$- \Phi_{\alpha\gamma}(\vec{L}-\vec{L}';\ell_3\kappa,\ell_3'\kappa')x_\beta(\vec{L}'\ell_3'\kappa')] = 0, \quad \forall\ \alpha,\beta,\gamma \tag{50}$$

(rotational invariance condition: RI).

Notice that the above conditions refer to rigid-body displacements and have nothing to do with the translational and point group symmetries of the lattice: they hold for any arbitrary array of atoms. The nature of the constraints due to the RI condition (49) on the force constants near the surface has been illustrated in a classical example due to LENGELER and LUDWIG [22]. In a model where the force constants are obtained by (45) from the crystal potential energy, whose phenomenological parameters fit the equilibrium condition of the crystal in the absence of external forces, the RI condition is automatically satisfied. Making use of (48) and (49), rewrite (50) in vector form:

$$\Sigma_{L'\ell_3'\kappa'}\ \Phi(\vec{L}-\vec{L}';\ell_3\kappa,\ell_3'\kappa') \times \vec{r} = 0\ . \tag{51}$$

Substituting in (51) the expression (47) for two-body potential, one has

$$I \times \sum_{L'\ell'\kappa'} \frac{\vec{r}}{r}\ \frac{\partial\phi_{\kappa\kappa'}}{\partial r} = 0. \tag{52}$$

This is true if and only if the components of the net force acting on the atom $(\vec{L},\ell_3\kappa)$, around which infinitesimal rotations are performed, are zero. Therefore, *RI conditions are equivalent to equilibrium conditions.*

The equation of motion for the harmonic lattice reads

$$M_\kappa(\ell_3)\ddot{u}_\alpha(\vec{L},\ell_3\kappa) = -\Sigma_{L'\ell_3'\kappa'\beta}\Phi_{\alpha\beta}(\vec{L}-\vec{L}';\ell_3\kappa,\ell_3'\kappa')\quad u_\beta(\vec{L}',\ell_3'\kappa')\ , \tag{53}$$

where $M_\kappa(\ell_3)$ are the atomic masses. The reduction of (53) to (1) in the long wave limit, and the connection between the RI condition and (2) are shown in [21]. Possible solutions of (53) are plane waves travelling along the surface of the form

$$\vec{u}(\vec{x},\ell_3\kappa) = [\hbar/2N_s\omega_j(\vec{K})M_\kappa(\ell_3)]^{1/2}\vec{e}(\ell_3\kappa|\vec{K}j)\quad \exp i[\vec{K}\cdot\vec{x}(\vec{L})-\omega_j(\vec{K})t]\ , \tag{54}$$

where N_s is the number of surface unit cells, i.e., the number of values taken by the vector index $\vec{L}$. The polarization vectors $\vec{e}(\ell_3\kappa|\vec{K}j)$ and the squared frequencies $\omega_j^2(\vec{K})$ are deduced from the secular equation

$$\Sigma_{\ell_3'\kappa'\beta}D_{\alpha\beta}(\ell_3\kappa,\ell_3'\kappa';\vec{K})e_\beta(\ell_3'\kappa'|\vec{K}j) = \omega_j(\vec{K})e_\alpha(\ell_3\kappa|\vec{K}j)\ , \tag{55}$$

where

$$D_{\alpha\beta}(\ell_3\kappa,\ell_3'\kappa';\vec{K}) = [M_\kappa(\ell_3)M_{\kappa'}(\ell_3')]^{-1/2} \tag{56}$$

$$\times\ \Sigma_{L'}\Phi_{\alpha\beta}(\vec{L}-\vec{L}';\ell_3\kappa,\ell_3'\kappa')\ \exp[i\vec{K}\cdot(\vec{x}(\vec{L})-\vec{x}(\vec{L}'))]$$

is the dynamical matrix, whose dimensions are $3sN_L \times 3sN_L$. The branch index $j = 1,2,\dots\ , 3sN_L$ labels the eigensolutions. The dimensionless vectors $\vec{e}(\ell_3\kappa|\vec{K}j)$ form a complete and orthogonal set, and are normalized to unity. They contain all the information about the spatial behaviour of the slab

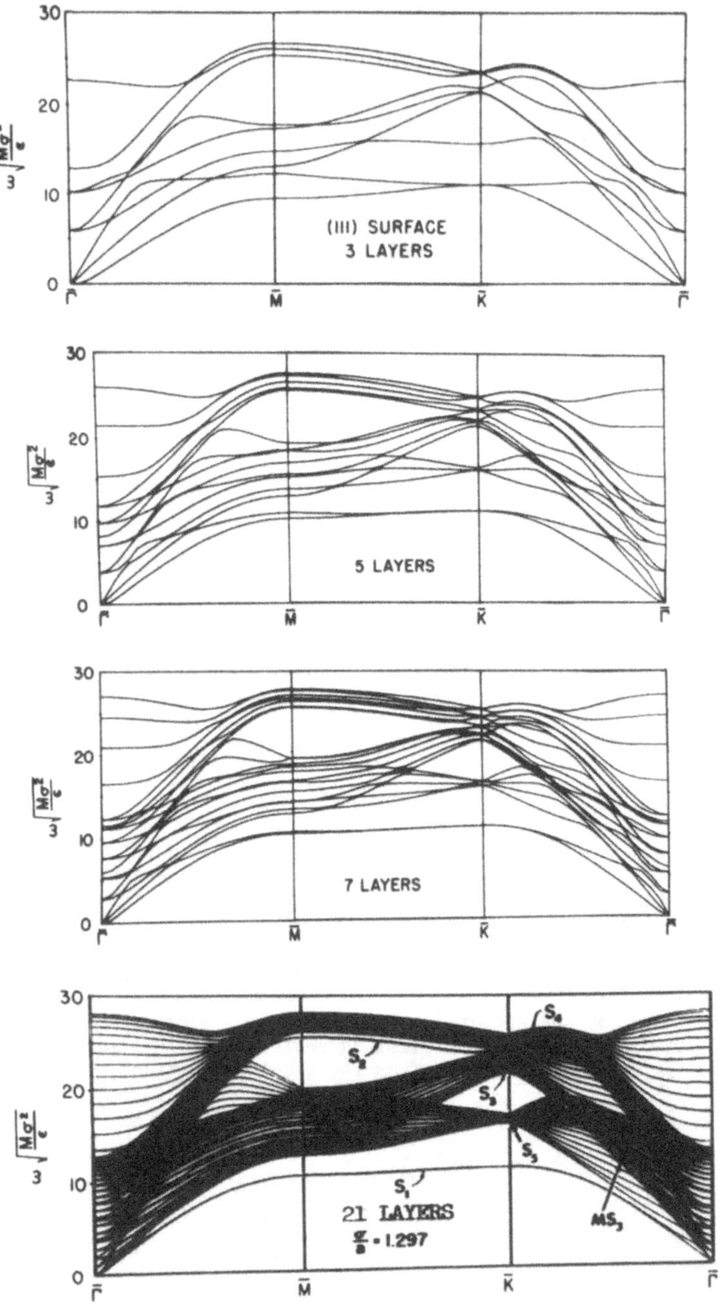

Fig.3 Dispersion curves of a monoatomic f.c.c. (111)-oriented slab along the boundaries of the irreducible part of the SBZ, for different thicknesses (from ALLEN, ALLDREDGE and DE WETTE, [18])

modes in the direction normal to the surfaces. For values of $\vec{K}$ along some symmetry directions of the SBZ the eigenvectors are perfectly polarized in the sagittal plane (sagittal polarization: $\perp$), or along the normal to the sagittal plane, namely parallel to the surface (parallel polarization: $\parallel$). In this case the set of equations (55) splits into two subsets for $\perp$ and $\parallel$ modes, and the secular determinant of (55) factorizes correspondingly.

Figure 3 shows an instructive example: the dispersion curves for a (111)-oriented monoatomic f.c.c. slab are plotted along the symmetry directions of the SBZ. The calculations, due to ALLEN, ALLDREDGE and DE WETTE [18], are based on a Lennard-Jones two-body potential, and are reported for increasing number of layers (N_L = 3,4,7,21). It appears that, as N_L increases, the bulk modes thicken forming a certain number of bands. These bands become continuous in the limit of an infinitely thick slab; in this limit the band edges and the gaps appearing between different bands for each $\vec{K}$ are exactly those of the distribution of the frequencies $\omega j(\vec{k})$ of the cyclic infinite lattice, where j = 1,2,...,3s, and $\vec{k} = (\vec{K},k_3)$ are the branch index and the 3D wavevector, when j and k_3 are allowed to assume all the possible values and $\vec{K}$ is kept constant. However, a finite number of slab modes, labeled S_1, S_2, etc., remain outside the bands. Since their eigenvectors are found to be large near the surface and rapidly decreasing with the increasing distance from the surface, they are interpreted as surface modes. Particularly the acoustic surface mode S_1 is localized below the acoustic band also in the long-wave limit. This mode is just the surface Rayleigh wave discussed in the previous section. The other surface modes have no analogue in the continuum theory, since they appear in the dispersive region of the Brillouin zone.

In the long-wave limit the monoatomic slab admits only acoustic surface waves, while the diatomic slab may have surface modes of optical character, whose frequencies do not vanish as $K \to 0$. In addition to the macroscopic FK modes, microscopic optical surface modes exist, such as the Lucas modes [17]. Figure 4 shows the dispersion curves for a 15-layer slab of NaCl with (001) orientation, calculated from a 11-parameter shell model by DE WETTE et al. [19]: the modes S_5 and S_4 are examples of Lucas modes with $\parallel$ and $\perp$ polarization, respectively. An interesting feature occurring for slabs with three-dimensional inversion symmetry (as in the case considered here)

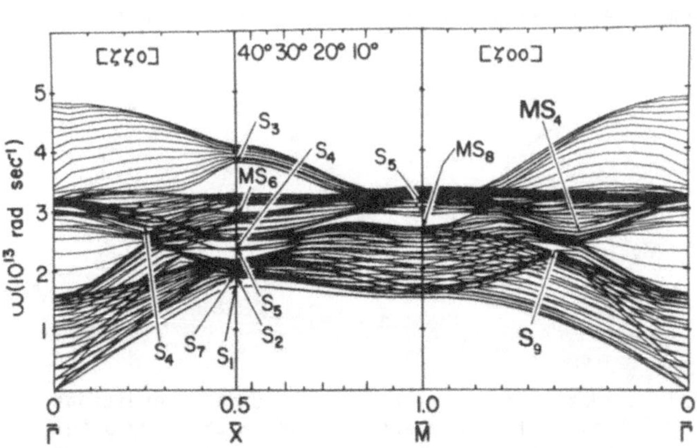

Fig.4 Dispersion curves of a 15-layer (001)-oriented NaCl slab (shell-model calculation from [19])

is that the surface modes occur in nearly degenerated pairs. The degeneracy occurs first in the short-wave region at the SBZ boundary, but, as N_L increases, it involves modes of larger and larger wavelength. This is due to the fact that each long-wave surface mode is deeply penetrating and involves the opposite surface. However, for $N_L = \infty$, all surface modes are degenerate everywhere. These modes should be regarded as single surface modes of a semiinfinite lattice. Actually, quite precise information on the vibrations of the semiinfinite lattice can be extrapolated from slab calculations with reasonably large values of N_L.

4.3 The Green's Function Method

4.3.1 The Free Surfaces as a Perturbation

The dynamical problem of a semiinfinite lattice can be approached by a perturbation method. We consider the infinitely thick slab to be originated by perturbing an infinite lattice with three-dimensional cyclic boundary conditions. The free surfaces are constructed by cutting the infinite lattice along an ideal plane Σ, as shown in Fig. 5. The resultant perturbation of the infinite lattice force constants $\phi^0_{\alpha\beta}(\vec{\ell}\kappa,\vec{\ell}'\kappa') = \phi^0_{\alpha\beta}(\vec{L}-\vec{L}'\ell_3-\ell_3',\kappa\kappa') \equiv \phi_0$ is then described by setting to zero all the interatomic force constants crossing the plane Σ. More precisely, we define the perturbation matrix Λ according to

$$\Lambda_{\alpha\beta}(\vec{L}-\vec{L}';\ell_3\kappa,\ell_3'\kappa') = \phi_{\alpha\beta}(\vec{L}-\vec{L}';\ell_3\kappa,\ell_3'\kappa') - \phi^0_{\alpha\beta}(\vec{L}-\vec{L}';\ell_3-\ell_3',\kappa\kappa') \qquad (57)$$

where both ϕ, the slab force constant matrix, and ϕ_0 are assumed to satisfy the TI and RI conditions; thus also Λ must satisfy the TI and RI conditions. In (57) it is intended that the slab force constant elements connecting atoms across Σ are zero. In practice, Λ works as a perturbation only when its nonzero elements are restricted within a small perturbation subspace σ (Fig. 5); for example if $\ell_3 = 1/2$ and $\ell_3 = -1/2$ denote the two surface layers, the subspace σ is defined by ℓ_3 and $\ell_3' = 1/2, 3/2, \ldots, N_p-1/2, -3/2, \ldots, -N_p+1/2$, where N_p, the number of atomic layers on each side of the slab involved in the cutting procedure, is "small". This requirement seems to rule out the applicability of the perturbation method to lattices with long-range forces, like ionic crystals. However, we will see that in the limit $N\to\infty$ it is possible to work with a small perturbation matrix also for ionic crystals.

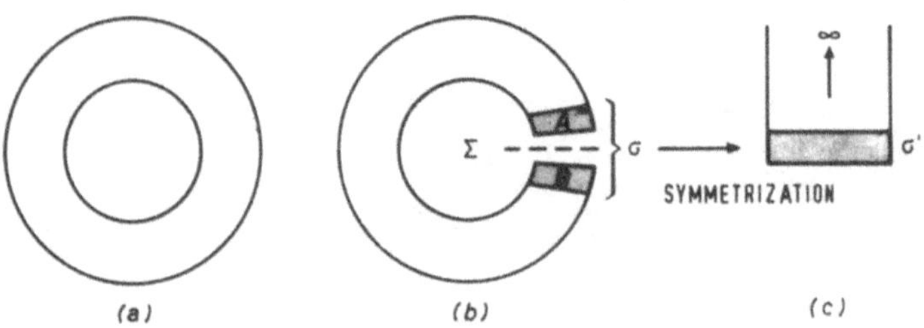

Fig.5 Representation of an ideal lattice with cyclic boundary conditions (a), and of the same lattice after two free surfaces are created by a cut along the plane Σ(b); σ denotes the perturbation subspace. In the representation of symmetrized coordinates we have a semiinfinite lattice with a perturbation subspace σ', restricted to a single surface (c).

Let us rewrite (53) for stationary states in a self-explanatory matrix notation

$$-M\omega^2\vec{u} = -\phi\vec{u} = -(\phi_o + \Lambda)\vec{u}, \tag{58}$$

where M is the mass matrix, and ω the variable frequency. We assume that (58) has been already Fourier transformed with respect to the coordinates along which translational symmetry is maintained, i.e. we work in the surface wave vector representation.

The eigenvalue spectrum of $D_o \equiv M^{-1/2}\phi_o M^{-1/2}$, the dynamical matrix of the infinite lattice, is represented for each wavevector $\vec{K}$ by a set of 3s continuous bands. First, we consider ω^2 outside the spectrum $\{D_o\}$ of D_o; in this case (58) can be written as

$$[I + (\phi_o - M\omega^2)^{-1}\Lambda]\vec{u} = 0 \qquad \omega^2 \notin \{D_o\} . \tag{59}$$

The advantage of (59) with respect to (58) is that now we have a secular determinant with the same dimension of the perturbation Λ, which in the K representation is $6sN_p x 6sN_p$. The values of ω^2 outside $\{D_o\}$ for which

$$\det|I + (\phi_o - M\omega^2)^{-1}\Lambda| = 0 \tag{60}$$

for any given $\vec{K}$ give the dispersion relations of the surface modes; they are localized at the surface since no frequency outside $\{D_o\}$ can propagate in the interior of the lattice.

When ω^2 falls into $\{D_o\}$, (58) is no longer reducible to the form (59) since $\phi_o - M\omega^2$ cannot be inverted. However, in this case we have a wave propagating in the bulk which is distorted by the surfaces through scattering processes. Again the problem can be reduced to the perturbation subspace, according to the standard theory of scattering [23], by inverting ϕ_o - Mz, where $z = \omega^2 + i0^+$ is the squared frequency removed from the real axis by an infinitesimal positive quantity. In this case (58) can be written as

$$\vec{u} = \alpha\vec{u}_o - (\phi_o - Mz)^{-1} \Lambda \vec{u}, \qquad \omega \in \{D_o\} \tag{61}$$

$$= \alpha\{I + (\phi_o - Mz)^{-1}\Lambda\}^{-1}\vec{u}_o, \tag{61'}$$

$$= \alpha\vec{u}_o - \alpha(\phi_o - Mz)^{-1} T\vec{u}_o, \tag{61''}$$

where $\vec{u}_o$ is any eigenvector of the infinite lattice for the frequency ω, and α is a normalization constant. Each solution of (61) for $\omega^2 \in \{D_o\}$ can be seen as a superposition of an incoming unperturbed wave and a scattered wave, represented by $-(\phi_o - Mz)^{-1}T\vec{u}_o$ in (61"), where

$$T = \Lambda \{I + (\phi_o - Mz)^{-1}\Lambda\}^{-1} \tag{62}$$

is the transition matrix. T has the same dimensions as Λ; thus only the elements of $(\phi_o - Mz)^{-1}$ in the subspace σ are involved. They form a $6sN_p$ x $6sN_p$ matrix

$$g \equiv (\phi_o - Mz)^{-1} \text{ projected onto } \sigma, \tag{63}$$

known as the unperturbed projected Green's function matrix. Resonant states, i.e. enhanced scattering waves, occur when

$$\mathrm{Re}\,\det|I + g\Lambda| = 0. \tag{64}$$

Since, for ω^2 outside $\{D_o\}$, g becomes just the real matrix $(\phi_o - M\omega^2)^{-1}$, (64) includes (60) as a particular case. Equation (64) should be regarded as the general condition for the existence of surface modes. For each wavevector $\vec{K}$, (64) has a certain number of solutions $\omega^2 = \omega_i^2(\vec{K})$; the solutions outside $\{D_o\}$ give surface-localized modes, while the in-band solutions, inside $\{D_o\}$, correspond to surface resonant modes, or pseudosurface waves. For an infinite lattice, the perturbative effects of the free surfaces on the continuous frequency bands are understood in terms of phonon densities; useful concepts are the unperturbed and the perturbed projected phonon densities

$$\rho(\vec{K},\omega) = (2\omega/\pi)\ \mathit{Tr}\ \mathrm{Im}\ g(\vec{K},\omega^2), \tag{65}$$

$$\tilde{\rho}(\vec{K},\omega) = (2\omega/\pi)\ \mathit{Tr}\ \mathrm{Im}\ \tilde{g}(\vec{K},\omega^2), \tag{65'}$$

respectively, where Tr denotes trace in the subspace σ of the perturbation, and

$$\tilde{g} \equiv (\phi - Mz)^{-1} \quad \text{projected onto } \sigma \tag{66}$$

$$= (I + g\Lambda)^{-1} g \tag{66'}$$

is the perturbed Green's function matrix. Along symmetry directions all modes have either sagittal ($\perp$) or parallel ($\parallel$) polarization, yielding the factorization

$$\{D_o\} = \{D_o\}_{\perp} \{D_o\}_{\parallel} \tag{67}$$

and the block diagonalization of all the above matrices. Having in mind (66'), the following cases for the values ω^2 fulfilling (64) occur:

i) *Surface modes:* ω^2 is outside $\{D_o\}$. The imaginary part of g is infinitesimal and $\tilde{\rho}(\vec{K},\omega)$ exhibits a δ peak.

ii) *Pseudo-surface modes:* ω^2 falls into a band of $\{D_o\}$, but the displacement vector $\vec{u}$ is orthogonal to all vectors $\vec{u}_o$ of that band; in (61) we must set $\alpha = 0$ and again we have a local mode, crossing a transparent bulk band. This occurs, e.g., when ω^2 belongs to $\{D_o\}_\perp$ but is outside $\{D_o\}_\parallel$ (or viceversa). Clearly, pseudo-surface modes exist only along symmetry directions.

iii) *Surface resonances:* ω^2 falls into a band of $\{D_o\}$ whose $\vec{u}_o$ are not orthogonal to $\vec{u}$ ($\alpha \neq 0$). In this case we have a resonance: $\tilde{\rho}(\vec{K},\omega)$ displays a Lorentian-shaped peak, whose width is proportional to Im g, which is finite. When deviating from a symmetry direction any pseudo-surface mode transforms into a resonance. However, surface resonance may exist also along symmetry directions in addition to local modes.

4.3.2 The Semiinfinite Lattice

In many cases the slabs created by cutting the cyclic lattice exhibit like surfaces and a mirror symmetry with respect to the plane Σ (Fig. 5b). Sometimes, the mirror symmetry is recovered by a shear translation of the surface B with respect to A, as for the (001) surfaces of a NaCl lattice. In these cases it is possible to replace the coordinate $|\ell_3)$ with a symmetrized coordinate $||\ell_3|,p)$ defined by the transformation

$$||\ell_3|,p) = 2^{-1/2}\{|\ell_3) + f(p)|-\ell_3)\}, \tag{68}$$

where $p = \pm$ is the parity index, and $f(p)$ is a suitable function, having the properties $f(-p) = -f(p)$ and $|f(p)|^2 = 1$, chosen in such a way that the matrices Λ, g and $\tilde{g}$ are diagonalized with respect to p into blocks $3sN_p \times 3sN_p$. Call these blocks Λ_p, g_p and

$$\tilde{g}_p = \{I + g_p \Lambda_p\}^{-1} g_p , \qquad p = \pm . \tag{69}$$

This enables us to work in a reduced $3sN_p$-dimensional subspace σ'. When $N_L \to \infty$, even- and odd-parity modes become degenerate: thus in subspace σ' we must have

$$\tilde{g}_+ = \tilde{g}_- \equiv \tilde{g}. \tag{70}$$

The $3sN_p$-dimensional matrix $\tilde{g}$ is interpreted as the perturbed projected Green's function for the single surface of the semiinfinite lattice. Equation (69) can be written as

$$\tilde{g} = (I + \bar{g}\,\bar{\Lambda})^{-1}\bar{g} , \tag{71}$$

where

$$\bar{g}^{-1} = \frac{1}{2}(g_+^{-1} + g_-^{-1}) \tag{72}$$

and

$$\bar{\Lambda} = \frac{1}{2}(\Lambda_+ + \Lambda_-) \tag{73}$$

are the inverse of the unperturbed projected Green's function for the semiinfinite lattice, and the pertaining perturbation, respectively. Notice that all the inversions are performed in the subspace σ'. The resonance condition, (64), in the subspace σ' becomes

$$\mathrm{Re}\,\det|I + \bar{g}\bar{\Lambda}| = 0 . \tag{74}$$

Besides dimensional reduction, the symmetrization (68) has the advantage that the symmetrized perturbation $\bar{\Lambda}$ takes a very simple form. Indeed, the symmetrized components of Λ are

$$\Lambda \equiv \Lambda_{\alpha\beta}(\vec{K};|\ell_3|\kappa p,|\ell_3'|\kappa' p) = \Lambda_{\alpha\beta}(\vec{K};|\ell_3|\kappa,|\ell_3'|\kappa') + f(p)\Lambda_{\alpha\beta}(\vec{K};|\ell_3|\kappa,-|\ell_3'|\kappa'). \tag{75}$$

According to (73) $\bar{\Lambda}$ is just the first term on the right-hand member of (75).

In the absence of surface elastic relaxation or reconstruction, the nondiagonal part of the perturbation concerns only pairs of atoms on opposite sides with respect to Σ. Thus all the elements of $\Lambda_{\alpha\beta}(\vec{K};|\ell_3|\kappa,|\ell_3'|\kappa')$ are zero, except those coming from the diagonal part of the force constant matrices (the Einstein part), which is obtained from the two-body force constants by imposing the TI condition. We call Λ^T such a diagonal part of $\bar{\Lambda}$.

Surface relaxation is caused by the unsaturated forces arising from the crystal fracture. They correspond to the perturbation in the first derivatives of the crystal potential, which are in turn related to the shear force constants (the terms proportional to $\phi'_{\kappa\kappa'}/r$ in (47)). Since the creation of a free surface destroys the inversion symmetry, the RI conditions relative to all the independent rotation axes parallel to the surface and lying within the reduced perturbation σ' yield non-trivial constraints on the shear force constants connecting atoms in σ'. Thus the shear force constants have to be perturbed within σ', and this contributes a non-diagonal part to $\bar{\Lambda}$. We argue that the independent non-diagonal elements of $\bar{\Lambda}$ are as numerous as the independent non-trivial RI conditions. These are in turn as numerous as the independent surface equilibrium conditions. We call Λ^R this non-diagonal part of $\bar{\Lambda}$. The role of Λ^R in surface dynamics is quite important, ensuring that the calculation refers to the surface in an equilibrium configuration. MARADUDIN et al. show in their book [21] that the RI condition is equivalent to the boundary condition in the elastic theory and is therefore to be fulfilled in order to have Rayleigh waves with the appropriate velocity.

We write the perturbation matrix as

$$\bar{\Lambda} = \Lambda^T + \Lambda^R, \tag{76}$$

by which we mean that $\bar{\Lambda}$ is fully defined by TI and RI conditions. The procedure actually used to derive the elements of Λ is based on the equality

$$\Lambda_p = \bar{\Lambda} + \frac{1}{2}(g_{-p}^{-1} - g_p^{-1}). \tag{77}$$

We apply now the sum rules associated with TI and RI conditions (49) and (50), which we formally rewrite respectively as

$$\mathcal{T}\,\phi = 0, \qquad\qquad \mathcal{R}\,\phi = 0 . \tag{78}$$

Since also $\mathcal{T}\,\Lambda_p = 0$ and $\mathcal{R}\,\Lambda_p = 0$, we have

$$\mathcal{T}\,\bar{\Lambda} = \frac{1}{2}\mathcal{T}(g_p^{-1} - g_{-p}^{-1})\,\vec{K} = 0 , \tag{79}$$

$$\mathcal{R}\,\bar{\Lambda} = \frac{1}{2}\mathcal{R}(g_p^{-1} - g_{-p}^{-1})\,\vec{K} = 0 ,$$

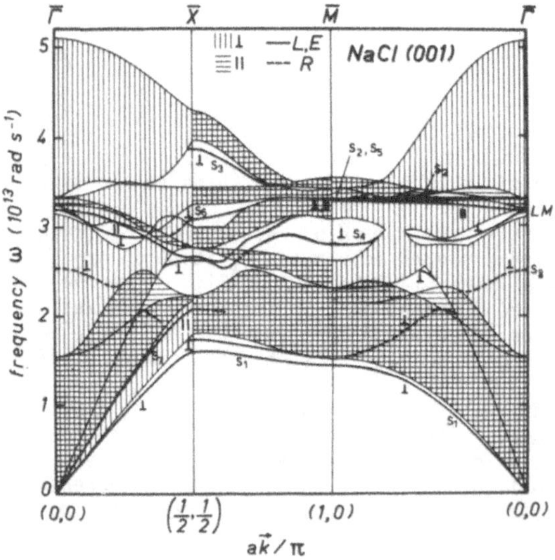

Fig.6 Surface dispersion curves of NaCl(001) calculated by the Green's function method.

where the summation over all lattice sites selects, in the $\vec{K}$ space, only the $\vec{K} = 0$ matrix elements of the inverse Green's functions. Equations (79) form an inhomogeneous linear system whose unknowns are the independent elements of Λ. Whatever is the assumed perturbation subspace σ', i.e., the number N_p of layers included in the perturbation, there are as many unknown elements of $\bar{\Lambda}$ are equations (79). Thus such elements are uniquely (and self-consistently) defined in terms of the $\vec{K} = 0$ elements of the unperturbed Green's function.

Furthermore, in ionic crystals the summation over the ion index κ implicit in the operators T and R produces in the right-hand members of (79) a cancellation of the long-range Coulomb contributions. Thus the diagonal elements of Λ as well as the non-diagonal ones, related to the surface elastic relaxation, decay in a very fast exponential way for increasing ℓ_3. It is therefore an excellent approximation to take only $\ell_3 = 1/2$ ($N_p = 1$), namely to restrict the perturbation to the first layer [24]. This receives an *a posteriori* support from the very good agreement obtained between Green's function and slab calculations for alkali halide surfaces. Such a comparison is done in the next paragraph. As long as the perturbation can be cut at the first layer, the Green's function method applied to alkali halide (001) surfaces deals with 6 x 6 matrices, which makes this technique rather convenient and practical.

4.4 Examples and Comparison with Experimental Data

I show now a few examples of calculated dispersion curves of surface modes in alkali halide crystals, obtained using the Green's function method. The technical aspects of the calculation, based on the breathing shell model (BSM) for the bulk lattice dynamics and on a perturbation model restricted to the first layer, are illustrated in [13]. In Fig. 6 I report the surface dispersion curves for the surface (001) of NaCl along the symmetry directions $\bar{\Gamma}\bar{X} = [\xi\xi 0]$, $\bar{X}\bar{M} = [1\xi 0]$ and $\bar{M}\bar{\Gamma} = [\xi 00]$. They should be compared with the slab calculations shown in Fig. 4, based on an 11-parameter shell model [19]. The main difference between the two calculations is that in the semiinfinite lattice the bulk bands become continuous (grey regions). Essentially all surface modes found in the slab calculation are also appearing in Fig. 6. There are of course small discrepancies due to the differences in the dynamical models and parameters, as well as to the fact the Green's function method is after all an approximation, but the overall agreement is very good. This method, however, has the advantage of giving the dispersion relations of resonances (broken lines): for instance, the resonance S_8 crossing the acoustic bands along both $\bar{\Gamma}\bar{X}$ and $\bar{\Gamma}\bar{M}$ directions. On the other hand no resonance corresponding to the Fuchs and Kliever macroscopic mode is found in the long wave limit. The deep penetration of this mode diverges as $K \to 0$, giving a vanishing amplitude at the first surface layer. Actually no resonant structure association with the FK mode occurs in the $K = 0$ surface-projected sagittal phonon density (Fig. 7).

Sagittal modes normally have elliptical polarization, but at the high-symmetry points $\bar{\Gamma}$ and $\bar{M}$ (C_{4v}) they become linearly polarized either along z or along x. But the x mode is indistinguishable from the $\|$ mode polarized along y, so that at $\bar{\Gamma}$ and $\bar{M}$ all $\|$ surface modes must be degenerate with some x-polarized $\perp$ mode, each mode pair transforming like the two components of the irreducible representation E. At $\bar{\Gamma}$ the degenerate pairs are the Lucas modes S_4 and S_5, and the acoustic modes S_1 (Rayleigh wave) and S_7 (the so-called shear-horizontal mode discovered by ALLDREDGE [25]). At $\bar{M}$ the $\|$ local mode S_5 meets the $\perp$ resonance S_2. Of course S_2 becomes a local mode at $\bar{M}$. This happens because the sagittal band which S_2 belongs to is polarized along

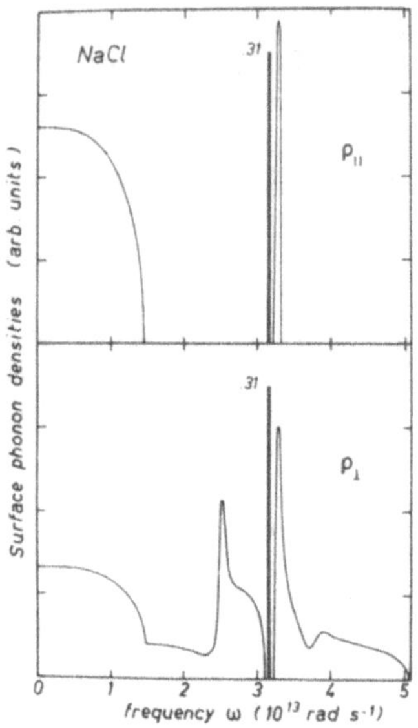

Fig.7 Surface-projected phonon densities of NaCl(001) at the symmetry point $\bar{\Gamma}$

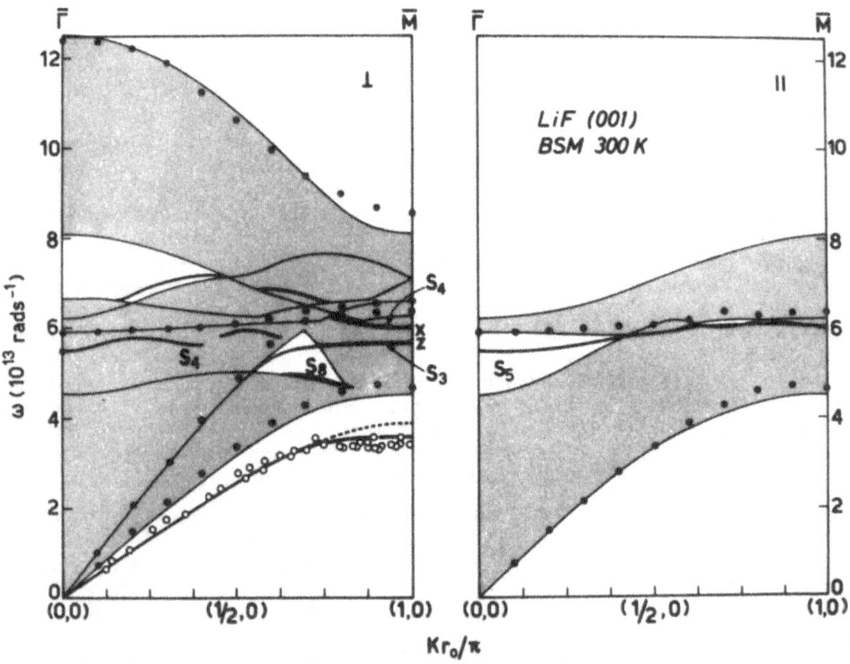

Fig.8 Surface phonon dispersion curves of LiF(001) along (100) calculated by the Green's function method (room temperature data)

z at the $\bar{M}$ point: thus we have a resonance which becomes a pseudo-surface mode not along a symmetry direction but only at isolated points of the SBZ [26].

The second example is LiF(001). The bulk bands (Fig. 8), now reproduced separately for $\perp$ and $\parallel$ polarizations only along $\bar{\Gamma}\bar{M}$, are delimited by certain bulk dispersion curves (thin lines). The comparison of these curves with the available neutron data (black points) illustrates the quality of the BSM, here used with room temperature data. Very recently, thanks to the great progress in the nozzle beam technique, BRUSDEYLINS, DOAK and TOENNIES [27] succeeded in measuring the dispersion curve of Rayleigh waves by means of inelastic scattering of He atoms. Rayleigh waves, more than any other surface modes, give sharp peaks in the time-of-flight spectra of the scattered atoms, which allows for a precise determination of phonon energy and momentum.

The experimental points (open circles) are in a quite good agreement with the dispersion curve calculated by either the Green's function method and BSM (heavy line) or the slab method and an 11-parameter shell model [19] (broken line). The only small discrepancy, more pronounced for the slab calculation, occurs at the $\bar{M}$ point, and is probably removed if the effects of the surface change in ionic polarizabilities and/or anharmonicity are taken into account [28].

4.4.1 The Energy-Loss Profile of Inelastic Atom Scattering

Besides the RW dispersion curve, time-of-flight (TOF) spectra of the scattered atoms show something similar to the projected density of sagittal modes $\tilde{\rho}_\perp(K,\omega)$ with K and ω related through the experimental kinematical condition. Obviously, the projected density is modulated by the scattering amplitude, which is different for the various modes, as well as by the K-dependent Debye-Waller and ω-dependent Bose factors. Consider an inelastic process casting an incident atom of mass m, initial energy $\hbar\omega_i$ and initial momentum $\hbar\vec{k}_i$ into a final state of energy $\hbar\omega_f \equiv \hbar(\omega_i-\omega)$ and momentum $\hbar\vec{k}_f \equiv \hbar(\vec{k}_i-\vec{k})$ oriented within a solid angle $d\Omega$ around the detector direction. The differential reflection coefficient is written as a Fourier transform of the time- and temperature-dependent correlation between dimensionless transition matrices [29]

$$\frac{d^2R}{d\omega d\Omega} = \frac{1}{2\pi}\,\frac{k_f}{|k_{iz}|}\int_{-\infty}^{+\infty} dt\; e^{-i\omega t}\langle T(\vec{k},0)T^+(\vec{k},t)\rangle \;. \qquad (80)$$

Here we define $T(\vec{k},t)$ as the ordinary T matrix divided by $2\pi\hbar^2/mL^2$, L being a quantization length. If the atom wavefunction $\psi(\vec{r},t)$, solving the Lippmann-Schwinger equation, is written as a superposition of incident and all final state waves

$$\psi(\vec{r},t) = e^{i\vec{k}_i\cdot\vec{r}}e^{-i\omega_i t} + \int d\vec{k}_f A(\vec{k},t)e^{i\vec{k}_f\cdot\vec{r}}e^{-i\omega_f t} \quad , \qquad (81)$$

the time-dependent perturbation coefficients $A(\vec{k},t)$ are given by

$$A(\vec{k},t) = \frac{i\hbar L}{2\pi m}\int_{-\infty}^{t} dt'\; e^{i\omega t'}\; T(\vec{k},t') \;. \qquad (82)$$

Of course $A(\vec{k},-\infty) = 0$, whereas $A(\vec{k},\infty)$ is proportional to the Fourier transform of $T(\vec{k},t)$, whose squared modulus gives the scattering cross sections.

Surface boundary conditions yield a linear equation for the coefficients, allowing in principle for their determination. The derivation, however, is complicated by the fact that the locus of turning points is oscillating in space and time as a consequence of static and phonon-induced corrugations.

Let us express $T(\vec{k},t)$ as a generalized Fourier transform of the scattering source function $f(\vec{R},t)$ [30]

$$T(\vec{k},t) = \frac{1}{2L}\int d\vec{R}\, \exp(i\vec{k}\cdot\vec{d})f(\vec{R},t) \ . \tag{83}$$

Here the 2D vector $\vec{R}$ denotes the surface position at which the collision occurs and

$$\vec{d} \equiv (\vec{R},\ D(\vec{R},t)) \tag{84}$$

is the 3D vector describing the locus of the turning points at time t. For a hard corrugated surface (HCS), $D(\vec{R},t)$ gives the surface profile, sum of the static corrugation $D_0(\vec{R})$ and of dynamic part $D_1(\vec{R},t)$ due to lattice vibrations. From the HCS boundary condition

$$\psi(\vec{d},t) = 0 \ , \qquad \forall\ \vec{R},t, \tag{85}$$

we have an integral equation for the source function

$$\int d\vec{R}'dt'M(\vec{R},\vec{R}',t,t')f(\vec{R}',t') = -1 \qquad \forall\ \vec{R},t \tag{86}$$

whose kernel

$$M(\vec{R},\vec{R}',t,t') = \frac{i\hbar}{4\pi m}\,\theta(t-t')\int d\vec{k}_f e^{i\vec{k}\cdot(\vec{d}-\vec{d}')}e^{-i\omega(t-t')} \tag{87}$$

is the atom Green's function at the surface boundary times $(2\pi)^2\exp[i\omega_i(t-t') - i\vec{k}_i\cdot(\vec{d}-\vec{d}')]$, and θ is heaviside step function.

Here the phonon-modulated potential works as a small perturbation when $|k_zD_1(\vec{R},t)| \ll 1$. In this case we are allowed to expand M and f in powers of k_zD_1 and keep only the linear terms (*one-phonon approximation*). Using matrix notations we write

$$M = M_0 + \mu \ , \quad f = f_0 + \phi \tag{88}$$

with

$$Mf = -1 \ , \qquad M_0f_0 = -1 \tag{89}$$

so that, to first order in D_1 ,

$$\phi = -M_0^{-1}\mu f_0 \ . \tag{90}$$

Here M_0 is given by (87) with $\vec{d}_0 = (\vec{R},\vec{D}_0(\vec{R}))$ replacing $\vec{d}$, and

$$\mu(\vec{R},\vec{R}',t,t') = i[D_1(\vec{R},t)-D_1(\vec{R}',t')]L_0(\vec{R},\vec{R}',t,t') \tag{91}$$

with

$$L_0 = \frac{i\hbar}{4\pi m}\theta(t-t')\int dk_f e^{i\vec{k}\cdot(\vec{d}_0-\vec{d}_0')}e^{-i\omega(t-t')}k_z \ . \tag{92}$$

The static source function $f_o(\vec{R})$ is assumed to be known. Thus the one-phonon contributions to $T(\vec{k},t)$ in (83) can be explicated, and their correlation can be written as

$$<...> = \frac{1}{(2L)^2} \int d\vec{R}d\vec{R}' e^{i\vec{K}\cdot(\vec{d}_0-\vec{d}_0')} \int d\vec{R}_1 d\vec{R}_1' dt_1 dt_1' e^{-W(\vec{R}_1)-W(\vec{R}_1')} \tag{93}$$

$$\times Z^*(\vec{R},\vec{R}_1,t-t_1) Z(\vec{R}',\vec{R}_1',t-t_1') <D_1(\vec{R}_1,t_1) D_1(\vec{R}_1',t_1')>$$

where the Debye-Waller factor $W(\vec{R}) = \frac{1}{2}k_z^2 <D_1(\vec{R},t)D_1(\vec{R},t)>$ is a periodic function of $\vec{R}$. The coupling matrix Z is given by

$$Z = k_z \tilde{f}_o - M_o^{-1}(\widetilde{L_o f_o}) + (M_o^{-1} L_o)\tilde{f}_o \quad , \tag{94}$$

where the vectors f_o and $L_o f_o$ have been replaced by the diagonal matrices

$$\begin{aligned} \tilde{f}_o &\equiv \delta(\vec{R}-\vec{R}')\delta(t-t')f_o(\vec{R}), \\ \widetilde{L_o f_o} &\equiv \delta(\vec{R}-\vec{R}')\delta(t-t') \int d\vec{R}'' dt'' L_o(\vec{R},\vec{R}'', t-t'') f_o(\vec{R}'') \ . \end{aligned} \tag{95}$$

In the limit of flat surface ($D_o \to 0$) and small phonon momentum transfer ($K \to 0$) the last two terms in (94) give contributions which cancel each other, whereas the first term keeps finite since $f_o \to -2ik_{iz}$ (eikonal approximation [29]). Thus the last two terms of (94) are argued to give a small contribution also at finite corrugations like in LiF, and can be omitted in the calculations. Note that BENEDEK and GARCIA, in the calculation reported in [28], have omitted these terms on the basis of the incorrect argument that they would involve, at the lowest order, only phonons of vanishing parallel momentum [30].

Now we represent $D_1(\vec{R},t)$ as a superposition of slab normal modes (54)

$$D_1(\vec{R},t) = \Sigma_{\vec{K}j} \left[\frac{\hbar}{2N_s\omega(\vec{K}j)}\right]^{1/2} \left[\Sigma_{\vec{\ell}\kappa} M_\kappa^{-1/2} \vec{e}_{\ell_3\kappa}(\vec{K}j) \cdot \frac{\partial D_o(\vec{R})}{\partial \vec{u}(\vec{\ell}\kappa)} e^{i\vec{K}\cdot\vec{X}_L}\right]$$

$$\times \left[b^+_{\vec{K}j} e^{i\omega(\vec{K}j)t} + b_{\vec{K}j} e^{-i\omega(\vec{K}j)t}\right] \tag{96}$$

where $b^+_{\vec{K}j}$ and $b_{\vec{K}j}$ are phonon creation and annihilation operators, respecttively, and $\partial D_o(\vec{R})/\partial \mu(\vec{\ell}\kappa)$ represents the deformation of the turning-point locus (corrugation in HCS model) under a unit displacement of the κ-th ion in the cell $\vec{\ell} \equiv (\vec{L},\ell_3)$ of the semiinfinite crystal.

Inserting (96) and $Z = k_z \tilde{f}_o$ into (93), performing the thermal average, and performing the Fourier transform in (80), we obtain

$$\frac{d^2R}{d\omega d\Omega} = \frac{1}{8\pi^2} \frac{k_f}{|k_{iz}|} \left|1 + \coth \frac{\hbar\omega}{2k_B T}\right|$$

$$\times \Sigma_{\ell_3\ell_3'\kappa\kappa'\alpha\beta} Z^*_{\ell_3\kappa\alpha}(\vec{K}) Z_{\ell_3'\kappa'\beta}(\vec{K}) \tilde{\rho}_{\ell_3\kappa\alpha,\ell_3'\kappa'\beta}(\vec{K},\omega), \tag{97}$$

where T is the surface temperature, $\vec{K}$ the parallel momentum transfer, namely $\vec{k} = (\vec{K}, k_z)$, $\tilde{\rho}(\vec{K},\omega)$ the surface phonon density matrix (65'), and

$$Z_{\ell_3\kappa\alpha}(\vec{K}) = \frac{k_z}{2L}\left(\frac{\hbar}{\omega M_\kappa}\right)^{1/2}\int d\vec{R} e^{-W(\vec{R}) + i\vec{k}\cdot\vec{d}_0(\vec{R})} f_0(\vec{R})\,\frac{\partial D_0(\vec{R})}{\partial\vec{u}(\vec{\ell},\kappa)} \quad . \qquad (98)$$

The transducer function $\partial D_0(\vec{R})/\partial\vec{u}(\vec{\ell}\kappa)$ is the interesting part of (98), which involves in principle a microscopic description of the surface potential and its modification under nuclear displacements. For metal and semiconductor surfaces the best way to know this function could be an extension of the calculations reported by HAMANN [31] for GaAs(110) and Ni(110)-H 2x1 static surfaces.

For closed-shell insulator surfaces like LiF(001) one can possibly use some simple model. BENEDEK and GARCIA [28] have considered only the effect of ions in the first layer ($\ell_3 = \ell_3^1 = 1/2$), and taken

$$\frac{\partial D_0(\vec{R})}{\partial\vec{u}(0\tfrac{1}{2}\kappa)} = \begin{cases} (-\partial D_0(\vec{R})/\partial\vec{R}, 1)\ , & \text{if } \vec{R} \text{ is inside } A_\kappa \\ 0 & \text{otherwise} \end{cases} \qquad (99)$$

where A_κ is an area surrounding the κ-th ion, proportional to the square of its ionic radius and such that $\Sigma_\kappa A_\kappa$ = unit cell area, and $D_0(\vec{R})$ is known from fitting the diffraction amplitudes.

The one-phonon energy-loss profiles calculated in this way compare quite well with the observed TOF spectra [28,32]. Figure 9 shows such a comparison with a spectrum taken in the 90° scattering configuration for a large inci-

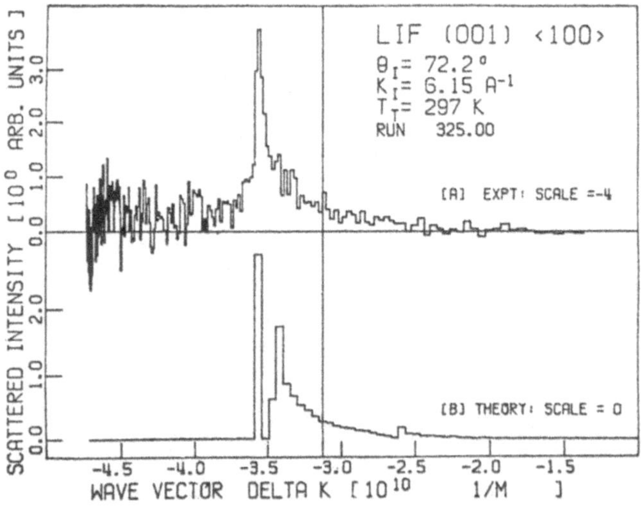

Fig.9 Time-of-flight spectrum of He^4 atoms scattered from LiF(001) in the (xz) plane with 90° geometry (incidence angle = 72.2°) (from [27] and [33]) compared with calculated energy-loss spectrum due to one-phonon inelastic processes.

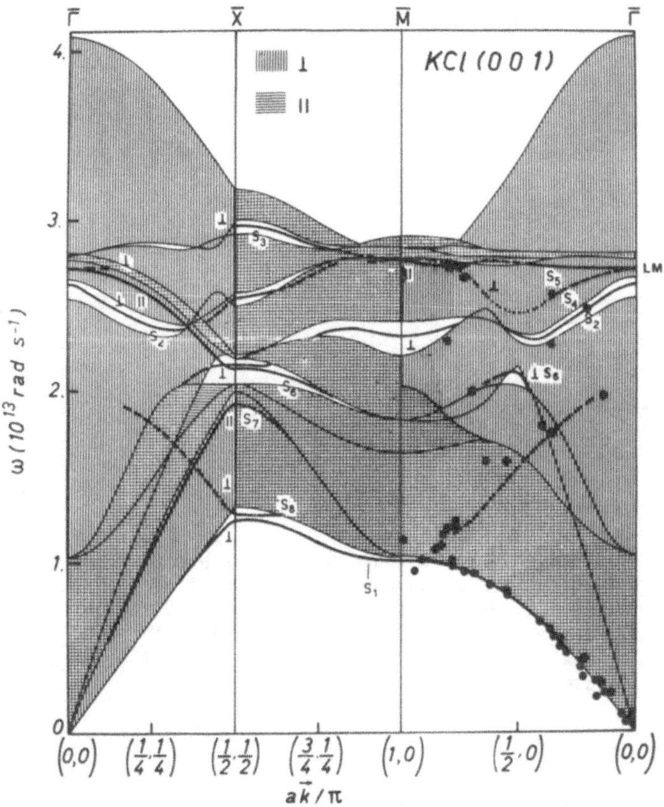

Fig.10 Calculated surface phonon dispersion curves of KCl(001) by the Green's function method using the breathing shell model and 0 K input data. Experimental points from [33].

dence angle (72.2°). Above the sharp peak corresponding to RW, there are other features coming from the projected density of bulk modes which compare quite well with the experiment.

Unfortunately LiF has too high frequencies compared to the energy of the incident atoms for the optical surface modes to be detected. For this reason Doak et al. [33] have started experiments on KCl(001), whose maximum frequency is of the order of the incident atom energy and the RW intensity is weak compared to LiF. Indeed some weaker structures found in TOF spectra correspond to energy transfers above the RW frequencies. The experimental points (black points in Fig.10) fall all quite close to sagittal dispersion curves. Quite interesting are the points along the dispersion curve of S_8 modes, which provide the first evidence of surface resonances. Indeed the resonance S_8 has quite a large intensity all over the SBZ. This appears from Fig. 11 which shows the sagittal phonon density at a point $(1/2,0)\pi/a$.

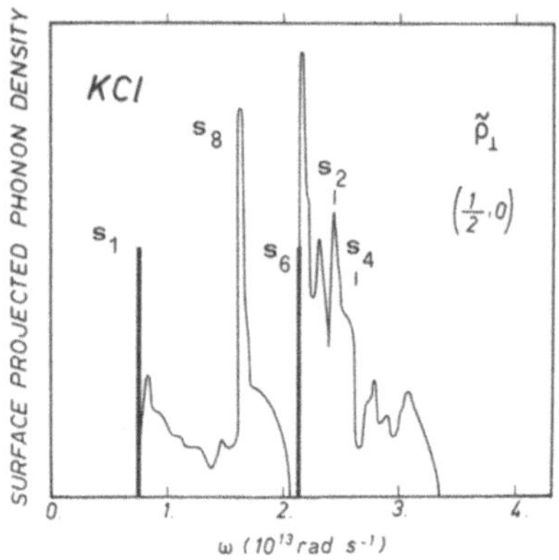

Fig.11 Surface-projected sagittal phonon density for K = (1/2,0) in KCl(001) showing the sharp resonance S_8 of sagittal polarization

Acknowledgement

I am much indebted to Prof. Peter Toennies, Dr. Bruce Doak and Dr. Guido Brusdeylins (Max-Planck-Institut für Stromungsforschung, Göttingen) for the continued information on their experimental results; to Prof. Vittorio Celli (University of Virginia) for several useful discussions and particularly his illuminating comments on equation (94); and to Mrs. Suzie Garrett for the accurate typing of the manuscript.

References

1. G.W. Farnell, Physical Acoustics, 6, 109 (1970).
2. A.A. Maradudin, R.F. Wallis and L. Dobrzinski, Handbook of Surfaces and Interfaces, Vol. 3: Surface Phonons and Polaritons (Garland STPM Press, New York, 1980).
3. A.A. Maradudin, *Surface Waves* in Festkörperprobleme-Advances in Solid State Physics (Plenum, New York 1981).
4. F. Garcia-Moliner, Ann. Phys. (Paris) 2, 179 (1977).
5. R. Fuchs and K.L. Kliever, Phys. Rev. 140, A2076 (1965).
6. K.L. Kliever and R. Fuchs, Phys. Rev. 144, 495 (1966); 150, 573 (1966).
7. T.S. Chen, F.W. de Wette and G.P. Alldredge, Phys. Rev. B 15, 1167 (1977).
8. A. Otto, Z. Phys. 216, 398 (1968).
9. V.V. Bryskin, Yu.M. Gerbshtein and D.N. Mirlin, Fiz. Tverd. Tela 13, 2125 (1972) (Sov. Phys. Sol. State, 13, 1779 (1972)); and *ibidem* 14, 543, 3368 (1972) (Sov. Phys. Sol. State, 14, 453, 2849 (1972)).
10. D.J. Evans, S. Ushioda and J.D. McMullen, Phys. Rev. Letters, 31, 369 (1973).
11. H. Boersch, J. Geiger and W. Stickel, Phys. Rev. Letters 17, 379 (1966).
12. H. Ibach, Phys. Rev. Letters, 24, 1416 (1970).
13. G. Benedek, Surface Sci. 61, 603 (1976).
14. R.F. Wallis, Rendiconti S.I.F., Course LII (Academic Press, New York and London, 1972).
15. T.E. Feuchtwang, Phys. Rev. 155, 731 (1967).
16. V. Bortolani, F. Nizzoli and G. Santoro, Proc. Int. Conf. on Lattice Dynamics, M. Balkanski, Ed. (Flammarion, 1978).

17. A.A. Lucas, J. Chem. Phys. 48, 3156 (1968).
18. R.E. Allen, G.P. Alldredge and F.W. de Wette, Phys. Rev. B4, 1648 (1971); 1661 (1971); 1682 (1971).
19. Ref. 7. See also T.S. Chen, G.P. Alldredge and F.W. de Wette, Sol. State Comm. 10, 941 (1972).
20. I.M. Lifshitz and L.M. Rozenzweig, Zh. Eksp. Teor. Fiz. 18, 1012 (1948); I.M. Lifshitz, Nuovo Cim. Suppl., 3, 732 (1956).
21. A.A. Maradudin, E.W. Montroll, G.H. Weiss and I.P. Ipatova, Solid State Physics Suppl. 3 (2nd edition, 1971).
22. W. Ludwig and B. Lengeler, Solid State Comm. 2, 83 (1964).
23. S. Doniach and E.H. Sondheimer, *Green's Functions for Solid State Physicists* (Benjamin Inc., 1974); E.N. Economou, *Green's Functions in Quantum Physics*, Spriner Series in Solid-State Sciences, Vol. 7 (Springer, Berlin, Heidelberg, New York 1979).
24. G. Benedek, Phys. Stat. Solidi B 58, 661 (1973).
25. G.P. Alldredge, Phys. Letters 41A, 281 (1972).
26. In my previous Green's function calculations (ref. 13) the $\bar{M}$ point degeneracy was not verified. Recently I could discover a computational error affecting the calculation in a small region around $\bar{M}$, and causing the misfit. An erratum is to be published on Surf. Science.
27. G. Brusdeylins, R.B. Doak, and J.P. Toennies, Phys. Rev. Letters, 46, 437 (1981).
28. G. Benedek and N. Garcia, Surf. Sci. 103, L143 (1981).
29. J.R. Manson and V. Celli, Surface Sci. 24, 495 (1971). This important work is based on the distorted wave Born approximation. The theory of inelastic processes in the eikonal approximation for a HCS is systematically treated by A.C. Levi, Nuovo Cim. B54, 357 (1979).
30. Here we work in direct space and time coordinates as in G. Benedek and N. Garcia, Surface Sci. 80, 543 (1979), using, however, different and more standard notation [23].
31. D.R. Hamann, Phys. Rev. Letters, 46, 1227 (1981).
32. G. Benedek, G. Brusdeylins, R.B. Doak and P.J. Toennies, Proc. Int. Conf. on Phonon Physics, Bloomington 1981, W.E. Bron ed. (to appear on J. Physique, Suppl.)
33. R.B. Doak, J.P. Toennies and G. Brusdeylins, private communication; R.B. Doak, Thesis (M.I.T., 1981; unpublished).

Inelastic Scattering of Neon from the (001) LiF Surface

L. Mattera, M. Rocca, C. Salvo, S. Terreni, F. Tommasini and U. Valbusa
Gruppo Nazionale di Struttura della Materia and
Istituto di Scienze Fisiche dell'Università, Viale Benedetto XV,5
I-16132 Genova, Italy

1. Introduction

Recent studies of inelastic He scattering from LiF [1] have shown that the interaction with Rayleigh phonons is dominant. However, to gain deeper knowledge of atom surface scattering it is interesting to use heavier incident surfaces probes, such as neon atoms, because of the larger collision time and the like interaction with the bulk modes.

In the present lecture we present some recent data on inelastic scattering of neon from LiF(001). The experiment has been carried out by measuring both time-of-flight (TOF) and angular distributions of the scattered particles. In Sect.1 we give a brief description of the apparatus used in the present experiment. In Sect.2 data are reported together with a description of the analysis performed to obtain information on the phonon dispersion curves. Section 3 is devoted to the discussion. There the Ne/LiF(001) interaction at low temperatures is shown to be characterized by a single predominant phonon. This conclusion follows from an analysis of the data based on accurate measurements of the Debye-Waller factor (DWF) and of the widths of the TOF distributions as a function of temperature.

1.1. Experimental Apparatus

The apparatus has been extensively described elsewhere [2]. Briefly it consists of four differentially pumped stages, denoted in Fig.1 by the letters A, B, C and D, where the last two are pumped by a liquid helium cryopump. Typical pressures are 10^{-4} and 10^{-7} torr in the first two stages, while in the others the vacuum depends on the gas used as probe; with neon we have about 10^{-10} torr. The supersonic nozzle beam source is placed in the first chamber, the chopper disc in the second and the target in the innermost. The quadrupole mass spectrometer detector is located in chamber D and it is followed by a TOF analysis system. The incident angle θ_i and the outgoing

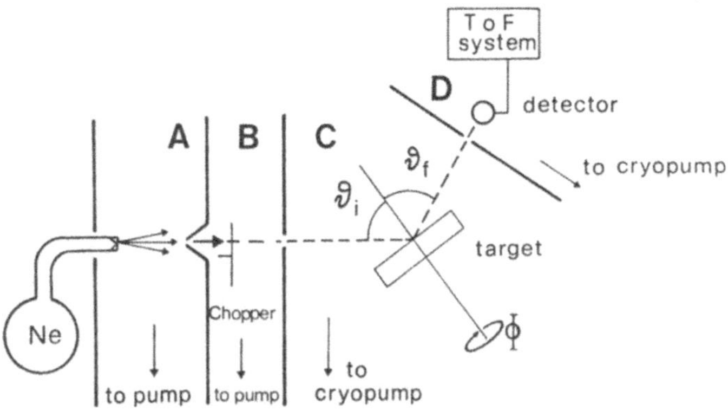

Fig.1. Experimental apparatus. ABCD refer to the four pumping stages. The detector is a quadrupole mass spectrometer followed by a time-of-flight analysis system

angle θ_f can both be varied independently, while other mechanisms allow tilting and rotating the crystal in order to optimize in-plane scattering and the azimuthal direction. The incident beam is chopped by a slotted disc with a duty cycle of 1/24 and a minimum gating time of 8 μs. The temperature of the crystal can be varied from 5 to 1000 K. The neon beam used has a most probable wave vector $\vec{k}_i$ = 26.4 Å^{-1}, an angular spread $\Delta\theta$ = 0.6° and an energy spread $\Delta E/E$ = 10% FWHM.

1.2. Measurements and Analysis

In Fig.2a we report some typical TOF spectra recorded around the diffraction peaks (4,4) and (5,5). The time scale refers to the time of flight of the neon atom from the chopper to the detector (flight path L = 309.1 mm). The arrows indicate the elastic TOF corresponding to the most probable velocity of the incident beam accurately measured as reported in [2]. The structures in the TOF patterns evidence loss or gain in energy of the neon atoms during the interaction.

The analysis of these measurements has been carried out by using the conservation of energy and momentum,

$$k_F^2 = k_i^2 + \sum_j \frac{2m}{\hbar^2} \omega_j$$

$$\vec{K}_F = \vec{K}_i + \sum_j \vec{Q}_j \qquad (1)$$

where ω_j and $\vec{Q}_j$ are the frequency and the momentum of the jth exchanged phonon and $\vec{k}_i = (\vec{K}_i, k_{iz})$, $\vec{k}_f = (\vec{K}_f, k_{fz})$ are the wave vectors of the inci-

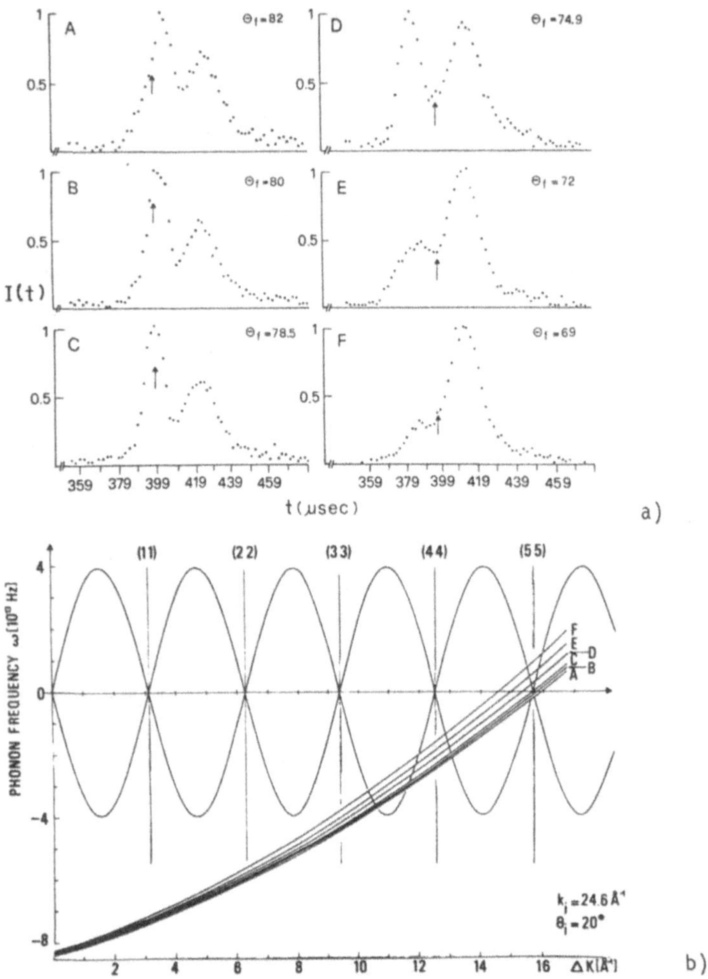

Fig.2. (a) Typical time-of-flight patterns. The structures correspond to phonon creation and annihilation. (b) Rayleigh dispersion curves in the sine approximation. The parabolae refer to the TOF patterns of Fig.2a

dent and outgoing atom of mass m. Considering the total energy and the total momentum exchanged with the surface $\omega = \sum_j \omega_j$ and $\Delta\vec{K} = \sum_j \vec{Q}_j$ we obtain from (1)

$$\omega = \frac{\hbar}{2m \sin^2\theta_f} (\Delta K)^2 + 2\Delta K\, k_i \sin\theta_i + k_i^2(\sin^2\theta_i - \sin^2\theta_f) \qquad (2)$$

which relates ω to ΔK for given scattering conditions (θ_i, θ_f, k_i). Equation (2) is a parabola in the ΔK, ω plane. Inelastic processes are possible whenever it crosses the phonon dispersion curves. In Figs.2a,b we see in fact that varying θ_f from 82° to 69° we pass from creation of phonons associated with G = (4,4) and G = (5,5) to annihilation associated with G = (5,5) and creation with G = (4,4).

The number of phonons participating in the interaction cannot be determined from the kinematic of the process. From data like those in Fig.2 it is possible to obtain only information about the total energy and the total momentum exchanged. A procedure to estimate the number of phonons involved in the process is described in Sect.1.3.

The structures of Fig.2 are also affected by the experimental resolution which broadens the TOF structures. We assume that the resolution is mainly determined by the velocity spread of the incident beam with a velocity distribution described by a Gauss function of 5% FWHM. The phonon interaction produces a broadening in the TOF patterns which can be also described by a Gauss function. In this framework the deconvolution of the experimental data is a simple deconvolution between Gauss functions as fully described in [3]. Such an analysis has been carried out over a large set of TOF measurements taken at different scattering angles ranging from $\theta_f = 10^o$ to $\theta_f = 82^o$. The results are summarized in Fig.3 where all the measured ω and ΔK are reported in the reduced Brillouin zone. The points refer to the ω, ΔK values corresponding to the time of flight of the most probable velocity; the lines refer to the FWHM of the distribution of the exchanged phonons which has been obtained by the previous described deconvolution procedures. From the figure it comes out that both bulk and Rayleigh phonons are relevant during the interaction, although we cannot tell them apart because of the resolution of the experiment. In fact every time the parabola of (2) crosses the two branches of the Rayleigh curves the experiment gives a value for ΔK close to zero while when only one branch is cut the values of ΔK are closer to it. The first case occurs because the total energy exchanged and the total momentum arise from the Rayleigh phonons moving in the backward and forward directions and from the bulk phonons which are in between. In the second case the total energy and the total momentum arise from Rayleigh phonons moving in only one direction and from bulk phonons.

1.3. Multiphonon Scattering

In order to proceed in the analysis of the data it is necessary to know the number of phonons participating in the interaction. Such information cannot be obtained by the previous measurements because they are sensitive only to the total energy and total momentum exchanged. The probability P_n of the n-phonon scattering process can be estimated as proposed in [4] making use

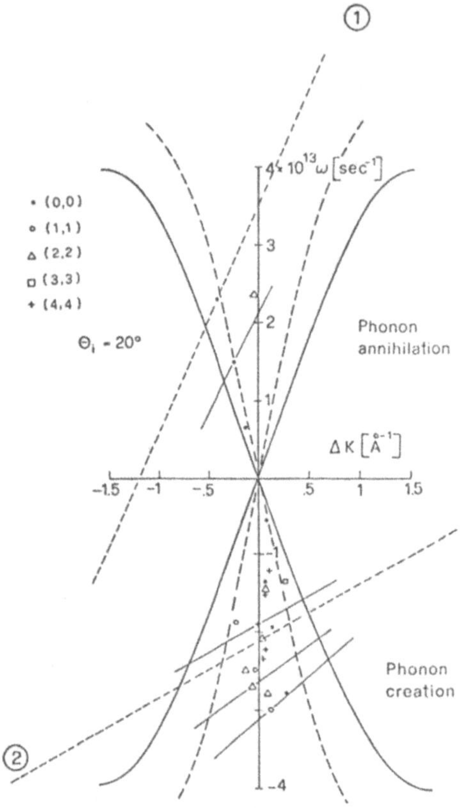

Fig.3. Reduced zone plot of the measured values of ω and ΔK. The solid curves represent the Rayleigh-phonon dispersion curves in the sine approximation. The dashed curves are the limits of the bulk band dispersion curves. Curves 1 and 2 indicate the parabola of (2) for $\theta_i = 20$, $k_i = 24.6$ Å^{-1} at $\theta_f = 17$ and $\theta_f = 42.5$

of the Debye-Waller factor e^{-2W} as

$$P_n = \frac{2^n}{n!} W^n e^{-2W} \quad . \tag{3}$$

We measured the DW factor as a function of temperature by an accurate experimental procedure [3]. By reporting these values in (3) we obtained the probability of the n-phonon scattering process for different temperatures of the crystal. The results are summarized in Fig.4. The figure shows that for the temperature of 78 K, where we carried out the reported measurements, the most probable process involves two or three phonons, while in order to have a one-phonon process the crystal must be maintained at 5 K. Therefore we cooled down the crystal to 5 K and we carried out measurements similar to those previously reported. Figure 5 shows one of the TOF patterns at 5 K compared to similar ones at higher crystall temperatures. Raising the crystal temperature the TOF patterns show broader structures as expected for a multi-phonon process. However there is no difference in the width of the TOF dis-

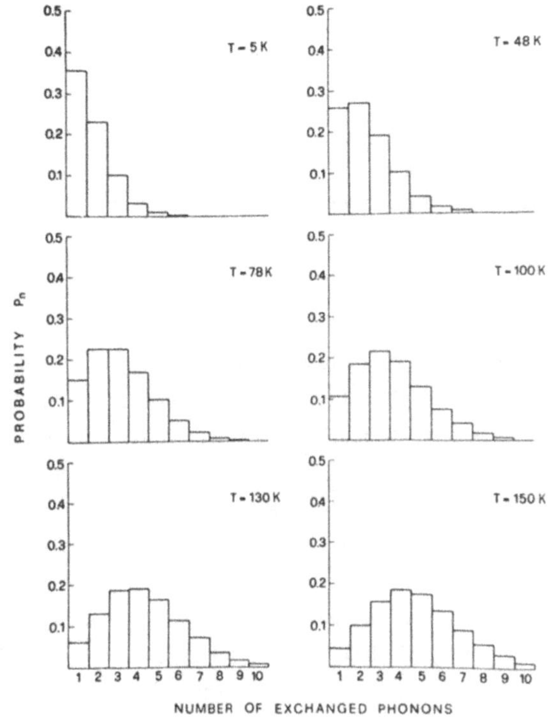

Fig.4. Probability of a n-phonon scattering process calculated by using (3) for different crystal temperatures

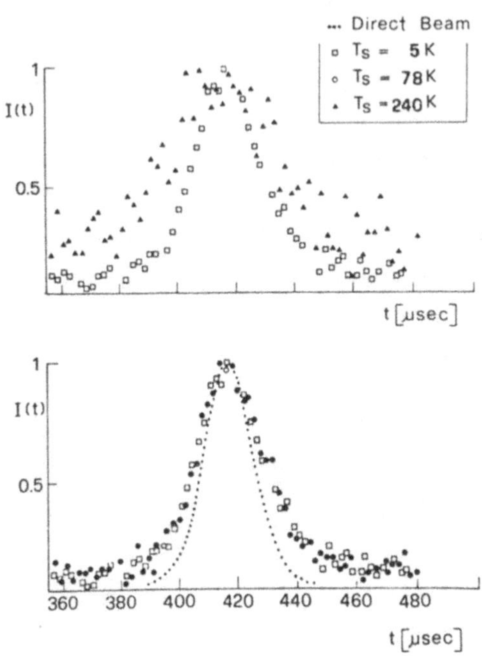

Fig.5. TOF patterns taken at three different crystal temperatures. For all of them $\theta_i = 20^\circ$ and $\theta_f = 23^\circ$

tribution in the temperature range 5 K to 78 K. This seems to contradict the result of (3) where at 78 K processes involving two or three phonons should be the most probable. This fact can be explained assuming that at this temperature whenever we have processes with two or three phonons only one of them is of high energy giving rise to a one-phonon-like process. This conclusion seems to be in agreement to the experiment of SEMERAD and HÖRL [5] who explained all their results by using a single-phonon model.

1.4. Conclusions

We have measured inelastic scattering of neon atoms from the (001) LiF surface at low temperatures. We have proved that the atoms exchange energy and momentum with both Rayleigh and bulk band modes, even though their contribution cannot be separated because of the energy resolution of the experiment. Debye-Waller factor measurements as a function of temperature allowed for an estimation of the probability of a n-phonon scattering process. We conclude that at the crystal temperature of 78 K the most probable processes are those occurring via 2-3 phonons while at 5 K the one-phonon processes are dominant. However by looking at the width of the TOF patterns at the same temperature we conclude that at 78 K, even though the dominant processes are those involving 2-3 phonons, only one of them has large energy, while the others have energy close to zero. The overall process is then one-phonon-like.

References

1 J.P. Toennies: In this book and references therein
2 E. Cavanna, A. Gussoni, L. Mattera, S. Terreni, F. Tommasini, U. Valbusa, C. Salvo: J. Vac. Sci. Technol. *19*, 161 (1981)
3 L. Mattera, C. Salvo, S. Terreni, F. Tommasini, M. Rocca, U. Valbusa: Surf. Sci. (to be published)
L. Mattera, C. Salvo, S. Terreni, F. Tommasini, U. Valbusa: Phys. Rev. B (to be published)
4 H.D. Meyer: Surf. Sci. *104*, 117 (1981)
5 E. Semerad, E.M. Hörl: Proceedings of the VIII Symposium on Molecular Beams, Cannes 1981

Inelastic Scattering from Metal Surfaces

B. Feuerbacher

Astronomy Division, Space Science Department of ESA ESTEC
NL-Noordwijk, The Netherlands

For a long time it has been recognized that the atom scattering technique has a remarkable potential for obtaining new and interesting information on solid surfaces. However, technical difficulties, in particular for inelastic measurements, have held up progress until recently. Elastic scattering measurements on insulators have now given detailed information on surface structure and interaction potentials. Recently those experiments were extended to the more difficult, but also more rewarding case of metal surfaces |1-2|, and shortly later to the exciting area of metal surfaces with well defined adsorbates |4-6|. In the realm of inelastic scattering experiments, a major breakthrough were the recent results of BRUSDEYLINS, DOAK, and TOENNIES |7,8|, which demonstrate the power of this method to study the low-energy elementary excitations of solid surfaces and showed sharp, detailed spectra for phonon excitation on a LiF crystal face. These exciting results call for an extension to metal surfaces. In fact, a small number of very recent results are available to date, which cannot be regarded conclusive as yet, but clearly demonstrate the power of this approach and indicate the directions for future experimentation. As it appears that we are right now at the threshold of a new and exciting era, the present paper will refrain from reviewing the pioneering earlier results that prepared the way for todays progress, and refer the reader to a number of detailed review articles |9-11|. The present discussion will focus exclusively on current results as well as the prospects and consequences arising from them.

1. What Is So Exciting About Metal Surfaces ?

All major breakthroughs in atom scattering from surfaces have been achieved with insulator surfaces, starting with the early measurements by Estermann and Stern |12| on LiF crystals. This is, by the way, in sharp contrast to other surface science techniques such as LEED, AES, UPS and similar spectroscopies, which are based on electron scattering and which are nearly exclusively used for the investigation of metal surfaces. The reason for this discrepancy rests in the fact that, first of all, the pronounced corrugation of ionic crystal faces offers an effective diffraction grating for the wavelengths represented by thermal atoms. Secondly, ionic crystals are easy to prepare as single crystal faces (by cleaving) and easy to clean on an atomic scale in vacuum (by heating). On the other hand, electron spectroscopies always suffer from charging problems when attempting to investigate insulator surfaces.

The basic reason for the pronounced interest in metal surfaces is the distinct role played by the conduction electrons. While their properties in the bulk are reasonably well understood, the lack of translational invariance at the surface imposes severe difficulties in their theoretical accessibility. This applies both to the spatial distribution, characterized by a rapid decrease of electron density normal to the surface |13| and to their excitation spectrum, which exhibits marked differences to the bulk behaviour |14|. A recent calculation |15| indicates that the scattering of light, low-energy atoms might serve as an effective probe for the electron density profile near surfaces. On the other hand, the role of the conduction electrons in the composition of the atom-surface interaction potential is by no means understood. This question leads on to the problem of energy and momentum transfer in a collision between a thermal atom and a metal surface, which is fundamental for the understanding of sticking, accommodation, and desorption processes.

An important aspect of metal surfaces is their ability to effectively absorb atoms and molecules, leading to chemisorption, surface reactions, or catalysis. The atom scattering technique holds promise of providing relevant information on various aspects of these processes. The ability to probe surface electron density profiles allows studying the rearrangement of surface conduction electrons upon adsorption. In conjunction with the observation of the dynamic properties of an adsorbate system by inelastic atom scattering, this might give insight into the microscopic processes relevant for the chemisorptive interaction. Further information might be gained from the study of the atom-surface interaction potential, in particular for the case of chemical interaction between the probing molecule and the surface, a case that is usually termed reactive surface scattering.

2. Basic Differences Between Metal and Insulator Surfaces

As a wide range of information is available on insulator surfaces, both from the theoretical and the experimental side, the question arises as to what fundamental differences are expected for metal surfaces, justifying the additional effort in the latter case. As mentioned above, the crucial differences arise from the presence of the conduction electrons, and they express themselves in the excitation spectrum and the spatial distribution at the surface.

A schematic comparison of the low-energy excitation spectrum of an insulator and a metal is given in Fig.1. An insulator is characterized by a gap for electronic excitations, which separates the occupied and unoccupied electron bands. There is no overlap between phonon excitations, which are in the range of the kinetic energies of thermal atoms, and electronic excitations. For metals, low-energy electron-hole pair excitations are possible across the Fermi level, so the phonon frequencies are embedded in an electron

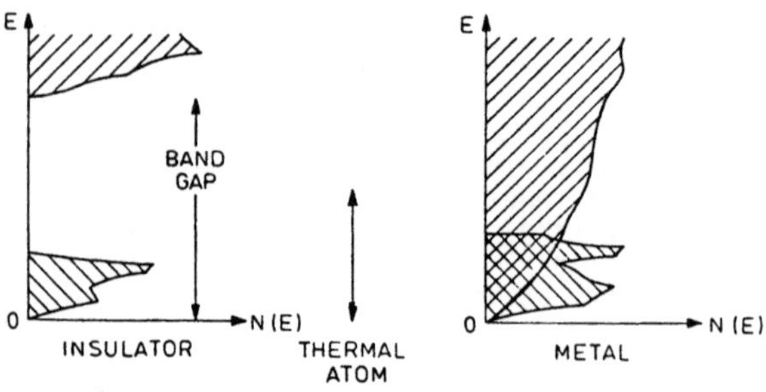

Fig.1 Schematic excitation spectrum of an insulator surface (left) and a metal surface (right). The lower band represents the phonon spectrum, while the upper arises from electronic excitations. The center arrow indicates the range of kinetic energies of light thermal atoms.

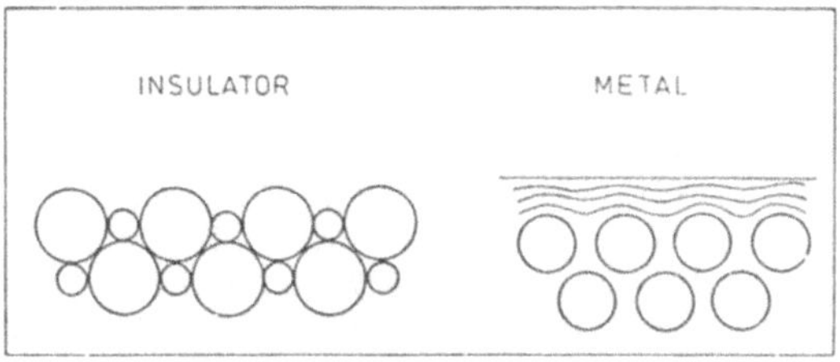

Fig.2 Schematic comparison of the electron density profile at an insulator (left) or metal (right) surface.

excitation continuum. The influence of this overlap is twofold. Firstly, electron-hole pairs may be directly excited in the atom-surface collision. This will be negligible for the case of nonreactive scattering (e.g. noble gas atoms), where the dominant interaction is of the van de Waals type and, as such non-dissipative. It can become an important energy transfer mechanism for atoms that show electronic interaction with the surface. A second effect arises from the decay possibility of phonon excitations into electron-hole pairs via electron-phonon coupling and expresses itself in a decreased phonon lifetime and such a broadening of the phonon-induced spectral features.

Fig.2 shows a comparison of the spatial charge distribution at an insulator and a metal surface. The insulator surface is best represented by an array of spheres having the ionic diameter of the constituents. As the gradient of the charge distribution is steep, a corrugated hard wall is a good description of the surface. This is in contrast to a metal surface, where the conduction electrons tend to smooth the potential variations. Therefore, the overall corrugation seen by an atom is small, and it depends on the penetration of the atom into the surface electron sheeth. One would expect an increasing corrugation seen by the atoms with increasing impact energy. The decreased corrugation of a metal surface compared to an insulator surface is illustrated by the fact that the first clear atom diffraction features on a clean, low index metal |1| were found only 46 years after the original discovery of atom diffraction on LiF |12|. An additional effect of the electron blanket on a metal surface is a distribution of interaction forces which may lead, under particular circumstances, to a breakdown of the pairwise additivity assumption usually employed to construct the surface-atom interaction potential.

3. Experimental Results

3.1 Light Probing Atoms: He on Cu

A recent experiment by MASON and WILLIAMS |16| has provided the first results on light atom scattering from a clean metal surface with a resolution that allows comparison to the earlier data on alkali halide crystals |7|. The experimental setup is somewhat unconventional in that it is based on a crystal diffractometer for energy analysis instead of the more widely used time-of-flight method. A sketch of the geometrical layout is given in Fig.3. A beam of He atoms at an energy of 22.6 meV is generated in a liquid nitrogen cooled nozzle source. The angle between beam source, sample, and analyzer is fixed at 130°, while the sample crystal can be rotated to vary incidence angle θ_i and detection angle θ_s simultaneously. A LiF(100) crystal serves as an energy analyzer with its (11) diffraction maximum. The resolution is approximately 1 meV.

The spectra of the He kinetic energy after scattering from a Cu(100) surface at 16°K are shown in Fig. 4 for three incidence angles. All spectra

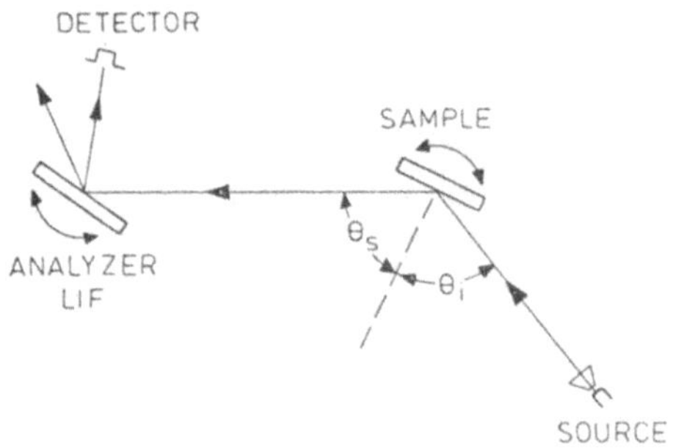

Fig.3 Geometric arrangement of the inelastic atom scattering spectrometer of MASON and WILLIAMS (Ref. 16). Incidence and scattering conditions are changed simultaneously by sample rotation. The analyzer is a rotatable LiF(100) crystal face used in the (11) diffraction condition.

are characterized by a sharp peak at the elastic energy $\Delta E = 0$, and a well separated continuum of energy losses. The interpretation of these data proceeds along the coupling diagram shown in Fig.5, which presents a plot of the excitation parabola P |17| in the ΔE (energy transfer) vs. ΔK (momentum transfer) plane. In the same figure, curves for Rayleigh phonon emission or absorption (R) are shown, together with the transverse bulk phonon branch (T), which bounds the continuum of bulk phonons (hatched).

The sharp peak at $\Delta E = 0$ occurs at the crossing of the coupling parabola P with the horizontal axis (heavy dot). This feature probably arises from diffuse scattering at surface irregularities and was also observed from LiF

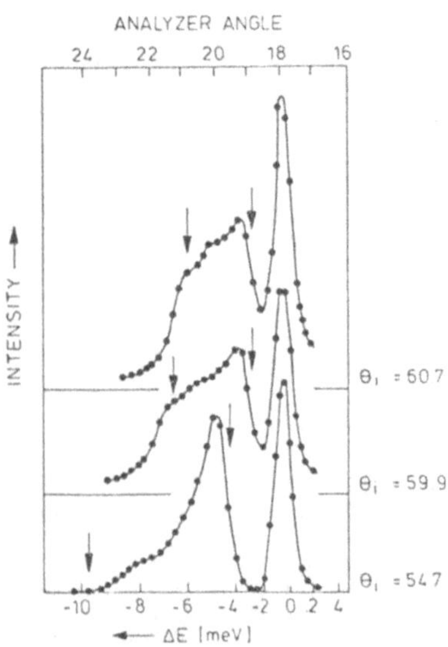

Fig.4 Kinetic energy spectra of 22.6 meV He atoms scattered from a low-temperature CU(100) surface. Incidence angles θ_i are shown on the right hand side. Vertical arrows indicate positions of expected Rayleigh wave interaction. (From Ref. 16).

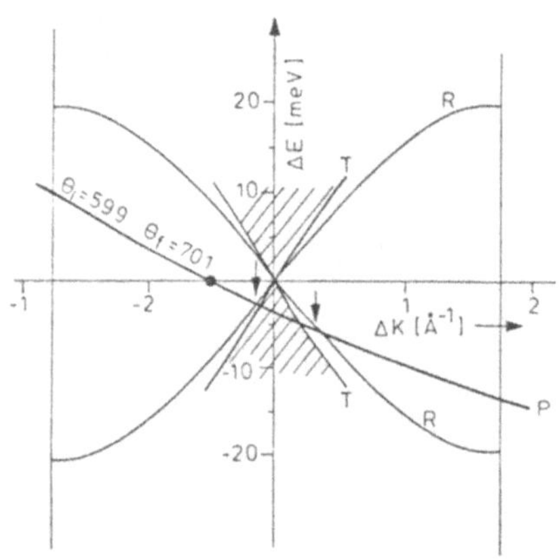

Fig.5 Coupling parabola P in the ΔE vs ΔK plane for scattering conditions as center curve in Fig.4. Curves for Rayleigh waves (R) and transverse bulk phonons (T) are included. The latter bound the bulk phonon continuum (hatched).

surfaces |7,8|. The vertical arrows mark the crossing with the Rayleigh phonon dispersion curves in both Fig.4 and Fig.5. A remarkable observation is the fact that the Rayleigh phonon position seems to bound the observed loss continuum approximately, but not exactly. This has led MASON and WILLIAMS to an interpretation which assigns the left hand structure (momentum loss) to surface phonons and the sharp peak on the right hand side (momentum gain) to the excitation of transverse bulk phonons.

While at present a satisfactory interpretation of these data is not yet at hand, there are a few remarkable features that should be noted. First of all, the spectra show continuum absorption by phonons (within the resolution available) in clear contrast to the LiF data that were dominated by discrete line absorption or emission. At first glance, this would indicate that the contribution of bulk phonons is much stronger for the metal surface. It should be kept in mind that the Cu (100) surface is non-diffractive, so rather simple spectra centered around $\Delta K = 0$ are expected, in contrast to the alkali halide data, where strong contributions from higher Brillouin zones are observed. While a clear distinction between single-phonon or multiphonon processes is not possible as yet, the relatively sharp structure and the limitation of absorption features to the region bounded by the Rayleigh dispersion curve seems to indicate dominating single-phonon coupling.

3.2 Heavy Probing Atoms: Ne on Ni

Quite different results are found if a relatively heavy noble gas atom like neon (M = 20) is scattered from a close packed NI(111) surface. The scattering geometry was similar to that described in Fig.3, with a fixed source-detector angle of 135°, however the source was a fast pulsed valve |18| and energy analysis was performed by standard time-of-flight techniques. Due to the pulsed source, the resolution was limited to about 8 meV at a beam energy of 65 meV.

A set of typical time-of-flight spectra for Ne scattered from Ni(111) is shown in Fig.6 for various incidence and detection angles θ_i and θ_s. The heavy curve in the center represents specular scattering. The spectra in-

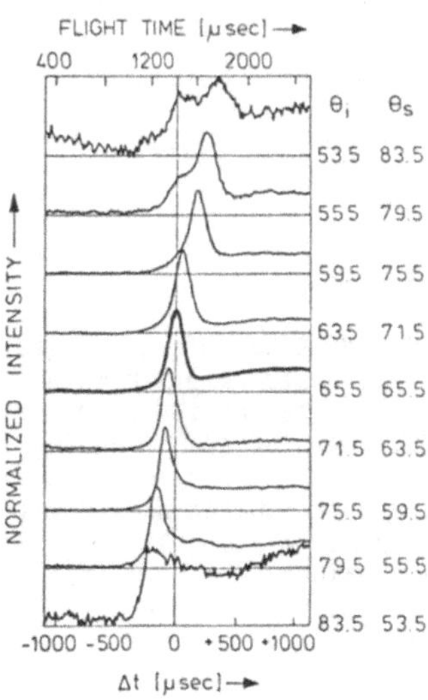

Fig.6 Time-of-flight spectra for 65 meV neon atoms scattered from a Ni(111) surface. Scattering conditions are given on the right hand side in terms of incidence angle and scattering angle (θ_i and θ_s)

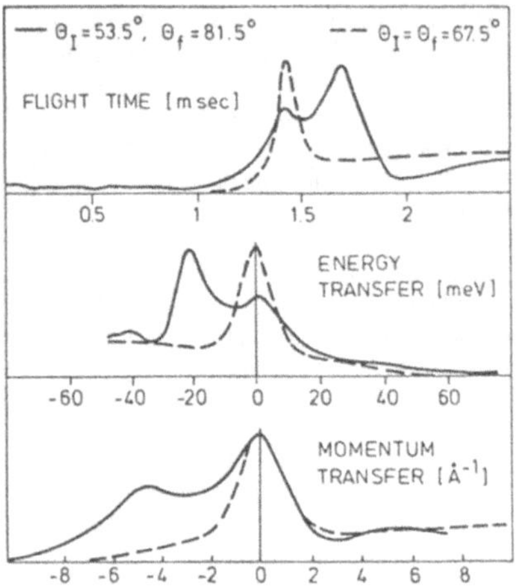

Fig.7 Ne scattered from Ni (111) at $\Delta\Theta_i = 53.5°$ compared to specular scattering (dashed). The top panel gives the measured time-of-flight spectra. The lower panels show the same spectra transformed into energy space or momentum space as indicated.

dicate energy gains or losses for scattering towards the surface normal or towards grazing scattering angles, respectively. An evaluation in terms of energy and momentum transfer is presented in Fig. 7. Here the top panel shows a time-of-flight spectrum $\Theta_i = 53.5°$, which has been transformed into energy space or momentum space in the lower two panels. Each panel contains the corresponding curve for specular scattering (dashed) for comparison. Two peaks are resolved in this spectrum, of which one is found at $\Delta E = 0$ and the other at $\Delta K = 0$.

A summary of the observed peak positions in a ΔE vs. ΔK diagram is shown in Fig. 8, together with a schematic phonon band structure along the $\bar{\Gamma}\bar{M}$ symmetry line of the Ni(111) face. The observed points fall into three distinctly different groups marked A, B and C in Fig.8. Group A occurs at $\Delta E = 0$ and has the same origin as the sharp peak of the He spectra in Fig.4. Group B shows a linear relationship between ΔE and ΔK which corresponds, within the accuracy of the measurements, to the continuum surface sound wave speed or equally, to the Rayleigh wave phonon dispersion, up to a critical value of transferred parallel momentum $\Delta K = K_C$. Higher energy transfers extend well beyond the maximum single-phonon energies (group C) and are characterized by vanishing parallel momentum transfers.

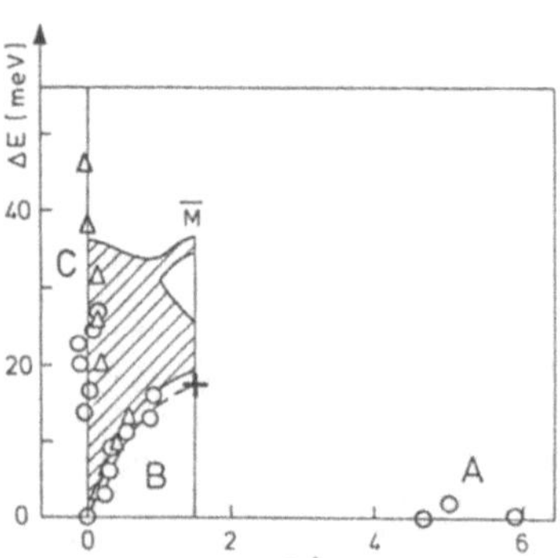

Fig.8 Absolute values of energy vs. parallel momentum transfer in Ne-Ni (111) scattering as derived from TOF spectra. Circles and triangles correspond to energy loss and gain, respectively. The shaded area gives the phonon spectrum along the $\bar{\Gamma}\bar{M}$ line. The straight line indicates the surface wave velocity in the continuum limit.

Sharp structures associated with vanishing parallel momentum transfer are not observed for the scattering of He from CU(100), and it should be noted that they are not found for He scattering from a Ni(111) surface |19| . This feature therefore calls for a different interpretation, which proceeds along an idea expressed by Beeby |20|. Heavy atoms move slow in the vicinity of the surface, and a large momentum is required to inverse their normal velocity component in the scattering process. Therefore more than a single collision with a surface atom may be required for reflection, leading to a multiphonon description of the collision process, as also suggested by the observation of energy transfers well beyond the single-phonon limit.

The picture that emerges for the interpretation of the above measurements therefore is as follows. As the Neon atom impacts, it interacts with several surface atoms that cooperate to provide the momentum to invert the normal momentum component of the Ne atom. This may lead to the excitation of continuum surface waves of the Rayleigh type, however only for wavelengths longer than the diameter of the distortion circle formed by the interacting surface atoms, giving rise to the linear dispersion branch B. The observed cutoff wavevector $K_c \cong 1$ Å^{-1} indicates that the distortion circle contains about 5 Ni atoms. For larger energy transfers, excitation of surface waves is not possible any more, so parallel momentum transfer becomes forbidden and the surface reacts as a flat reflector. This is, by the way, the limit assumed by the cubes models |20|, which have been so successful in describing the lobular scattering of heavy gas atoms from metal surfaces. These models have one common basic assumption, which implies that the surface does not take up parallel momentum from the beam. The present results serve to support this assumption only for the limit of large energy transfers, while parallel momentum transfer is shown to be possible for wavelengths larger than the distortion circle (or the size of the "cube") via the excitation of elastic surface waves. It should be noted in this context that the mass ratio for Ne on Ni ($\mu = 0.34$) is quite close to that of the He-LiF system ($\mu = 0.24$), where distinct single-phonon processes are clearly observed. In generalizing this observation, one may conclude that one of the actions of the conduction electrons in a metal is to increase the sensitivity to multiphonon interaction. Experiments of Ne on LiF |22| also show spectra indicative of multiphonon effects, and their analysis leads to vanishing momentum transfer parallel to the surface.

3.3 Adsorbate Layers

One of the exciting features of inelastic atom scattering spectroscopy is its capability to observe the dynamics of surface adsorbate layers. Vibrational spectra of adsorbates have been observed in infrared reflection spectroscopy and by low-energy electron scattering. Both these techniques are limited in their range of accessibility of momentum transfers parallel to the surface. Atom scattering holds promise of providing complete dispersion relations for adsorbates and thus of giving access to the lateral interactions within an adsorbate layer.

The first measurement of an adsorbate vibration by inelastic atom scattering spectroscopy has been performed by MASON and WILLIAMS |16|. Using the instrument described above (see Fig.3), they have deposited an ordered layer of xenon atoms on a low temperature Cu(100) surface. The spectrum of He atoms scattered from this layer is shown in Fig. 9. A set of well defined, sharp peaks is observed, which have a constant separation (on an energy scale) of 2.5 meV and are found both for the atom energy gain and loss. Variation of the scattering angle does not result in a shift of the energetic position of the sharp structures.

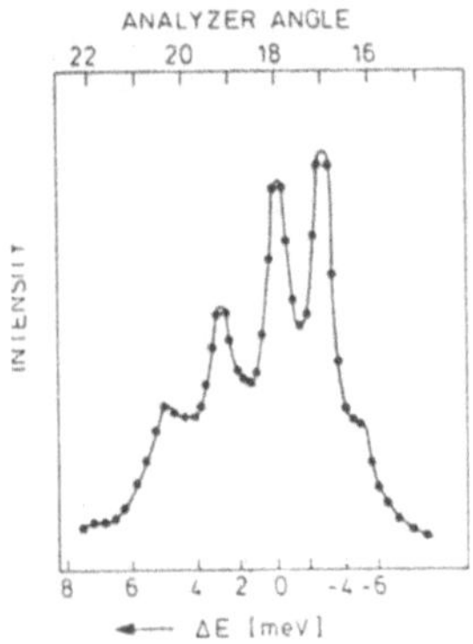

Fig.9 Inelastic spectrum of 22.6 meV helium atoms scattered from a Cu(100) surface covered with an ordered monolayer of Xe. Incidence angle is 59.9°. The sharp peaks arise from excitation or annihilation of single or multiple vibrations of Xe atoms against the Cu surface.

The results indicate a normal vibration eigenmode for Xe physisorbed on Cu(100) at 2.5 meV. This mode is dispersionless, pointing to lateral interactions between the Xe atoms negligible compared to the Xe-Cu interaction forces, as expected in such a system. An interesting feature is the high probability for multiple excitations for both the Stokes and the anti-Stokes lines, which is in contrast to electron spectroscopy results, where multiple excitations are found with very small probabilities.

No results have as yet been reported on the observation of vibrational excitation of chemisorbed species by inelastic atom scattering. However, elastic scattering measurements on such systems have provided deep insight into the structural arrangement of electron density profiles at metal surfaces covered with ordered chemisorption layers |5|. An example of the sensitivity of atom scattering to long-range ordering is presented in Fig.10. Here an ordered p(2 x 2) layer of oxygen atoms has been formed on a Ni(111) surface, and the intensity of He atoms diffracted into the $(\frac{1}{2},\frac{1}{2})$ beam has

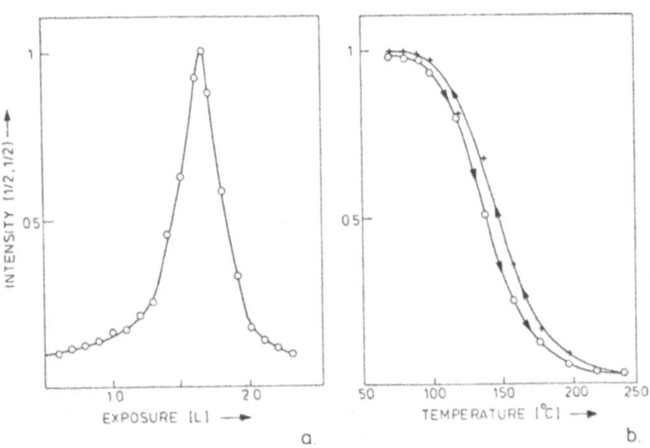

Fig.10 Intensity of the half order diffraction peak of 65 meV He atoms scattered from a Ni(111) surface with a p(2 x 2) oxygen overlayer, a) as a function of surface exposure to O_2, b) after formation of the superstructure, as a function of surface temperature. The arrows indicate the direction of temperature change (heating or cooling).

been observed. Figure 10a shows behaviour of the diffraction intensity during deposition of the layer as a function of surface exposure to oxygen gas at room temperature. The intensity rises sharply as the surface coverage for an ordered superstructure is approached, and decreases rapidly as additional oxygen atoms destroy the long range order. At elevated temperatures, the oxygen p(2 x 2) superstructure undergoes a reversible order-disorder phase transition. This is illustrated in Fig.10b, which shows the intensity of the half order He diffraction beam as a function of surface temperature. As the temperature is increased, the diffraction intensity decreases rapidly over a range of about 100°C. Upon cooling, the intensity returns to the same value as measured with the deposited film.

4. Conclusions

The present paper has reviewed a number of recent experimental results that demonstrate the feasibility to observe dynamic processes on metal surfaces by means of inelastic atom scattering spectroscopy. This conclusion is by no means self-evident, even after similar results have been available from insulator surfaces. Even though the resolution of experiments on metals is not yet sufficient for direct comparison to the spectra from insulators, a few preliminary conclusions can be drawn. It appears, that both bulk and surface features are observable, however the systematics of their appearance are by no means clear. At first glance, bulk features in the spectra seem to be more important, relative to the surface features, than has been found for insulator surfaces. There are still a good deal of spectral features that require interpretation, and the specific role of the conduction electrons is not yet obvious.

Heavy probing gases seem to emphasize multiphonon effects much more than observed on insulators. Surface waves can be excited by atom impact, however only for wavelengths longer than the size of the impact distortion circle.

Adsorbate vibrations are observable by inelastic atom scattering. For an ordered physisorbed layer with negligible dispersion of the phonon modes, multiple vibrational excitation was observed both for the Stokes and the anti-Stokes features. The high sensitivity of atom diffraction to long range order makes a reversible order-disorder phase transition of an adsorbate observable.

References

1. G. Boato, P. Cantini, and R. Tatarek, J. Phys. F.: Metal Phys. 6, L237 (1976).
2. J.M. Horne, S.C. Yerkes, and D.R. Miller, Surface Sci. 93, 47 (1980).
3. J. Lapoujoulade, Y. Le Cruër, M. Lefort, Y. Lejay, and E. Maurel, Surface Sci. 103, L85 (1981).
4. K.H. Rieder and T. Engel, Phys. Rev. Letters 43, 373 (1979).
5. K.H. Rieder and T. Engel, Phys. Rev. Letters 45, 824 (1980).
6. J. Lapoujoulade, Y. Le Cruër, M. Lefort, Y. Lejay, and E. Maurel, Phys. Rev. B 22, 5740 (1980).
7. G. Brusdeylins, R.B. Doak, and J.P. Toennies, Phys. Rev. Letters 44, 1417 (1980).
8. G. Brusdeylins, R.B. Doak, and J.P. Toennies, Phys. Rev. Letters 46, 437 (1981).
9. J.P. Toennies, Appl. Phys. 3, 91 (1974).
10. See also: F.O. Goodman and H.Y. Wachman, Dynamics of Gas-Surface Scattering , Academic Press, New York 1976.

11. B. Feuerbacher, in Vibrational Spectroscopy of Adsorbates , ed. by R.F. Willis, Springer Series in Chemical Physics, Vol. 15. Springer, Berlin, Heidelberg, New York 1980, p. 91.
12. I. Estermann and O. Stern, Z. Physik 61, 95 (1930).
13. N.D. Lang and W. Kohn, Phys. Rev. B 1, 4555 (1970).
14. See, e.g. F. Forstmann in Photoemission and the Electronic Properties of Surfaces , ed. by B. Feuerbacher, F.R. Willis, and B. Fitton, Wiley, Chichester 1978, p. 193.
15. N. Esbjerg and J.K. Nørskov, Phys. Rev. Letters 45, 807 (1980).
16. B.F. Mason and B.R. Williams, Phys. Rev. Letters 46, 1138 (1981).
17. G. Benedek, Phys. Rev. Letters 35, 234 (1975).
18. M.R. Adrianes, W. Allison and B. Feuerbacher, to be published in J. Phys. E: Scientific Instruments.
19. B. Feuerbacher, unpublished.
20. J.L. Beeby, in Dynamic Aspects of Surface Physics , ed. by F.O. Goodman, Editrice Compository, Bologna 1974, p. 751.
21. See, e.g. Ref. 10, p. 108 ff.
22. L. Mattera, C. Salvo, S. Terreni, F. Tommasini, and U. Valbusa, Proc. IV ECOSS, Le Vide, les Couches Minces 201, Suppl. II, 838 (1980). See also paper by the same authors in this volume.

Bound State Resonance in the Inelastic Scattering of He-Graphite

P. Cantini

Gruppo Nazionale di Struttura della Materia del CNR and
Instituto di Scienze Fisiche dell'Universitá di Genova, Viale Benedetto XV,5
I-16132 Genova, Italy

1. Introduction

Helium atom scattering is a powerful technique to study surface lattice dynamics [1]. Although highly monochromatized atomic beam and energy analysis of the scattered intensity can be necessary to yield precise measurement of the surface phonon dispersion relation, in special cases useful and sufficiently detailed information on lattice dynamics can be obtained through the study of only the inelastic bound state resonance angular behaviour (inelastic selective adsorption). The selective adsorption process was extensively studied in recent years as one of the most relevant phenomena in the diffraction of atoms from surfaces. The present paper will be devoted to illustrating the selective adsorption mechanism in the inelastic scattering, a process that was studied in detail for He/graphite. The results of this study were manyfold; several different ways of forming a bound state have been observed and a general theory for resonant inelastic scattering was proposed [2] to explain the experimental data.

I recall that the kinematic conditions for a scattering process involving the exchange of a single phonon are

$$k^2 = k_o^2 \pm 2M\omega_q/\hbar \qquad (1)$$

$$\vec{K}_G = \vec{K}_o + \vec{G} \pm \vec{Q} \qquad (2)$$

where $\vec{k} = (\vec{K}, k_z)$ is the wave vector of the gas atom of mass M while $\hbar\omega_q$ and $\vec{Q}$ are the energy and the parallel wave vector of the exchanged phonon; $\vec{G}$ is a surface reciprocal lattice vector. The gas atom may undergo a phonon-assisted resonant transition in a bound state, if its perpendicular energy corresponds to an energy level ε_j of the laterally averaged gas-surface potential $V_0(z)$, while the atom travels in a nearly free particle state parallel to the surface after the exchange of a phonon. Due to the discrete nature of the bound state eigenvalues ε_j, the phonon-assisted resonance selects, at any incident angle and energy, a well identified family of phonons. Their frequencies $\omega_{N,j}(\vec{Q})$ are given, through the kinematic conditions (1) and (2), by

$$k_o^2 \pm 2M\omega_{N,j}(\vec{Q})/\hbar - (\vec{K}_o + \vec{N} \pm \vec{Q})^2 = 2M\varepsilon_j/\hbar^2 \qquad (3)$$

where $\vec{N}$ is a closed-channel reciprocal vector. On the other hand, when a scanning of the inelastic angular distribution is taken, each final scattering angle ϑ_f corresponds to inelastic processes which select a family of phonons $\omega_f(\vec{Q})$ given, for in-plane scattering, by the parabolic equation

$$k_o^2 \pm 2M\omega_f(\vec{Q})/\hbar - (\vec{K}_o + \vec{F} \pm \vec{Q})^2/\sin^2\vartheta_f = 0. \qquad (4)$$

In general the family of phonons yielding a resonant contribution (3) and that selected by the kinematic condition (4) do not coincide; under this condition the position $(\vartheta_o, \vartheta_f^*)$ of a resonant structure observed in the tail of a diffraction peak will fix, through (3) and (4), both the energy $\hbar\omega_q^*$ and the parallel momentum $\vec{Q}^*$ of the phonon involved in the resonance. This procedure, repeated for a set of incident angles ϑ_o, allows a phonon dispersion curve to be obtained.

2. Experimental Observations

The inelastic selective adsorption was studied for He/graphite scattering in several different experimental conditions. I will report hereafter some examples of the relevant mechanisms observed in the inelastic resonances.

2.1 Phonon-Assisted Bound State Resonance

The resonant transition in a bound state after the exchange of one phonon with the surface was observed in several experimental conditions. The in-plane scattered intensity was measured for a large number of incident angles, both in the ϕ=0° and ϕ=30° azimuthal plane. Structures in the angular distribution were associated with inelastic bound state resonances. A criterion for this assignement is the following: a typical behaviour of inelastic resonance is that the angular position of the resonant structure changes little as the incident angle is changed, therefore both the reciprocal vector $\vec{N}$ and the energy level ε_j of the resonance can be identified with the help of the corresponding structure found in the elastic peak intensity. In Fig.1 some typical measured angular distributions are reported; the azimuth is ϕ=0° while k_o = 11.05 Å^{-1}. The minimal which appear are associated with the $\vec{N}$= (01)-(1$\bar{1}$) closed channels and with the

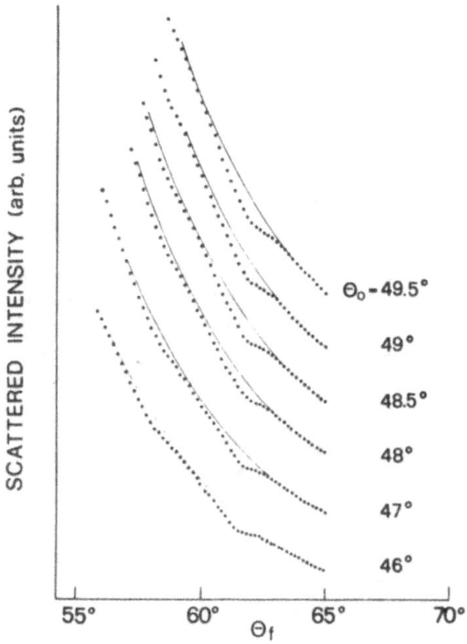

Fig.1 In-plane angular distribution of the inelastic tail of the specular peak at different incident angles ϑ_o. Resonant minima, associated with the ε_2 and ε_1 levels and with the $\vec{N}=(01)-(1\bar{1})$ closed channel are present (k_o=11.05Å^{-1} and ϕ =0°)

$\vec{F}$ = (00) final channel; the energy levels involved are ε_1= -6.33 and ε_2= -2.85 meV, known with great accuracy for He/graphite system [3,4] .

2.2 Specular Phonon-Assisted Bound State Resonance

The resonant transition in a bound state, reached after the exchange of one phonon, can also occur without exchange of a reciprocal lattice vector, say with $\vec{N}$ = (00). This process can be called "specular" phonon-assisted resonance. It is possible with phonon creation alone, to select the $\omega_{N,j}(\vec{Q})$ of Eq.(3); the scattering condition Eq.(4) will correspond to the $\vec{F}$ = $(\bar{1}0)$ final channel. The specular inelastic resonances observed at different incident angles, with k_o = 11.05 Å^{-1} and ϕ =0° are shown in Fig.2. Two maxima clearly appear near $\vartheta_f \simeq$ 50° and $\vartheta_f \simeq$ 54°, associated with the energy levels ε_1 = -6.33 and ε_o = -11.98 meV; resonances associated to the higher levels ε_2, ε_3 and ε_4 give rise to a small shoulder around $\vartheta_f \simeq$ 47.5°. The identification of the observed resonances as "specular" inelastic was confirmed by the angular position ϑ_f and the use of (3) and (4) to obtain the $\hbar\omega_q^*$ and $\vec{Q}^*$ of the involved phonons. The solution which involves phonons belonging to the first Brillouin zone corresponds to the $\vec{N}$ = (00) closed channel and $\vec{F}$ = $(\bar{1}0)$ final channel.

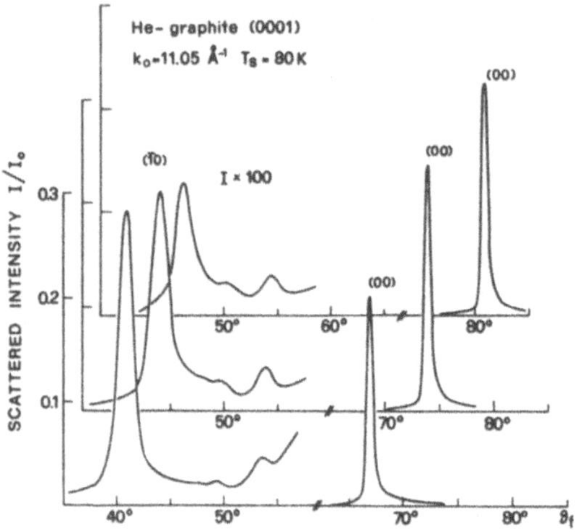

Fig.2 In-plane angular distribution of the scattered intensity at three different incident angles $\vartheta_o = 67°$, $\vartheta_o = 74°$ and $\vartheta_o = 81°$. Specular inelastic resonances appear as maxima near $\vartheta_f = 50°$ and $\vartheta_f = 54°$

We studied also the behavior of such resonances at different surface temperatures T_S. The trend observed at different temperatures is shown in Fig.3, where the specular resonance is reported in an enlarged scale.

An inelastic background for $\vartheta_f > 50°$ mainly given by the phonon annihilation process associated with the specular peak is apparent; due to the Bose statistics of phonons the background increases strongly with the temperature. On the other hand the specular inelastic resonance peak is related to phonon creation and associated with the $\vec{F}=(10)$; therefore the Bose factor is slowly increasing with surface temperatures T_S, while the Debye-Waller factor strongly attenuates the resonant intensity.

2.3 Double Bound State Resonance

As is well known, at selected incidence conditions the incoming can go elastically in a bound state $(\varepsilon_j - \vec{N})$; at the same time it

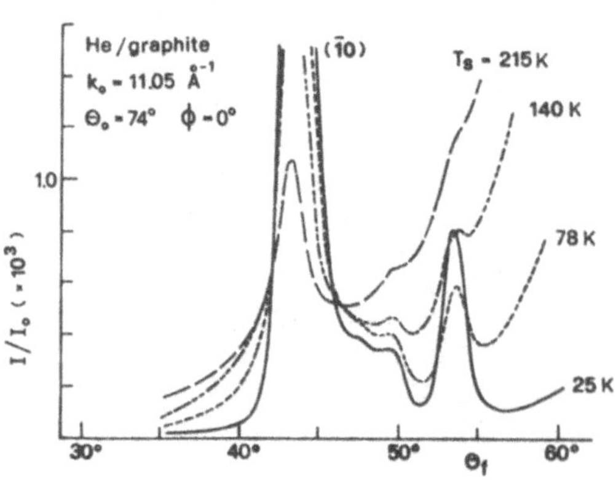

Fig.3 In-plane angular distribution of the inelastic tails of the $\vec{F}=(\bar{1}0)$ elastic peak observed at different surface temperatures

can go also inelastically in a different bound state $(\varepsilon_{j'}-\vec{N})$. In this incident condition a third resonant transition is possible, with an inelastic jump from the first $(\varepsilon_j-\vec{N})$ to the second $(\varepsilon_{j'}-\vec{N})$ bound state. This third resonant transition selects the same family of phonons $\omega_{Nj'}(\vec{Q})$ of the inelastic resonance $(\varepsilon_{j'}-\vec{N})$, therefore its effect is a large interference contribution to the inelastic selective adsorption at the incident angles where the elastic bound state resonance occurs. A typical example is shown in Fig.4, where the inelastic tail of the specular peak is reported for different incident angles. The angular distributions were taken at $k_o = 9.07$ Å^{-1} ($E_o = 42.96$ meV) in the azimuth $\phi = 30°$. The inelastic resonance reported is the $(\varepsilon_o-\vec{N}_{10})$-$-(\varepsilon_o-\vec{N}_{01})$ crossing, while the elastic resonance associated with the $(\varepsilon_1-\vec{N}_{10})-(\varepsilon_1-\vec{N}_{01})$ crossing interferes at $\vartheta_o \simeq 51.2°$, and a strong maximum takes the place of the minimum.

3. Calculation of the Specular Phonon-Assisted Resonance

Following the theoretical formalism recently proposed [2] , the resonant contribution to the inelastic scattering can be easily evaluated for single-phonon processes in the eikonal approximation. As an example of the proposed theory I will show the calculation performed to describe the specular phonon-assisted resonances observed for He/graphite scattering and previously shown

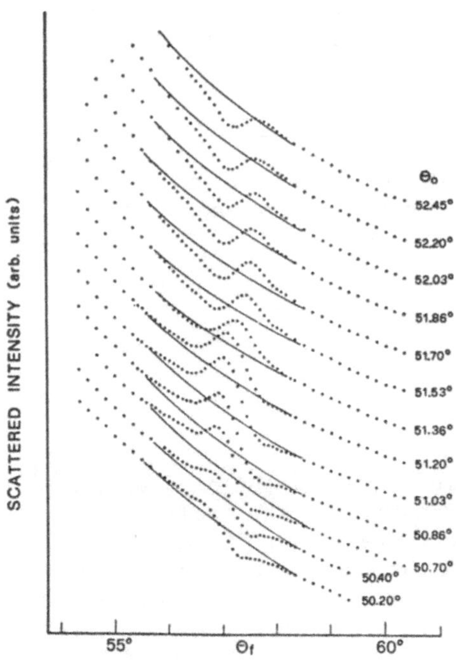

Fig.4 Angular distributions of the inelastic tails of the specular peak. The maximum at $\vartheta_o \simeq 51.2°$ corresponds to the double resonance condition, with elastic transition in the ε_1 level and the inelastic transition in the ε_o level ($k_o = 9.07$ Å^{-1}, $\phi = 30°$ and $\vec{N} = (01)-(10)$)

in Fig.3. Details of this calculation are published elsewhere [2,5]. The inelastic scattering probability was calculated by assuming a simplified surface phonon spectrum, with Rayleigh phonons alone, which are taken to be identical with the $TA_\perp$ mode in the ΓM direction of the bulk [7]. For each energy level ε_j we considered only resonant processes with creation of backward phonons ($\vec{Q}$ opposite to $\vec{K}_o$). Furthermore the Debye-Waller attenuation was calculated in the usual form, with $2W_{G,G'}=(p_{G-Q}+p_{G'})^2\langle u_\perp^2\rangle$, where the mean square displacement of surface atoms $\langle u_\perp^2\rangle$ was that experimentally obtained from the thermal attenuation of the specular peak [6]. Because the described inelastic resonance is related to a scattering process involving a non zero reciprocal vector ($\vec{F}=(\bar{1}0)$), different incoming energies give different angles for the resonant structure; the effect of the energy spread of the beam was therefore taken into account. The resulting angular distributions calculated at three different temperatures are displayed in Fig.5. The agreement with the experimental curves plotted in Fig.3 is quite satisfactory, the resonances associated with the higher levels ε_2, ε_3 and ε_4 give rise to a small shoulder around $\vartheta_f \simeq 47.5°$, while only the resonance associated with the ε_1 and ε_o levels survive as well resolved maxima. However the calculated structures, even at the lowest temperature T_S=25 K, are too sharp; this may in part be due to the assumed simplified phonon spectrum. The discrepancies with experiment however increase with surface temperature, and this can lead to the conclusions that for the He/graphite system the one phonon approximation loses its validity at quite low temperatures, this being an effect of the quasi-two-dimensional phonon spectrum [6].

4. Information on Surface Phonon Dispersion Relation

As was shown in the Introduction, the position of a given inelastic resonance ϑ_f^* observed at a given incident angle ϑ_o de-

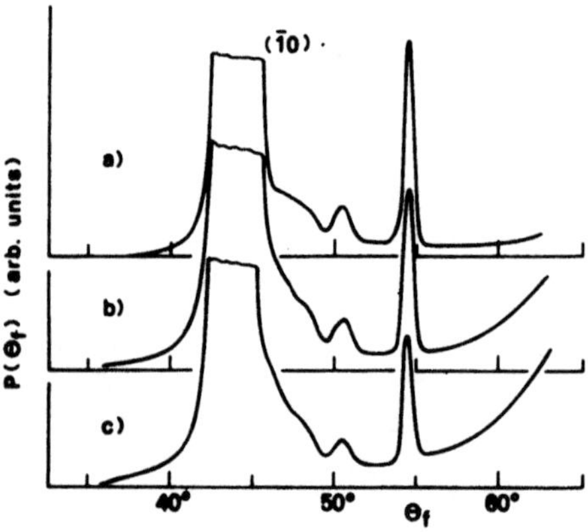

Fig.5 Calculated angular distribution of the inelastically scattered intensity at ϑ_o=74°, in the experimental condition of Fig.3. Curve (a) corresponds to T_S=25 K, curve (b) to T_S=78K and curve (c) to T_S=140 K

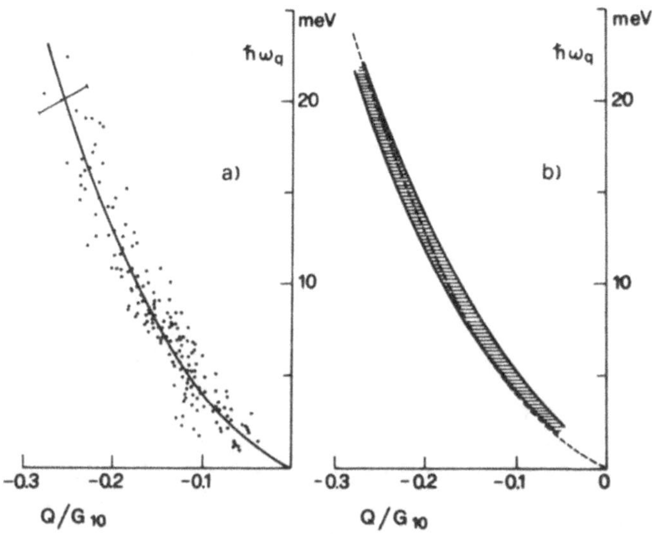

Fig.6 Mean energy $\hbar\omega_q^*$ and parallel momentum $\vec{Q}^*$ of phonons which give inelastic selective adsorption (a) and most probable dispersion relation of surface phonons (b) compared with the $TA_\perp$ mode of bulk phonons along ΓM

termines, through (3) and (4), the energy $\hbar\omega_q^*$ and the parallel momentum $\vec{Q}^*$ of the phonon involved in the resonance; thus the study of inelastic resonances observed at different incidence conditions allows obtaining a phonon dispersion curve. For He/graphite we analyzed about 120 angular distributions, where about 180 resonance structures were identified. All the resonances were observed at crystal temperature T_S=80 K, to avoid multiphonon processes. We discarded also the incident angles where the double-resonance process occurs, because the determination of the resonance position involves too large an uncertainty in this process, as appears in Fig.4. The values of $\hbar\omega_q^*$ and $\vec{Q}^*$ for the phonons involved in the selected measurements are reported in Fig. 6-a. All the observed events refer to backward phonons, as forward phonons are not likelay to be favored by Bose tics. The error bar shown for one experimental point represents the estimated uncertainty of ±4' in the angular position of the resonance. It can be noted that all points belong to a limited region of the (ω,Q) space, which contains the dispersion relation of the $TA_\perp$ transverse acoustic mode along (0001) for bulk graphite, as determined by neutron inelastic diffraction measurements [7]. This last experiment allowed an accurate determination of the lower branches of phonon spectrum; a relevant result of this work was to show clear evidence for a quadratic dispersion relation for the $TA_\perp$ branch, an effect which is typical of layered materials having strong intraplanar and weak interplanar forces. In bulk graphite the effect is well described by the dispersion relation

$$(\hbar\omega_{\perp})^2 = \alpha(Q/G_{10})^2 + \beta(Q/G_{10})^4 \qquad (5)$$

with $\alpha = 6.6\ 10^2$ [meV]^2 and $\beta = 8.5\ 10^4$ [meV]^2. The effect was theoretically predicted [8] while the calculated values of α and β [9] are in agreement with the neutron data. The dispersion relation (5) is expected to be approximately valid also for surface Rayleigh phonons propagating along the basal plane, since graphite layers are bound through weak forces. This assumption was recently confirmed by calculation of the surface modes of graphite slabs [10]. We tried to fit the experimental points of Fig.6-a with a curve having the form of Eq.(5). The most probable dispersion relation of surface phonons obtained from inelastic resonances is reported in Fig.6-b as a dashed region. It nearly coincides with the $TA_{\perp}$ bulk mode. The confidence in the measured phonon energy is about ± 1 meV. The discrepancy at low phonon energy may be connected with the complexity of the surface projected phonon spectrum in this range and with the presence of multi-phonon processes.

5. Conclusions

The bound state resonance process, as effect of multiple reflections inside the attractive well, still remains a very important process also responsible of many structures observed in the inelastic tails of elastic peaks. The inelastic resonance can be simply described in the eikonal approximation, as a modulation of the inelastic scattering. The effect is similar to the modulation observed in the elastic probability, even if several different mechanisms for going in a bound state are possible. In spite of these difficulties, the study of the angular positions of inelastic resonances allows obtaining reliable information on the surface phonon dispersion relation, without energy analysis of the scattered atoms.

REFERENCES

1 G. Brusdeylins, R. Bruce Doak and J.P. Toennies, Phys.Rev. Lett. 44 (1980) 1417; 46 (1981) 437.
2 P. Cantini and R. Tatarek, Phys.Rev. B 23 (1981) 3030.
3 G. Boato, P. Cantini, C. Guidi, R. Tatarek and G.P. Felcher, Phys.Rev. B 20 (1979) 3957.
4 G. Derry, D. Wesner, W. Carlos and D.R. Frankl, Surf.Sci. 87 (1979) 629.

5 P. Cantini and R. Tatarek, to be published.
6 G. Boato, P. Cantini, C. Salvo, R. Tatarek and S. Terreni, to be published.
7 R. Nicklow, N. Wakabayashi and H.G. Smith, Phys.Rev. B 5 (1972) 4951.
8 K. Komatsu, J.Phys.Soc. Japan 10 (1955) 346.
9 K.K. Many and R. Ramani, Phys.Stat.Sol. (b) 61 (1974) 659.
10 E. de Rouffignac, G.P. Alldredge and F.W. de Wette, Phys. Rev. B 23 (1981) 4208.

Index of Contributors

GPSR Compliance

The European Union's (EU) General Product Safety Regulation (GPSR) is a set of rules that requires consumer products to be safe and our obligations to ensure this.

If you have any concerns about our products, you can contact us on ProductSafety@springernature.com

In case Publisher is established outside the EU, the EU authorized representative is:

Springer Nature Customer Service Center GmbH
Europaplatz 3
69115 Heidelberg, Germany

Batch number: 10620143

Printed by Printforce, the Netherlands